251
Topics in Current Chemistry

Topics in Current Chemistry

Recently Published and Forthcoming Volumes

Chalcogenocarboxylic Acid Derivatives

Volume Editor: Shinzi Kato

With contributions by
S.-I. Fujiwara · A. Ishii · N. Kambe · N. Kano · S. Kato · T. Kawashima
T. Murai · J. Nakayama · O. Niyomura

 Springer

The series *Topics in Current Chemistry* presents critical reviews of the present and future trends in modern chemical research. The scope of coverage includes all areas of chemical science including the interfaces with related disciplines such as biology, medicine and materials science. The goal of each thematic volume is to give the nonspecialist reader, whether at the university or in industry, a comprehensive overview of an area where new insights are emerging that are of interest to a larger scientific audience.

As a rule, contributions are specially commissioned. The editors and publishers will, however, always be pleased to receive suggestions and supplementary information. Papers are accepted for *Topics in Current Chemistry* in English.

In references *Topics in Current Chemistry* is abbreviated Top Curr Chem and is cited as a journal.

Visit the TCC content at springerlink.com

Library of Congress Control Number: 2004111708

ISSN 0340-1022
ISBN 3-540-23012-2 **Springer Berlin Heidelberg New York**
DOI 10.1007/b98140

Springer is a part of Springer Science+Business Media
springeronline.com
© Springer-Verlag Berlin Heidelberg 2005
Printed in The Netherlands

The use of general descriptive names, registered names, trademarks, etc. in this publication does not imply, even in the absence of a specific statement, that such names are exempt from the relevant protective laws and regulations and therefore free for general use.

Cover design: KünkelLopka, Heidelberg/design & production GmbH, Heidelberg
Typesetting: Fotosatz-Service Köhler GmbH, Würzburg

Printed on acid-free paper 02/3141/xv – 5 4 3 2 1 0

Topics in Current Chemistry
Also Available Electronically

For all customers who have a standing order to Topics in Current Chemistry, we offer the electronic version via SpringerLink free of charge. Please contact your librarian who can receive a password for free access to the full articles by registration at:

springerlink.com

If you do not have a subscription, you can still view the tables of contents of the volumes and the abstract of each article by going to the SpringerLink Homepage, clicking on "Browse by Online Libraries", then "Chemical Sciences", and finally choose Topics in Current Chemistry.

You will find information about the

– Editorial Board
– Aims and Scope
– Instructions for Authors
– Sample Contribution

at springeronline.com using the search function.

Preface

Chalcogenocarboxylic acid derivatives are a large class of compounds including more than one chalcogenocarboxyl group in which one or two oxygen atoms of the carboxyl group are replaced by sulfur, selenium or tellurium atoms. There are 15 kinds of compounds of chalcogenocarboxylic acid (Table 1). As the hydrogen atom of hydrochalcogenoxide (EH) group in RCOEH (E=S, Se, Te) can also be replaced by all of the element in the periodic Table and in addition, Group 2 to 16 elements can formally bind with more than two chalcogenocarboxyl groups, the number of the types increases to over 10000, even limited to the case where R=methyl group. The chemistry of metal chalcogenocarboxylates has not been explored extensively as that of carboxylates and dithiocarbamates. This volume presents a comprehensive overview of the syntheses and their limitations, structures and reactions of chalcogenocarboxylic acid derivatives, by emphasizing the developments in organic and inorganic chalcogen chemistry over the last 5 to 20 years.

Takayuki Kawashima and Naokazu Kano wrote Chapter 3, Juzo Nakayama and Akio Ishii Chapters 4 and 5, Nobuaki Kambe and Shin-ichi Fujiwara contributed Chapter 6, and Toshiaki Murai submitted Chapter 7 and Osamu Niyomura and I presented Chapters 1 and 2.

Finally I thanks Professors Hisashi Yamamoto of Chicago University and Tamejiro Hiyama of Kyoto University for suggesting we write this book and the encouragement they gave us.

Nagoya, November 2004

Shinzi Kato

Contents

Contents of Volume 231
Elemental Sulfur and Sulfur-Rich Compounds II

Volume Editor: R. Steudel
ISBN 3-540-40378-7

Top Curr Chem (2005) 251: 1–12
DOI 10.1007/b101005
© Springer-Verlag Berlin Heidelberg 2005

Chalcogenocarboxylic Acids

Osamu Niyomura[1] · Shinzi Kato[2] (✉)

[1] Division of Chemistry, Center for Natural Sciences, College of Liberal Arts and Sciences, Kitasato University, Kitasato, Sagamihara, Kanagawa 228–8555, Japan
niyomura@clas.kitasato-u.ac.jp
[2] Department of Applied Chemistry, School of Engineering, Chubu University, Matsumoto-Cho, Kasugai, Aichi, 487–8501, Japan
kshinzi@nifty.com

Abstract Although thio- and dithio-carboxylic acids have been extensively studied for some time now, research into other chalcogenocarboxylic acids – containing selenium and tellurium – has only blossomed over the last decade. Monochalcogenocarboxylic acids exist as fast tautomeric equilibrium mixtures of chalcogenol and chalcogenoxo forms. The chalcogenol form is the predominant species in solid state and nonpolar solvents. In contrast, in polar solvents at low temperature, the acids predominantly exist in the chalcogenoxo form. Syntheses of heavier dichalcogenocarboxylic acids have only been attempted very recently. This chapter presents the results from recent studies of chalcogenocarboxylic acids, their syntheses, structures and reactions.

Keywords Chalcogenocarboxylic acids · Chalcogenol form · Chalcogenoxo form · Carboxylic acids · Chalcogens

1
Introduction

Carboxylic acids are one of the most fundamental and important groups of compounds in organic chemistry. *Chalcogenocarboxylic acids*, RCEE′H (E, E′=O, S, Se, Te), are derived from carboxylic acids by replacing one or both oxygen atoms of the carboxyl group with S, Se or Te.

As shown in Scheme 1, there are 15 kinds of chalcogenocarboxylic acid. There are six monochalcogenocarboxylic acids, that have oxygen and chalcogen atoms (**A**), as well as their tautomers (**A′**). There are also the dichalcogenocarboxylic acids (**B** and **C**), while the dichalcogenocarboxylic acids with two different chalcogen atoms (of which there are three kinds, **B**) also possess tautomers (**B′**).

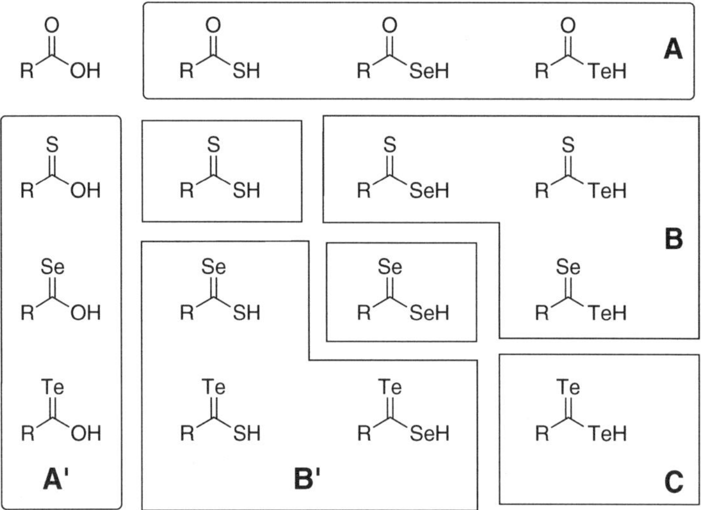

Scheme 1

Historically, the first chalcogenocarboxylic acid discovered was thiocarboxylic acid – thioacetic acid – reported by Kekulé in 1854 [1]. Since then, chalcogenocarboxylic acids, and particularly numerous thio- and dithiocarboxylic acid esters, have been synthesized and summarized in several review articles [2–9]. In contrast, until very recently, little has been known about the chemistry of the congeners containing heavier chalcogen atoms, such as selenium and tellurium, probably due to their instability and the handling difficulties associated with them. In this chapter, the chemistry of chalcogenocarboxylic acids, their syntheses, structures, spectral features and reactions are reviewed.

2
Syntheses

2.1
Monochalcogenocarboxylic Acids

As mentioned above, the first isolation of a monochalcogenocarboxylic acid –
thioacetic acid – was reported in the middle of the 1850s, having been synthesized from the reaction of thioacetic acid with P_4S_{10} [1]. This method cannot
applied to the synthesis of other thiocarboxylic acids, due to very low yields,
although reactions in the presence of a catalytic amount of Ph_3SbO have been
found to give good yields of aromatic thioacids [10]. Acidolysis of thiocarboxylic
acid alkali and alkaline earth metal or ammonium salts with hydrogen chloride
has proved to be the most convenient method to prepare the thiocarboxylic
acids 1 and 1′ (E=S) (Scheme 2) [5, 6, 11, 12]. The first example of selenocarboxylic acid was confirmed spectroscopically in 1972 by Jensen et al., who synthesized selenobenzoic acid by reacting benzoyl chloride with H_2Se in the
presence of pyridine, followed by sulfuric acid [13]. Isolations of a series of
selenocarboxylic acids 1 and 1′ (E=Se) were gained through HCl-acidolysis of
the corresponding sodium [14] or potassium salts (Scheme 2) [15]. Formation
of the tellurocarboxylic acids 1 and 1′ (E=Te) (purple for the aliphatic compounds and blue for the aromatic compounds in THF solution) by similar HCl
acidolysis of the corresponding cesium (or other alkali metals) tellurocarboxylates has also been observed spectroscopically (Scheme 2) and by conversion into
the acyl carbamoyl telluride [16]. However, these are too air-sensitive to isolate.

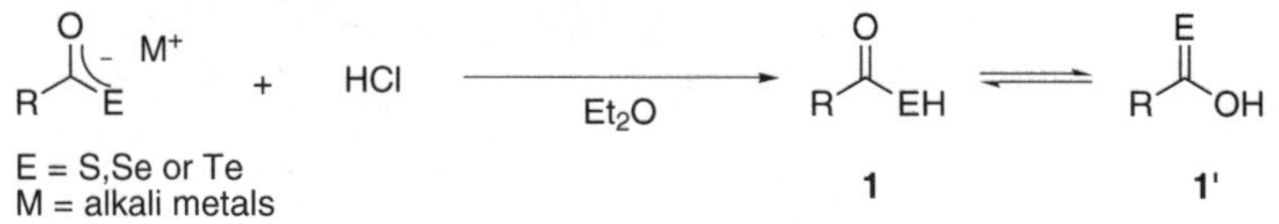

$$R\overset{O}{\underset{E}{\|}}{}^{-}\,M^{+} \;+\; HCl \xrightarrow{\;Et_2O\;} R\overset{O}{\underset{EH}{\|}} \;\rightleftharpoons\; R\overset{E}{\underset{OH}{\|}}$$

E = S, Se or Te
M = alkali metals
                               1   1'

Scheme 2

2.2
Dichalcogenocarboxylic Acids

Dithiocarboxylic acids can be readily obtained by HCl-acidolysis of the corresponding magnesium halide, prepared by the reaction of a Grignard reagent
with carbon disulfide [17] or ammonium salts [18]. Aromatic dithiocarboxylic
acids cannot, in general, be distilled. The formation of green to blue selen-
othiocarboxylic acid 2 or 2′ [$(CH_2=CHCH_2)_3CCSSeH$] in ether by treating the
corresponding tetraalkyl ammonium salts with hydrogen chloride has been
observed spectroscopically (Scheme 3) [19]. Nakayama et al. have found that
treating diselenocarboxylic acid inner salts with acids such as CH_3CO_2H,

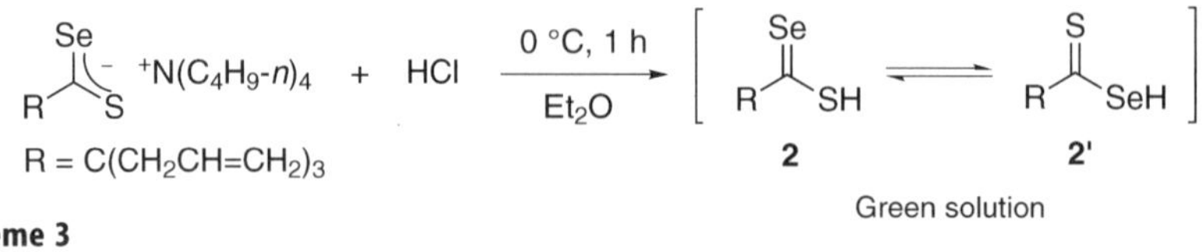

Scheme 3

R = C(CH₂CH=CH₂)₃

2 **2'**

Green solution

CF₃CO₂H, CF₃SO₃H, HBF₄, and H₂SO₄ (Scheme 4) results in the formation of diselenocarboxylic acid **3** [20]. The formation of 2-methylbenzenecarbodiselenoic acid **4** (green in ether) by acidolysis of the corresponding tetramethyl ammonium salt with CF₃SO₃H has also been reported [21]. The compound **4** reacts with methyl vinyl ketone to give γ-oxabutyl diselenoester **5** (Scheme 5) [21].

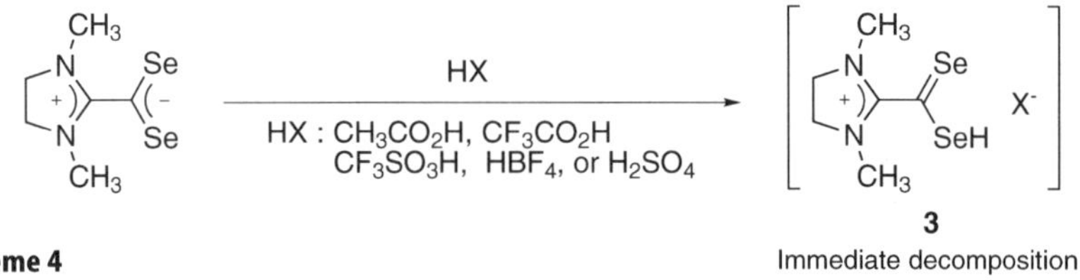

Scheme 4

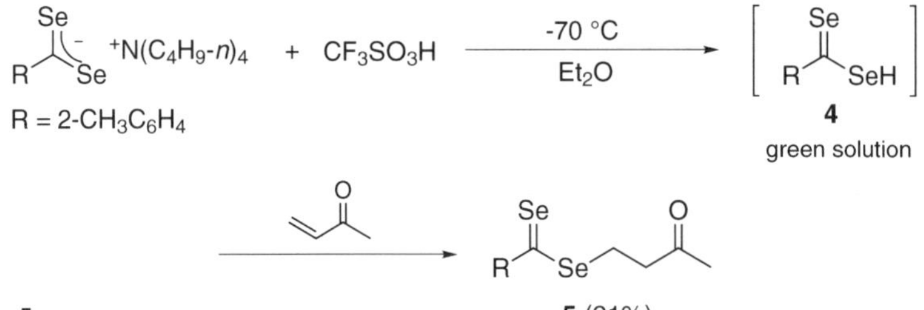

HX : CH₃CO₂H, CF₃CO₂H
CF₃SO₃H, HBF₄, or H₂SO₄

3

Immediate decomposition

R = 2-CH₃C₆H₄

4

green solution

Scheme 5   **5** (21%)

There is no reported example of a tellurium-containing dichalcogenocarboxlic acid (a tellurothio-, selenotelluro- or ditelluro-carboxylic acid) thus far.

3
Structures and Physical Properties

3.1
Spectroscopic Studies

The structures of thio- and dithiocarboxylic acids have been studied extensively using IR, UV/Vis and NMR spectra, and molecular orbital calculations for many decades now [3, 7–9]. Thiocarboxylic acids are considered to exist as fast tautomeric equilibrium mixtures of thiol (**I**) and thioxo (**II**) forms

(Scheme 6). Similarly, for seleno- and tellurocarboxylic acids, selenol and selenoxo, and tellurol and telluroxo forms may co-exist, respectively. Previous studies concerning the tautomerism of thiocarboxylic acids indicate that the thiol form (**I**) with the *s-cis* (*syn*) conformation is the predominant species.

Scheme 6   **I**   **II**

In 1996, spectroscopic observations were reported that indicated that the thioxo form of thiocarboxylic acid predominates in polar solvents at low temperature [16]. Since then, several theoretical studies of tautomerism in chalcogenocarboxylic acids have been reported. Spectroscopic experiments have shown that monochalcogenocarboxylic acids (RCOSH, RCOSeH, RCOTeH) exist in chalcogenol forms in nonpolar solvents and the solid state [14, 16]. The IR spectra (neat or Nujol) of selenocarboxylic acids show Se–H and C=O stretching frequencies at 2290–2324 and 1680–1720 cm^{-1}. In the ^{1}H and ^{13}C NMR spectra (in CDCl$_3$ solution), signals due to SeH and C=O are observed at δ=2.3–4.7 and δ=190–207, respectively. In contrast, in polar solvents such as tetrahydrofuran (THF) and acetone methanol, tautomeric equilibria between chalcogenol and chalcogenoxo forms have been observed, and the chalcogenoxo forms are the predominant species at low temperatures (below –90 °C) [14, 16]. In the IR spectra of 4-methoxybenzenecarboselenoic acid in THF at room temperature, the intensities of C=O stretching frequencies at 1682 cm^{-1} were observed to be less than those seen in the solid state. The resonance from ^{13}C=Se is observed at δ=222.2, and the peak from Se^{1}H occurs at δ=15.3, indicating the likely existence of hydrogen bonding with the oxygen atom of THF (Table 1). In

Table 1 NMR spectra for monochalcogenocarboxylic acids

E=S, Se, Te R=4–CH$_3$OC$_6$H$_4$		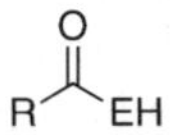 chalcogenol form		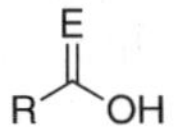 chalcogenoxo form
E=S	NMR (^{1}H)	SH	4.48^a	OH	14.52^c
	(^{13}C)	C=O	188.5^a	C=S	212.3^c
E=Se	NMR (^{1}H)	SeH	2.59^a	OH	15.3^c
	(^{13}C)	C=O	189.6^a	C=Se	222.2^c
	(^{77}Se)		427.5^a		753.9^c
E=Te	NMR (^{1}H)	TeH	–	OH	16.02^c
	(^{13}C)	C=O	–	C=Te	222.2^c
	(^{125}Te)		535^b		952^c

^{a} In CDCl$_3$ at rt; ^{b} in toluene-d_8 at –90 °C; ^{c} in THF-d_4 at –90 °C.

the UV/Vis spectra of 4-methoxybenzene-substituted chalcogenoxo acids, the absorption maxima attributed to the n-π^* transitions of the C=S, C=Se and C=Te groups were observed at 413, 502 and 652 nm, respectively.

It has been reported that dithiocarboxylic acids exist as monomers in dilute solution and as hydrogen-bonded dimers in concentrated solution. NMR spectra for neat dithioacetic acid **6** revealed that reversible covalent associations exist, resulting in dimer **7** and cyclic trimer **8** [22]. At 19.5 °C, the ratios of these spices are 61% monomer, 38% **7** and ~1% **8**.

The acidities of thio- and dithiocarboxylic acids have been discussed in an earlier review [3]. For example, the pK_a values of PhCOOH, PhCOSH and PhCSSH are 4.20, 2.48 and 1.92, respectively. The acidities of some chalcogenocarboxylic acids have been estimated through theoretical studies. The gas-phase acidities (ΔH) of formic acid and its sulfur congeners are 342.1 for HC(O)OH, 332.2 for HC(O)SH, 328.8 for HC(S)OH and 325.8 kcal/mol for HC(S)SH. Therefore, the acidity appears to increase roughly in proportion to the number of sulfur atoms present [23]. For selenocarboxylic acids, gas-phase acidities are 340.4 for HC(O)OH, 327.6 for HC(O)SeH, and 321.9 kcal/mol for HC(Se)OH, indicating that selenocarboxylic acids are more acidic than their parent carboxylic acids [24]. It was also predicted that selenocarboxylic acids may be more acidic than their corresponding thioic acids [14].

3.2
X-ray Structural Analyses

Little crystallographic information is available on chalcogenocarboxylic acids. Few (if any) thio- [25, 26] and dithiocarboxylic acids [25–28] are known, and no congeners containing selenium and tellurium have been found.

The substituted thiobenzoic acids **9** (R=2-HOC$_6$H$_4$) [25] and **10** (R=4-CH$_3$C$_6$H$_4$) [26] exist as cyclic dimers, with both molecules connected via intermolecular hydrogen bonds (S-H···O=C) in the crystal state, as in their parent benzoic acids [25, 29]. The bond distances of the thiocarboxylic acids indicate that the molecules exist in the thiol form in crystals. In the same way, in the solid state 4-methyldithiobenzoic acid **11** forms dimers through hydrogen bonds (S-H···S=C) [26]. Intramolecular C=S···HO and intermolecular S-H···O(H)-C hydrogen bonds are observed in 2-hydroxydithiobenzoic acid **12** [25].

3.3
Theoretical Studies

Semiempirical and ab initio theory molecular orbital calculations have been carried out on chalcogenocarboxylic acids, especially simple thiocarboxylic acids [8]. Some model compounds have been investigated recently using higher basis sets and theory levels [23, 30–34]. Further calculations for heavier congeners containing selenium and tellurium have also been carried out [24, 30, 32, 35].

Jemmis et al. reported theoretical calculations for tautomeric rearrangements in mono- and dichalcogen congeners of formic acid HC(E)E′H (E,E′=O, S, Se, Te) at the HF, MP2 and B3LYP levels (Scheme 7) [30]. The relative energies of the minima (**13a** and **13b**) and the transition states (**TS**) showed that the barrier to tautomerism reduced as the electronegativity of the chalcogens decreased. For monochalcogenocarboxylic acids (HCOEH), the relative energy values show a thermodynamic preference (~5–8 kcal/mol at the HF level) for the keto moiety (chalcogenol forms) more than the enol moiety (chalcogenoxo forms). The stabilization due to solvation (hydrogen bonding between HCOEH and a polar solvent, such as THF and acetonitrile) is more for the chalcogenoxo forms than for the chalcogenol forms. But the greater stabilities of the chalcogenoxo forms are not sufficient to reverse the thermodynamic stability. For thio-, seleno- and telluroacetic acids, the chalcogenol forms are still preferred, in both the gas phase and the solvent model (self-consistent reaction field, SCRF method) studies.

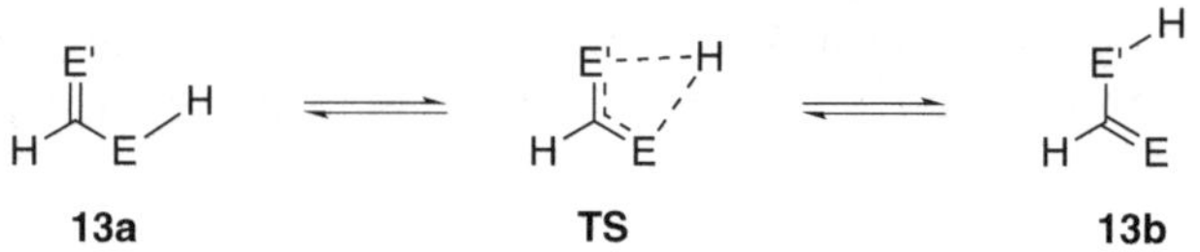

Scheme 7   13a   TS   13b

Hadad et al. found that XC(=O)SH (X=Me, NH$_2$, OH, F) is preferred over XC(=S)OH by 5–14 kcal/mol using ab initio calculations [23]. The C–O bond dissociation energy is greater than the C–S energy by ~30 kcal/mol and the C=O bond is significantly stronger than the C=S bond by about 40 kcal/mol. The C=O bond has more polar character (C$^{\delta+}$–O$^{\delta-}$) than the C=S bond, so that the bond order of the C=O bond is about 1.2 while the C=S bond order is ~2.0. For selenocarboxylic acids, density functional theory (DFT) calculation results suggest that the *syn* selenol form is more stable than the other forms [24].

On the other hand, a more detailed study into the tautomerism of thioformic acid with solvent interactions suggested that the thioxo form was favored [23]. For monomeric thioformic acid complexed with dimethyl ether, the thioxo isomer **14b** becomes ~5 kcal/mol more stable than the thiol isomer **14a** (Scheme 8). The reason for the stabilization of the thioxo isomer is the stronger hydrogen bond interactions (O-H···O type) in complex **14b** compared to those in **14a** (S-H···O type). These results are in agreement with experimental observations, as described infra [16].

Scheme 8   **14a**   **14b**

4
Reactions

Chalcogenocarboxylic acids, especially thio- and dithiocarboxylic acids, have been used as starting materials for synthesizing corresponding chalcogeno-carboxylic acid derivatives, and this topic is discussed in greater detail in Chapters 2 and 3. However, at this point we will review some of the most distinctive reactions reported over the last ten years.

4.1
Thiocarboxylic Acids

Additions of thiocarboxylic acids to unsaturated compounds to give thioesters have been extensively studied. Thiocarboxylic acids react smoothly with conjugated azoalkenes **15** to form the hydrazone 1,4-adducts **16** in high yields (Scheme 9) [36]. In the case of R^2=alkoxycarbonyl groups (-COOR′), the adducts exhibit tautomerism into the corresponding enamino forms **16′** until the ratio of the two tautomers almost reaches unity. On the other hand, when R^2 is a carbamoyl group (-CONHR′), only the hydrazono forms **16** are detected.

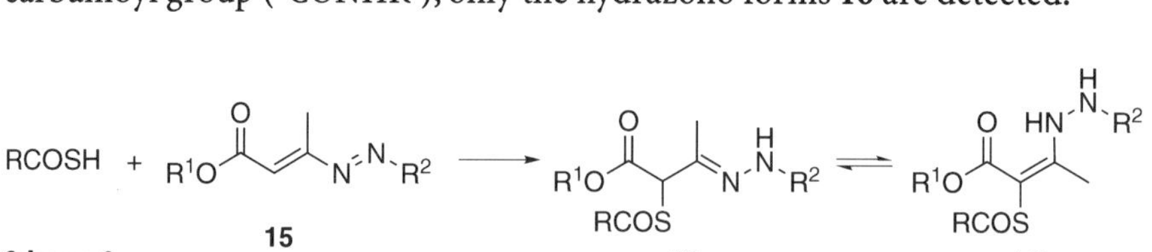

Scheme 9   **15**   **16**   **16'**

Reacting thiobenzoic acids with olefins in the presence of a clay catalyst (Montmorillonite K 10) gives the Markovnikov product **17** regioselectively (Scheme 10) [37]. In the absence of a clay catalyst, the addition proceeds in an anti-Markovnikov manner to afford the thioester **18**.

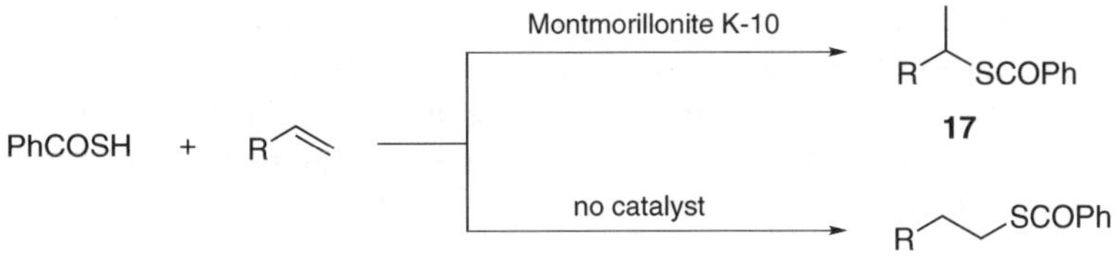

Scheme 10

A three-component reaction for β-aminothiocarboxylic acids **19** has been developed [38]. The thioacids **19** react with aldehydes then β-dimethylamino-2-isocyanoacrylates under mild conditions to afford substituted 1-thiazole-2-ylmethyl-azetidine-2-ones **20** (Scheme 11).

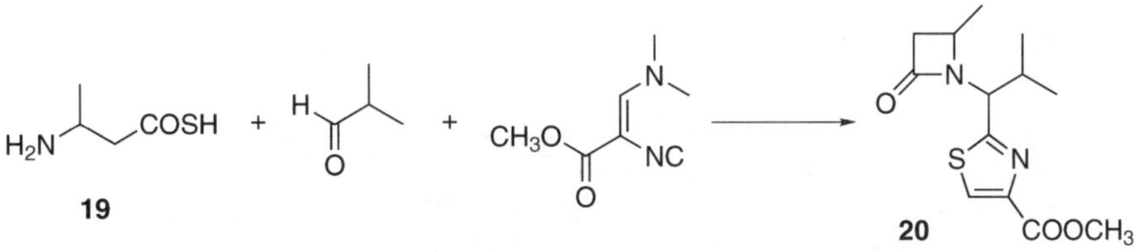

Scheme 11

Wang et al. investigated the chemoselectivities of $Rh_2(CH_3COO)_4$- or $BF_3 \cdot O(C_2H_5)_2$-mediated reactions of thioacetic acid with α-diazocarbonyl compounds [39]. The reaction of thioacetic acid with alkyl diazoacetates in the presence of $Rh_2(CH_3COO)_4$ affords S-ester **21** in high yields (Scheme 12). A formation mechanism where the Rh(II)-carbene intermediate inserts into the S–H bond has been proposed. In contrast, the reaction catalyzed by $Rh_2(CH_3COO)_4$ forms O-ester **22**. In this case, the $BF_3 \cdot O(C_2H_5)_2$ unit complexes with the carbonyl group of the diazo compound, followed by nucleophilic attack of the carbonyl oxygen of thioacetic acid on the complex.

Thiocarboxylic acids react with phosphazenes (R^1–N=PPh$_3$, Staudinger intermediate) generated from alkyl azides and electron-deficient triarylphos-

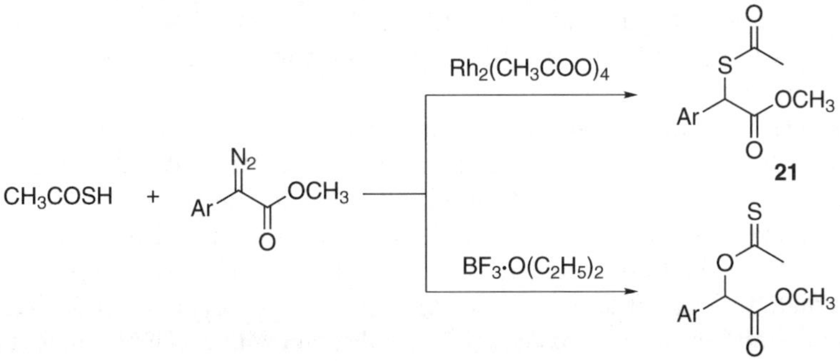

Scheme 12

phines to yield amides (Scheme 13) [40]. Another conventional method for synthesizing amides using thiocarboxylic acids and azides has been reported [41]. Reactions with various azides in the presence of 2,6-lutidine in methanol, chloroform or even water as a solvent give the corresponding amides (Scheme 13). Further, the use of $RuCl_3$ promotes reactions with less reactive azides in milder conditions [42].

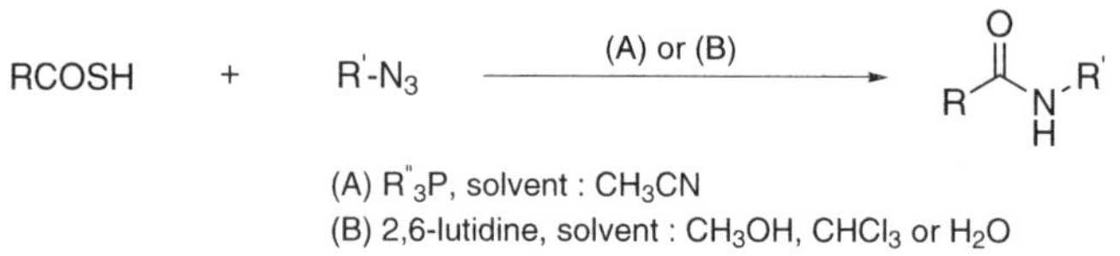

$$RCOSH \quad + \quad R'-N_3 \quad \xrightarrow{\text{(A) or (B)}} \quad R\overset{O}{\underset{}{\text{C}}}\text{-N(H)-}R'$$

(A) R''_3P, solvent : CH_3CN

Scheme 13   (B) 2,6-lutidine, solvent : CH_3OH, $CHCl_3$ or H_2O

4.2
Dithiocarboxylic Acids

Dithiobenzoic acid reacts with equimolar amounts of 2,2'-dithiobis(benzothiazole) **23** and triphenylphosphine to give 2-benzothiazolyl dithiobenzoate **24** in good yields (Scheme 14) [43]. The product **24** is an effective thiobenzoylation reagent in reactions with amines or alcohols.

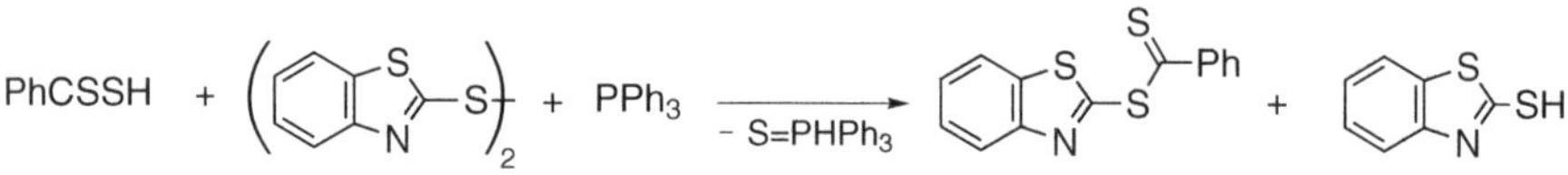

$$PhCSSH \quad + \quad (\text{benzothiazolyl-S})_2 \quad + \quad PPh_3 \quad \xrightarrow{-\ S=PHPh_3} \quad \text{24} \quad + \quad \text{benzothiazolyl-SH}$$

Scheme 14   **23**   **24**

4.3
Seleno- and Tellurocarboxylic Acids

In 1972, Jensen et al. reported that selenobenzoic acid reacts with *m*-nitrobenzaldehyde to afford *Se*-[hydroxyl(*m*-nitrophenyl)-methyl] selenobenzoate. The treatment of the acid in $NaHCO_3$ with *p*-nitrobenzyl bromide gave *Se*-(*p*-nitrobenzyl) selenobenzoate [13]. In 1987, Hirabayashi et al. reported that 4-biphenylselenocarboxylic acid reacts with methanol, phenyl isocyanate and pyridine to give ArC(O)OMe, ArC(O)SeC(O)NHPh and ArC(O)NHPh, respectively (Ar=4-biphenyl) [15].

In 1994, successful isolations of various alkyl and aryl selenocarboxylic acids [14] enabled them to undergo exact reactions. Reactions of selenocarboxylic acids with dicyclohexylcarbodiimide (DCC) yield the corresponding diacyl selenides **25** and selenourea **26** quantitatively (Scheme 15) [14], as in reactions of thio- [44] and dithiocarboxylic acids [45, 46]. In air, selenocarboxylic acids are immediately oxidized to afford the corresponding diacyl diselenides [13, 14]. Also, thio- and dithio-carboxylic acids readily react with aryl isocyanates to give acyl carbamoyl [47, 48] and thioacyl carbamoyl sulfides [49, 50], respectively.

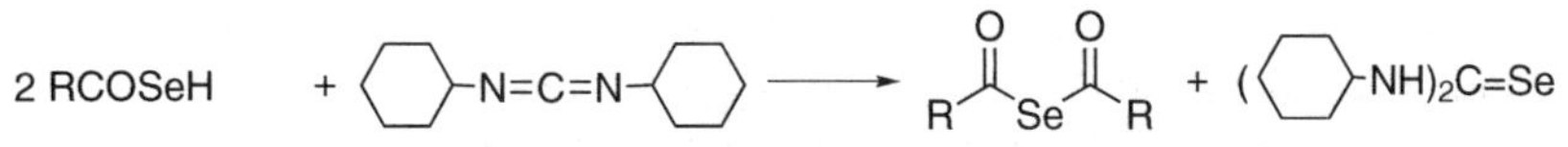

Scheme 15

$$2\ RCOSeH\ +\ \langle\text{aryl}\rangle\text{-}N{=}C{=}N\text{-}\langle\text{aryl}\rangle\ \longrightarrow\ R\underset{Se}{\overset{O}{\diagup}}\underset{}{\overset{O}{\diagdown}}R\ +\ (\langle\text{aryl}\rangle\text{-}NH)_2C{=}Se$$

25   **26**

Seleno- and tellurocarboxylic acids also add to aryl isocyanates to give the corresponding acyl carbamoyl selenides and tellurides **27** which are stabilized by intramolecular hydrogen bonding between carbonyl oxygens and NH hydrogen atoms (Scheme 16) [16, 51]. Selenocarboxylic acids react with alkenes or alkynes to give the corresponding adducts in high yields [52]. Reactions with styrenes mainly afford anti-Markovnikov adducts, whereas additions onto vinyl ethers proceed in a Markovnikov manner.

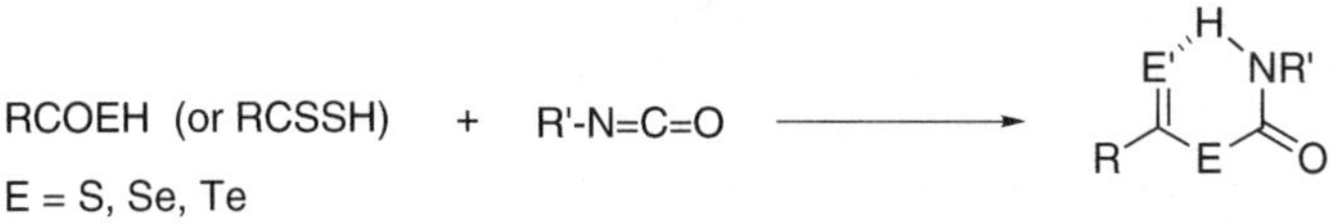

$$RCOEH\ (or\ RCSSH)\ +\ R'\text{-}N{=}C{=}O\ \longrightarrow$$

E = S, Se, Te

Scheme 16

27

E' = O, S

References

1. Kekulé A (1854) Ann Chem Pharm 90:309
2. Scheithauer S, Mayer R (1979) In: Senning A (ed) Topics in sulfur chemistry, vol 4. Georg Thieme, Stuttgart
3. Voss J (1979) In: Patai S (ed) Supplement B: The chemistry of carboxyric acid derivatives. Wiley, New York, p 1021
4. Reid DH (1970–1981) Organic sulphur, selenium and tellurium, vols 1–6. Royal Society of Chemistry, London
5. Mayer R, Scheithauer S (1985) In: Falbe J (ed) Methoden der organischen Chemie. Georg Thieme, Stuttugart, Band 5, Teil 2, p 891
6. Mayer R, Scheithauer S (1985) In: Falbe J (ed) Methoden der organischen Chemie. Georg Thieme, Stuttgart, Band 5, Teil 1, p 785
7. Ogawa A, Sonoda N (1995) In: Katritzky AR, Meth-Cohn O, Rees CW (eds) Comprehensive organic functional group transformation, vol 5. Pergamon, Oxford, p 231
8. Ishii A, Nakayama J (1995) In: Katritzky AR, Meth-Cohn O, Rees CW (eds) Comprehensive organic functional group transformation, vol 5. Pergamon, Oxford, p 505
9. Murai T, Kato S (1995) In: Katritzky AR, Meth-Cohn O, Rees CW (eds) Comprehensive organic functional group transformation, vol 5. Pergamon, Oxford, p 545
10. Nomura R, Miyazaki S, Nakano T, Matsuda H (1990) Chem Ber 123:2081
11. Engelhardt A, Latschinoff P, Malyschew R (1868) Z Chem 358:455
12. Noble P Jr, Tarbell DS (1963) Org Synth Coll Vol IV:924
13. Jensen KA, Boje L, Henriksen L (1972) Acta Chem Scand 26:1465
14. Kageyama H, Murai T, Kanda T, Kato S (1994) J Am Chem Soc 116:2195
15. Hirabayashi Y, Ishihara H, Echigo M (1987) Nippon Kagaku Kaishi 1430
16. Kato S, Kawahara Y, Kageyama H, Yamada R, Niyomura O, Murai T, Kanda T (1996) J Am Chem Soc 118:1262
17. Houben J (1906) Ber Deut Chem Ges 39:3219

18. Kato S, Mitani T, Mizuta M (1973) Int J Sulfur Chem 8:359
19. Murai T, Kamoto T, Kato S (2000) J Am Chem Soc 122:9850
20. Nakayama J, Kitahara T, Sugihara Y, Sakamoto A, Ishii A (2000) J Am Chem Soc 122:9120
21. Tani K, Murai T, Kato S (2002) J Am Chem Soc 124:5960
22. Rajab AM, Noe EA (1990) J Am Chem Soc 112:4697
23. Hadad CM, Rablen PR, Wiberg KB (1998) J Org Chem 63:8668
24. Remko M, Rode BM (1999) J Phy Chem A 103:431
25. Mikenda W, Steinwender E, Mereiter K (1995) Monatsh Chem 126:495
26. Kato S, Tani K (unpublished results)
27. Miyamae H, Oikawa T (1985) Acta Cryst C41:1489
28. Behr A, Keim W, Kipshagen W, Vogt D (1988) J Organomet Chem 344:C15
29. Wilson CC, Shankland N, Florence AJ (1996) J Chem Soc Farad T 92:5051
30. Jemmis ED, Giju KT, Leszczynski J (1997) J Phys Chem A 101:7389
31. Delaere D, Raspoet G, Nguyen MT (1999) J Phys Chem A 103:171
32. Clabo DA Jr, Dickson HD, Nelson TL (2000) J Mol Model 6:341
33. Romano RM, Della Védova CO, Downs AJ (2002) J Phy Chem A 106:7235
34. Fausto R, Teixeira-Dias JJC, Carey PR (1991) J Am Chem Soc 113:2471
35. González AI, Mó O, Yáñez M (1999) J Phy Chem A 103:1662
36. Attanasi OA, Buratti S, Filippone P, Fiorucci C, Foresti E, Giovagnoli D (1996) Tetrahedron 52:1579
37. Kanagasabapathy S, Sudalai A, Benicewicz BC (2001) Tetrahedron Lett 42:3791
38. Kolb J, Beck B, Dömling A (2002) Tetrahedron Lett 43:6897
39. Yao W, Liao M, Zhang X, Xu H, Wang J (2003) Eur J Org Chem 1784
40. Park SD, Oh JH, Lim D (2002) Tetrahedron Lett 43:6309
41. Shangguan N, Katukojvala S, Greenberg R, Williams LJ (2003) J Am Chem Soc 125:7754
42. Fazio F, Wong CH (2003) Tetrahedron Lett 44:9083
43. Yeo SK, Choi BG, Kim JD, Lee JH (2002) B Korean Chem Soc 23:1029
44. Mikolajczyk M, Kielbasinski P, Schiebel HM (1976) J Chem Soc Perkin T 1 564
45. Kato S, Katada T, Mizuta M (1976) Angew Chem Int Edit Jpn 15:766
46. Kato S, Shibahasi T, Katada T, Takgai T, Noda I, Mizuta M, Goto M (1982) Liebigs Ann Chem 1229
47. Motoki S, Saito T, Kagami H (1974) B Chem Soc Jpn 47:775
48. Katada T, Nishida M, Kato S, Mizuta M (1977) J Organomet Chem 129:189
49. Kato S, Mitani T, Mizuta M (1972) B Chem Soc Jpn 45:3653
50. Kato S, Wakamatsu M, Mizuta M (1974) J Organomet Chem 78:405
51. Kageyama H, Tani K, Kato S, Kanda T (2001) Heteroatom Chem 12:250
52. Kato S (unpublished results)

Top Curr Chem (2005) 251: 13–85
DOI 10.1007/b101006
© Springer-Verlag Berlin Heidelberg 2005

Group 1–17 Element (Except Carbon) Derivatives of Thio-, Seleno- and Telluro-Carboxylic Acids

Shinzi Kato[1] (✉) · Osamu Niyomura[2]

[1] Department of Applied Chemistry, School of Engineering, Chubu University
1-1 Matsumoto-Cho, Kasugai, Aichi 487-8501, Japan
kshinzi@nifty.com

[2] Division of Chemistry, Center for Natural Sciences, College of Liberal Arts and Sciences,
Kitasato University, Kitasato, Sagamihara, Kanagawa 228-8555, Japan
niyomura@clas.kitasato-u.ac.jp

Abstract Over the last two decades, our understanding of the chemistry of Group 1–17 element derivatives of chalcogenocarboxylic acids has improved considerably. In recent years, the structures of these compounds have been revealed using X-ray diffraction. The present review surveys the syntheses, structures and reactions of Group 1–17 element derivatives of thio-, seleno- and telluro-carboxylic acids, except for when carbon is the Group 1–17 element attached directly to the chalcogen atom.

Keywords Monochalcogenocarboxylates · Metal chalcogenocarboxylates · Metal thiocarboxylates · Metal selenocarboxylates · Metal tellurocarboxylates

1
Introduction

A monochalcogenocarboxylic acid (RCOEH, E=S, Se, Te) is a compound in which one of the two oxygen atoms of the carboxylic acid is replaced by sulfur, selenium or tellurium. There are six types of monochalcogenocarboxylic acid, as shown in Scheme 1. Since the hydrogen atom of the acid's hydroxy or hydro-chalcogenoxy (EH) group can be formally replaced by any elements in the periodic table, and substituting the hydrogen for any element from Groups 2 to 16 can enable connections between two or more chalcogenocarboxyl groups, there are a vast number of potential monochalcogenocarboxylic acid derivatives (at least >3000), even if we limit ourselves to R=CH$_3$. The first monochalco-genocarboxylic acid derivatives were reported in the middle of the nineteenth century by Kekulé, who synthesized dibenzoyl disulfide [178, 179]. However, prior to the beginning of the 1970s, the chemistry of monochalcogenocarboxylic acid derivatives (except for thio- and dithio-esters) was unknown, probably due to the difficulties involved in synthesizing their starting compounds (such as monochalcogenocarboxylic acids and their alkali metal salts). Despite the growing interest in the field, investigations into monochalcogenocarboxylic acid derivatives are (somewhat surprisingly) yet to be reviewed, although some discussion on them can be found sporadically in review articles concerning thio- [13, 119, 251, 263, 325], dithio- [119, 158, 217, 274] and seleno-esters [119, 153, 251], and metal dithiocarbamates [43, 44, 88, 234], and xanthates [329], as well as in advanced books concerning organosulfur [119, 217, 251, 264], selenium [111, 119, 121, 251, 263] and tellurium compounds [111, 119, 151, 253]. In this

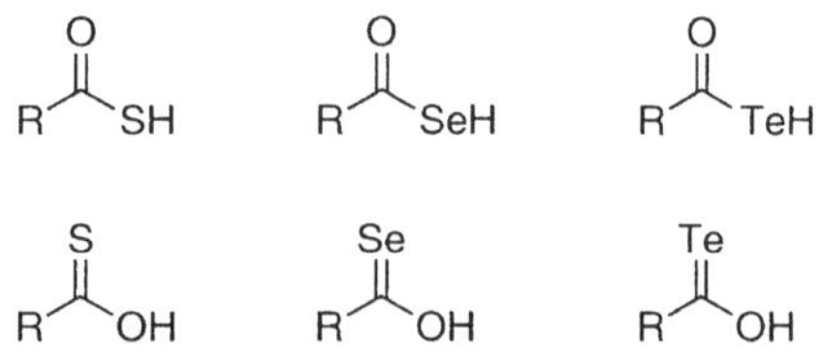

Scheme 1 Monothiocarboxylic acids

chapter, we review the synthetic methods and structures of Group 1–17 element derivatives [RC(O)EZ or RC(E)OZ (E=S, Se, Te; Z=Group 1–17 elements)] of monochalcogenocarboxylic acids. We will survey them by progressing from Group 1 element derivatives, through the groups of the periodic table, to Group 17 derivatives, ignoring chalcogenoesters RCOER′ and RCEOR′ (E=S, Se, Te) except when we need to compare their physical properties to other derivatives of monochalcogenocarboxylic acids. We believe that our literature coverage is comprehensive up to the end of March 2004 (we regret any pertinent references that may have been overlooked). Finally, it should be emphasized that we present a survey here rather than a critical evaluation of the literature..

2
Synthesis

2.1
Alkali and Alkali Earth Metal Salts

2.1.1
Group 1 Element Compounds (Li, Na, K, Rb, Cs)

2.1.1.1
Lithium

Purifying the lithium thiocarboxylates **1** is a difficult task, due to their strong hygroscopicities and high solubilities in solvents such as benzene and ether. Anhydrous salts of aromatic thiocarboxylic acid were synthesized by the reaction of thiocarboxylic acid with lithium hydride in a mixed solvent of hexane/ether (5:1) [150]. Banister and his co-workers [6] isolated lithium thiobenzoate **2** using butyllithium instead of LiH in the presence of tetramethylethylenediamine (TMEDA) and derived its X-ray molecular structure. Lithium seleno- **3** (E=Se) [187] and telluro-carboxylates **3** (E=Te) have been obtained as the LiCl-containing salts from the reaction of acyl chlorides with the corresponding lithium chalcogenides [164]. The reaction of phenyllithium with COSe does not give lithium selenobenzoate [72].

$$M(RCOS) \qquad [Li(PhCOS)(TMEDA)]_2 \qquad Li(ArCOE)(LiCl)$$

$$\textbf{1} \qquad\qquad \textbf{2} \qquad\qquad \textbf{3}$$

$$M = Li, Na, K \qquad\qquad\qquad E = Se, Te$$
$$Rb, Cs \qquad TMEDA: [(CH_3)_2N)]_2(CH_2)_2$$

$$M(RCOSe) \qquad\qquad M(RCOTe)$$

$$\textbf{4} \qquad\qquad\qquad \textbf{5}$$

$$M = Li, Na, K \qquad\qquad M = Li, Na, K$$
$$Rb, Cs \qquad\qquad\qquad Rb, Cs$$

2.1.1.2
Sodium

Sodium thiocarboxylates have been synthesized using the following three methods: (1) $(PhCOS)_2+C_2H_5ONa$ [70]; (2) $RCOCl+NaHS_x \cdot (H_2O)$ [103]; (3) $RCOSH+NaOH$ [68] or sodium carbonate [271]. The sodium salts are hygroscopic and removal of water is very difficult. Their anhydrous salts are obtained by treatment of thiocarboxylic acid with sodium hydride [150]. Sodium selenocarboxylates **4** (M=Na) have been prepared by the two reactions: (1) $(RCO)_2Se+C_2H_5ONa$ [161]; (2) $RCOCl+Na_2Se$ [161, 162]. Sodium tellurocarboxylates **5** (M=Na) can be isolated by the reaction of acyl chlorides with sodium tellurides [162, 167, 174].

2.1.1.3
Potassium

The first potassium thiocarboxylate **1** (M=K) to be synthesized was potassium thiobenzoate, which Engelhardt et al. [62] obtained by reacting thiobenzoic acid with potassium hydroxide. Since then, four routes to the salts have been developed: (1) $KS_xH+RCOCl$ [62, 195, 248]; (2) $KS_xH+(PhCOS)_2$ [62]; (3) $K_2CO_3+RCOSH$ [29]; (4) alcoholic $KS_xH+PhCO)_2O$ [62, 129, 213]; 5) PhCOCl + sulfur in the presence of KOH [69]. Among these routes, the method (1) is the most practical and useful [248]. Potassium trithiooxalate K_2(SOC-CSS) is synthesized using phenyl trichloroacetate and K_2S/KSH in ethanol [224]. In 1976 Hirabayashi and his coworkers reported the synthesis of potassium selenocarboxylates K(ArCOSe)/Se from the reaction of aromatic diacyl selenides with KOH [113]. Aliphatic compounds **4** (M=K, R=allyl) can be prepared using CH_3OK instead of potassium hydroxide [131]. The reactions of acyl chlorides with potassium selenide [246] and of *O*-trimethylsilyl selenocarboxylates with KF [172] yield elemental selenium-free potassium salts in good yields. Treatment of *Se,Se′*-diphenyl selenooxalate with K_2Se leads to the formation of potassium selenoxalate K_2(SeOC-COSe)·(KCl) [216]. Potassium tellurocarboxylate **5** (E=Te) can be synthesized from reaction of diacyl telluride with C_2H_5OK [133] and the reactions of acyl chlorides with potassium telluride [168].

2.1.1.4
Rubidium

A series of rubidium thio- **1** (M=Rb) [150], seleno- **4** (M=Rb) [171, 326] and telluro-carboxylates **5** (M=Rb) [174] have been isolated in moderate to good yields from the reactions of the corresponding chalcogenocarboxylic acids or *O*-trimethylsilyl esters with rubidium acetate or fluorides.

2.1.1.5
Cesium

Similarly to the rubidium salts, cesium thio- **1** [150], seleno- **4** [171, 326] and telluro-carboxylates **5** (E=Cs) [173, 174] are obtainable from the corresponding chalcogenocarboxylic acids or *O*-trimethylsilyl thioesters.

2.1.1.6
Structures

Selected bond distances of alkali metal chalcogenocarboxylates are collected in Table 1. The molecular structure of lithium benzoate **2** is a centrosymmetric dimer with an eight-membered ring composed of coplanar C, O, and S atoms and tetrahedrally-coordinated Li atoms above and below this plane, where the C–O [1.246(2) Å] and C–S distances [1.704(2) Å] show considerable multiple bond character [6]. From an ab initio optimization calculation on a dimeric lithium formate (HCOSLi)$_2$, the PhCOS unit in **1–2** has been predicted to be predominantly ionic [6].

The two anion moieties in both potassium 1,2-dithio-oxalate **6**, K$_2$[SOC-COS] [213], and potassium trithio-oxalate **7**, K$_2$[S$_2$C-C(O)S]·(KCl), possess mutually perpendicular thiocarboxyl groups, respectively [224]. In contrast, those in the 1,2-diselenooxalate **8**, K$_2$[Se(O)C-C(O)Se] [216], are in a planar *trans*-confor-

K$_2$[SOC–COS]   K$_2$[S$_2$C–C(O)S]•(KCl)   K$_2$[Se(O)C–C(O)Se]

6   **7**   **8**

mation. Relatively long C–O [1.227–1.239(4) Å] and short C–S [1.712(3) and 1.697(3) Å] or C–Se distances [1.87(1) Å] suggest highly ionic character. Potassium thioacetate K(CH$_3$COS) [30] and aromatic sulfur {K(C$_6$H$_5$COS) [245], K(2-CH$_3$OC$_6$H$_4$COS) [245], K(4-CH$_3$OC$_6$H$_4$COS) [245], K(2-CF$_3$OC$_6$H$_4$COS) [245]} and selenium isologues K(2-CH$_3$OC$_6$H$_4$COSe) [245] also have been characterized by X-ray structural analysis. The coordination number of the potassium in these salts is seven: three or four oxygens and three or four sulfurs (or selenium) atoms are attached to the K [30, 245]. Fundamentally, the structural units of these potassium salts [K(RCOE), R=aryl; E=S, Se, Te] are dimers in a head-to-head mode (**9** and **10**), where the oxygen and/or sulfur (selenium) atoms associate with the metal of the opposite molecule, and in addition the

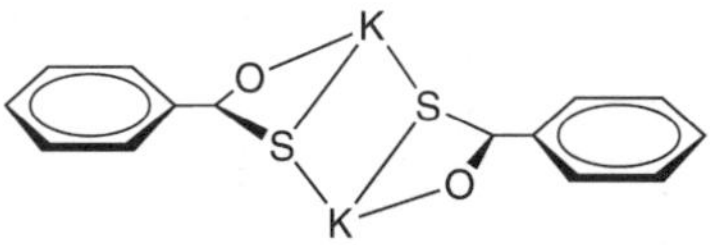

9 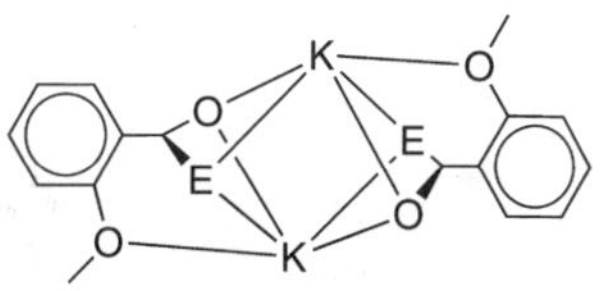   **10** (E = S, Se)

Table 1 Selected bond distances of alkali and alkaline earth metal monochalcogenocarboxylates

M(RCOE)	M	E	Distance [Å]				References
			$M{\cdots}O^a$	M–E	C=O	C–E	
$(C_6H_5COSLi)_2(TMEDA)$	Li	S	1.88(4)	2.478(4)	1.46(2)	1.704(2)	[6]
$K_2(SOC\text{-}COS)$	K	S	2.832 2.730	3.101 3.208	1.227 1.239	1.712 1.697	[214]
$K_2(SOC\text{-}CSS)(KCl)$	K	S	2.77(2) 2.94(2) 2.76(2)		1.26(3)	1.70(2) 1.71(2) 1.62(3)	[225]
$K(CH_3COS)$	K	S	2.860(8) 2.687(11)	3.316(5) 3.425(5)	1.231(19)	1.703(16)	[30]
$K(C_6H_5COS)$	K	S	2.699(2) 2.658(2)	3.483(1) 3.361(1)	1.242(3)	1.715(3)	[246]
$K(4\text{-}CH_3OC_6H_4COS)$	K	S	2.738(4) 2.724(4) 2.796(4)	3.267(1) 3.238(2) 3.251(2)	1.245(5)	1.708(2)	[246]
$K(2\text{-}CH_3OC_6H_4COS$	K	S	2.831(2) 3.091(2) 2.715(2) 2.819(2)	3.254(7) 3.492(2)	1.235(3)	1.708(2)	[246]
$K_2(COSe)_2$	K	Se			1.20(2)	1.87(1)	[217]
$K(2\text{-}CH_3OC_6H_4COSe)$	K	Se	2.810(4) 2.892(4) 2.750(4) 3.050(4)	3.309(1) 3.597(1) 3.625(2)	1.225(6)	1.861(5)	[246]
$Rb(CH_3COS)$	Rb	S	2.80(2) 2.98(1)	3.494(8) 3.442(8)	1.32(4)	1.69(4)	[32]

Table 1 (continued)

M(RCOE)	M	E	Distance [Å]				References
			$M\cdots O^a$	M–E	C=O	C–E	
$Rb(2\text{-}CH_3OC_6H_4COS)$	Rb	S	2.955(3) 3.070(4) 2.868(4) 3.039(4)	3.366(2) 3.410(2) 3.572(2)	1.228(6)	1.718(5)	[246]
$Rb(2\text{-}CH_3OC_6H_4COTe)$	Rb	Te	2.86(1) 3.02(1)	3.791(2) 3.833(7)	1.25(2)	2.12(2)	[246]
$Cs(2\text{-}CH_3OC_6H_4COS)$	Cs	S	3.142(2) 3.136(3) 2.997(3)	3.628(1) 3.799(1) 3.882(1)	1.231(4)	1.712(3)	[175]
$Cs(2\text{-}CH_3OC_6H_4COSe)$	Cs	Se	3.239(3) 3.124(3) 3.019(3)	3.7468(7) 3.914(7) 3.988(7)	1.226(5)	1.868(5)	[175]
$Cs(2\text{-}CH_3OC_6H_4COTe)$	Cs	Te	3.103(4) 3.181(4) 3.207(4)	3.902(1) 4.2159(8)	1.235(6) 1.23(2)	2.109(6) 1.73(2)	[175]
$Ca(CH_3COS)(CH_3COS)(H_2O)$	Ca	S	2.32(1) –2.55(1)				[33]
$Ca(CH_3COS)_2(15\text{-crown-5})$	Ca	S	2.490(6) –2.282(3)	3.224(4)			[195]
$Sr(CH_3COS)(CH_3COS)(H_2O)_4$	Sr	S	2.69(1)	3.224(4)	1.264(12)	1.696(16)	[31]
$Ba(CH_3COS)_2(H_2O)_3$	Ba	S		2.359(1) 2.382(1)	1.237(6) 1.239(6)	1.709(5) 1.722(5)	[35]

^{a} Intramolecular $M\cdots C{=}C$.

potassium atoms exist above and below the plane involving the chalcogeno-carboxyl groups [245]. The long C–O (1.22–1.25 Å) and short C–E (E=S, Se, Te) bond distances [1.70–1.72 Å for the C–S, 1.861(5) Å for the C–Se, 2.12 Å for the C–Te] of the COE groups suggest that the negative charge localized on the chalcogenocarboxyl group may be somewhat more localized on the less electronegative chalcogen atom, regardless of the alkali metal [245]. In the *para*-methoxy derivative 1 (M=K; R=4-CH$_3$OC$_6$H$_4$), one thiocarboxyl group chelates to potassium [245].

The structures of Rb(2-CH$_3$OC$_6$H$_4$COS) 1 (M=Rb) and Rb(2-CH$_3$OC$_6$H$_4$COTe) 5 (M=Rb) resemble those of the corresponding potassium salt 10 [245]. They exist in a dimeric structure in which the methoxy oxygen atoms are associated with the metal of the opposite molecule. The coordination number of the Rb metal is eight. The structure of Rb(CH$_3$COS) has been found [32]. X-ray molecular structure analyses of cesium 2-methoxybenzenecarbothioate [245], -selenoate [174] and -telluroate [174, 245] have also been carried out. Their structural units are essentially dimers, as observed for the potassium salts. It is worth noting that the Cs ion in Cs(2-CH$_3$OC$_6$H$_4$COTe) bonds with the benzene ring carbons of a neighboring molecule in the η^5-mode, including the *ipso*-carbon of the 2-CH$_3$OC$_6$H$_4$ group [245].

2.1.1.7
Reactions

Alkali metal chalcogenocarboxylates readily react with a variety of electrophiles, such as alkyl halides, acid chlorides and typical organo element halides to give corresponding chalcogen-substituted chalcogenoesters [113, 114, 130–133, 136, 150, 167, 171, 173, 174, 245, 187, 246]. Aromatic selenocarboxylic acid potassium salts react with nitriles to give primary selenoamides in moderate to good yields, whereas the reaction with acrylonitrile in similar conditions leads to Michael adducts as the major product [118].

2.1.2
Group 2 Element Compounds (Be, Mg, Ca, Sr, Ba)

2.1.2.1
Beryllium

No synthesis of beryllium chalcogenocarboxylates has been reported.

2.1.2.2
Magnesium

It is well-known that Grignard reagents such as RMgBr react with COS to form thiocarboxylic acid magnesium salts RCOSMgBr 11 [328]. Very little is known about the coordination properties of magnesium thiocarboxylates. Reactions

RCOSMgX (X=Br, I)   Ca(CH$_3$COS)(CH$_3$COO)·3H$_2$O
11   **12**

Sr(CH$_3$COO)(CH$_3$COS)·4H$_2$O   Ba(CH$_3$COO)(CH$_3$COS)·3H$_2$O
13   **14**

of Grignard reagents with COSe do not lead to the formation of the expected magnesium salts RCOSeMgBr [72].

2.1.2.3
Calcium

A synthesis of calcium thioacetate was described in Ulrich's report in 1859 [314], but the synthetic mechanism and its structure remained unclear. Bernard and Borel [17] found that calcium carbonate reacts with thioacetic acid in a mixed solvent of CH$_3$OH/H$_2$O to give the calcium complex **12** with ligands of both acetato and thioacetato in good yield, where the coordination number of the central Ca metal is seven [16]. The reaction of thioacetic acid with the corresponding metal hydride or carbonate in the presence of 12-crown-4 ether gives polyether complexes in which the thioacetato ligand binds to the central Ca metal as a monodentate through the oxygen atom [194].

2.1.2.4
Strontium

Through a similar ligand-exchange reaction between Sr(CH$_3$COO)$_2$ and thioacetic acid, the acetato/thioacetato Sr complex **13** is obtained [16]. In the presence of 15-crown-5 ether, the reaction of SrCO$_3$ with thioacetic acid yields a polyether Sr complex with both acetato and thioacetato ligands [194].

2.1.2.5
Barium

Similar to the strontium complex, the ligand-exchange reaction of Ba(CH$_3$COO)$_2$ with CH$_3$COSH and the reaction of BaCO$_3$ with thioactic acid in the presence of 15-crown-5 produce the acetato/thioacetato complex **14** Ba(CH$_3$COO)-(CH$_3$COS)(H$_2$O) [16] and a polyether complex of barium thioacetate [194], respectively.

These alkali earth metal complexes **12–14** are thermally stable and decompose above 300 °C to the corresponding metal acetate and carbonate MCO$_3$ and the sulfides MS$_x$ (M=Ca, Sr, Ba) [18].

2.1.2.6
Structure

Borel and Ledesert performed X-ray molecular structural analysis on **12–14** and revealed that the coordination number of each Ca, Sr and Ba metal is seven, nine and nine, respectively [28, 31, 33, 35] (see the selected bond distances in Table 1). On the other hand, the coordination of the polyether complex of calcium is eight, from binding to all of the oxygen atoms of the 15-crown-5 ring, the oxygen and sulfur atoms of the bidentate thioacetato ligand, and the oxygen of the other monodentate thioacetate ligand [194].

2.2
Transition Metal Complexes

2.2.1
Group 3 Element Compounds (Sc, Y, Lantanoids and Actinoids)

2.2.1.1
Scandium

Chalcogenocarboxylato scandium complexes have not been prepared.

2.2.1.2
Yttrium

The synthesis of chalcogenocarboxylato yttrium complexes has not been reported.

2.2.1.3
Lantanoids and Actinoids

Stoichiometric reaction of $SmCl_3$ with thiocarboxlic acid potassium salts in tetrahydrofuran (THF) gives the tris(thiocarboxylato) samarium complex $Sm(4-CH_3C_6H_4COS)_3(THF)_2$ **15**, in which the three thiocarboxylato ligands coordinate in a bidentate fashion (for bond distances, see Table 2) [140]. Only two chalcogenocarboxylato actinoid complexes complexes $Cp_2U(RCOS)_2$ (R=CH_3, Ph) **16** [6] and $UO_2(PhCOS)_2$ **17** [297] have been synthesized, by treating $Cp_2U(CH_3)_2N)_2$ with thiocarboxylic acid, and UO_2Cl_2 with sodium thiobenzoate, respectively. The complex **16** (R=Ph) has a pseudo-*cis*-octahedral geometry, where the thiocar-

$$Sm(4-CH_3C_6H_4COS)_3(THF)_2 \qquad Cp_2U(RCOS)_2$$

15   **16**

$$UO_2(RCOS)_2$$

17

Table 2 Selected bond distances of monochalcogenocarboxylato transition metal complexes

M(RCOE)x	M	E	Distance [Å]				References
			$M\cdots O^a$	M–E	C=O	C–E	
$Sm[4-CH_3C_6H_4C(O)S]_3\cdot(thf)_2$	Sm	S	2.408(3) 2.444(3)	2.877(2)	1.254(6)	1.705(5)	[140]
$Ta[C_6H_5C(O)S]_2(Cp)_2$	Ta	S	2.408(3) 2.444(3)	2.516(3)	1.203(15)	1.798(13)	[110]
$W(CH_3COS)(CO)_5$	W	S		2.548(5)	1.222(7)	1.654(20)	[52]
$Rh_2(CH_3COS)_4$	Rh	S		2.539(2)	1.364(19)	1.770(6)	[9]
$Fe(Cp)(CO)_2(2-NO_2C_6H_4COS)$	Fe	S		2.266(1)	1209(5)	1.734(4)	[58]
$Ru(tert-C_4H_9Cp)(CO)_2-(3-NO_2C_6H_4COS)$	Ru	S		2.388(1)	1.220(3)	1.742(2)	[123]
$Ru(tert-C_4H_9Cp)(CO)_2-(3,5-NO_2C_6H_4COS)$	Ru	S		2.498(1)	1.218(5)	1.897(4)	[123]
$Os_3H(CO)_{10}(C_6H_5COS)$	Os	S		2.405(8) 2.438(9)	1.199(7)	1.910(38)	[2]
$[Mn(CO)_4(C_6H_5COSe)]_2$	Mn	Se	2.214(4)	2.4816(9) 2.5192(8)	1.197(6)	1.989(4)	[56]
$(Ph_4P)^+[Mn(C_6H_5COS)_3]^-$	Mn	S	2.139(3)	2.576(2)	1.264(6)	1706(5)	[53]
$(Ph_4P)^+[Co(C_6H_5COS)_3]^-$	Co	S		2.468(1)	1.241(5)	1.941(4)	[53]
$[Ni(CH_3COS)_2]_2\cdot0.5C_2H_5OH$	Ni	S	2.010(7) –2.056(7)	2.221(5) –2.230(5)	1.244(12) –1.262(12)	1.701(10) –1.718(10)	[221]

^{a} Intramolecular $M\cdots O=C$.

Table 2 (continued)

M(RCOE)x	M	E	Distance [Å]				References
			M···O[a]	M–E	C=O	C–E	
$Ni(CH_3COS)_2(Py)_2$	Ni	S	2.140(6) 2.128(6)	2.454(2) 2.442(2) 2.080(6)(Ni–Ni) 2.069(6)(Ni–Ni)	1.24(1) 1.23(1)	1.704(9) 1.699(9)	[35]
$(Ph_4P)^+[Ni(C_6H_5COS)_3]^-$	Ni	S	2.100(2)	2.419(1)	1.260(4)	1.705(4)	[53]
$Ni(4\text{-}CH_3C_6H_4COSe)_2[(C_2H_5)_3P]_2$	Ni	Se	3.177(5)	2.3426(7)	1.209(6)	1.910(6)	[175]
$Pd(4\text{-}CH_3C_6H_4COSe)_2[(C_2H_5)_3P]_2$	Pd	Se	3.270(7)	2.456(1)	1.203(8)	1.906(8))	[175]
$Pt(4\text{-}CH_3C_6H_4COS)_2[(C_2H_5)_3P]_2$	Pt	S	3.256(6)	2.331(2)	1.219(7)	1.753(8)	[168]
$Pt(4\text{-}CH_3C_6H_4COSe)_2[(C_2H_5)_3P]_2$	Pt	Se	3.362(5)	2.452(1)	1.212(6)	1.913(6)	[175]
$Pt(4\text{-}CH_3C_6H_4COTe)_2[(C_2H_5)_3P]_2$	Pt	Te	3.50(2)	2.632(2)	1.23(2)	2.15(2)	[168]
$(Ph_3P)Cu(m\text{-}PhCOS)_2Cu(PPh_3)_2$	Cu	S		2.309(1) –2.424(1)		1.770(6)	[294]
$[Ph_4P][Cu(CH_3COS)_2]$	Cu	S	2.998	2.151(3)	1.210(12)	1.716(12)	[268]
$[Ph_4P][Ag(CH_3COS)_2]$	Ag	S	3.082	2.359(1)		1.730(4)	[268]
$[(C_2H_5)_3NH][Ag(PhCOS)_2]$	Ag	S		2.359(1) 2.382(1)	1.213(4) 1.232(4)	1.727(4) 1.729(3)	[289]
$[(Cp_2MoH_2)Ag(PhCOS)_3]_2$	Ag	S	3.098 3.729	2.559(2) 2.693(1) 2.828(4)(Ag–Ag)	1.221(3)	1.751(2)	[38]
$(CH_3)_3PAu[CCl_3COS]$	Au	S	3.722(1)	2.315(4)	1.22(2)	1.77(2)	[261]
$(2\text{-}CH_3C_6H_4)_3PAu[CCl_3COS]$	Au	S	3.293(7)	2.3104(13)	1.195(7)	1.723(6)	[261]
$Zn(PhCOS)_2(H_2O)_2$	Zn	S	2.031(6)[a]	2.278(2)	1.250(7)	1.729(8)	[27]

Table 2 (continued)

M(RCOE)x	M	E	Distance [Å]				References
			M$\cdots$O[a]	M–E	C=O	C–E	
[Ph$_4$P][Zn(PhCOS)$_3$]	Zn	S	2.328(4) −2.376(2)	2.304(2) −1.263(6)	1.215(6) −1.769(5)	1.737(5)	[321]
Zn(CH$_3$COS)$_2$(Lut)$_2$	Zn	S	2.312(1)	2.309(1) −1.237(8)	1.214(6) −1.749(4)	1.746(4)	[249]
Zn[(CH$_3$)$_3$CCOS]$_2$(Lut)$_2$	Zn	S		2.294(4)		1.738(7) −1.743(6)	[249]
[Ph$_4$P][Cd(PhCOS)$_3$] (m-form)	Cd	S	2.557(2) −3019(4)	2.453(2) −2.545(2)	1.214(6) −1.237(8)	1.712(5) −1.736(6)	[48]
[Ph$_4$As][Cd(PhCOS)$_3$]	Cd	S	2.557(2) −3019(4)	2.453(2) −2.545(2)	1.214(6) −1.237(8)	1.712(5) −1.736(6)	[321]
[Cd(PhCOS)$_2$(C$_2$H$_5$OH)]$_2$	Cd	S	2.54(3) −2.68(3)	2.53(1) −2.58(1)	1.21(4) −1.39(5)	1.73(4) −1.77(4)	[236]
Cd(CH$_3$COS)$_2$(Lut)$_2$	Cd	S		2.487(1) 2.496(1)	1.217(4) −1.221(4)	1.732(4) −1.745(3)	[249]
Cd[(CH$_3$)$_3$CCOS]$_2$(Lut)$_2$	Cd	S		2.489(1) 2.313(1)	1.215(6)	1.719(5) −1.743(6)	[249]
[(CH$_3$)$_4$N]{Na[Cd(PhCOS)$_3$]$_2$}	Cd	S	2.659(5) 2.659(5)	2.485(2) −2.521(2)	1.226(7) −1.237(8)	1.725(2) −1.737(8)	[319]
[(C$_2$H$_5$)$_3$NH][Cd(PhCOS)$_3$]	Cd	S	2.537(2) −3.111(2)	2.505(7) −2.5920(8)	1.229(3) −1.239(3)	1.723(2) −1.743(2)	[323]
[Ph$_4$P][Hg(PhCOS)$_3$]	Hg	S	2.858(6) −3.364(6)	2.402(2) 2.498(1)	−1.222(9)	1.714(9) −1.761(7)	[321]
[(CH$_3$)$_4$N][Hg(PhCOS)$_3$]	Hg	S	2.767 −3.051(9)	2.432(3) −2.555(3)	1.197(13) −1.234(12)	1.705(11) −1.737(11)	[322]

boxylato ligands are bidentate [6]. In general, thiocarboxylato transition metal complexes hardly react with alkyl halides. In contrast, the samarium complex **15** exhibits the corresponding alkali metal salt-like reactions such as *S*-esterification and an insertion reaction of C=N into the Sm–S (Scheme 2) [140].

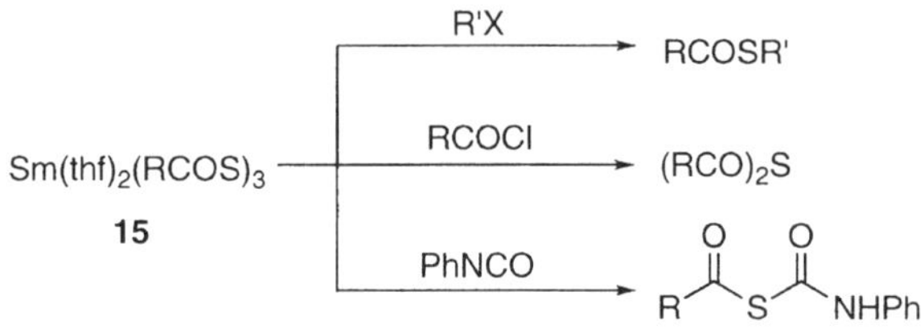

Scheme 2 Reactions of samarium tris(thiocarboxylate) **15**

2.2.2
Group 4 Element Compounds (Ti, Zr, Hf)

2.2.2.1
Titanium

A few chalcogenocarboxylato complexes of titanium are known, R$'_2$Ti(RCOS) [8, 42] and Cp$'_2$Ti(RCOS)$_2$ [Cp$'$=(CH$_3$)C$_5$H$_4$] [7], which can be synthesized by reacting R$_2$TiCl (Cp or indenyl) with sodium thiocarboxylate or Cp$'_2$TiCl$_2$ with sodium thiocarboxylate, respectively. Their detailed structures are not yet fully known.

2.2.2.2
Zirconium

No chalcogenocarboxylato zirconium complexes are known.

2.2.2.3
Hafnium

There are no previous reports concerning the synthesis of chalcogenocarboxylic acid hafnium complexes.

2.2.3
Group 5 Element Compounds (V, Nb, Ta)

2.2.3.1
Vanadium

Vanadium oxodichloride VOCl$_2$ reacts with sodium thiobenzoate and potassium 1,1-bis(thiooxalate) in an aqueous solution to give the corresponding complexes **18** [135] and VO(SOC-COS) [184], respectively. The ligands of the former complex have been investigated by ESR spectroscopy [134].

2.2.3.2
Niobium

Bis(thiocarboxylato) niobium **18** Nb(RCOS)$_2$ [207] and bis(η^5-cyclopenta-di-enyl)bis(thiobenzoato)niobium **19** [110] have been synthesized by stoichio-metric reactions of NbCl$_2$ with sodium thiocarboxylates and of Nb(Cp)$_2$X$_2$ (X=Cl, Br, I) with thallium thiobenzoate, respectively. The structure of **19** is considered to be analogous to the tantalum complex **20**, in which the thiocar-boxylato ligands are monodentate through the sulfur atom [110].

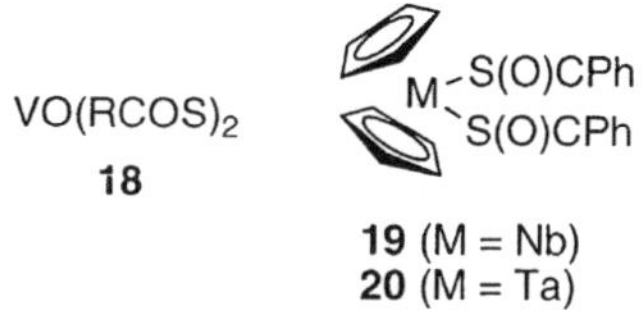

2.2.3.3
Tantalum

There has been only one example of the synthesis of (η^5-cyclopentadienyl)-bis(thiobenzoato)tantalum **20**, which was achieved by the reaction of Ta(Cp)$_2$X$_2$ (X=Cl, Br, I) with thallium thiobenzoate [110]. The molecular structure of **20** is a normal, bent sandwich geometry where the two thiocarboxylato ligands are planar and are monodentate through the sulfur atom (for bond distances, see Table 2) [110].

2.2.4
Group 6 Element Compounds (Cr, Mo, W)

2.2.4.1
Chromium

Only the green crystals of the chromium tris(thiobenzoato) complex Cr(PhCOS)$_3$ **21** have been obtained, by the stoichiometric reaction of CrCl$_3$·6H$_2$O with sodium thiobenzoate in aqueous conditions [73, 271]. Like many other Cr(III) complexes, the complex **21** is a monomer in which the three thiobenzoato ligands bind to the metal in chelate mode.

2.2.4.2
Molybdenum

Dinuclear thiocarboxylato molybdenum complexes **22** have been synthesized by the ligand exchange reaction of $Mo_2(CH_3COO)_4$ with thiocarboxylic acids and their ammonium salts in methanol [107]. On the basis of infrared and electron spectra and magnetic measurement, the dimeric structure **23** has been deduced, in which the thiocarboxylate ligands bridge the two Mo metals [107, 271]. The preparation of the Mo(II) complexes, including the thio- or seleno-ester moieties have also been documented, although the thio- and seleno-carboxyl groups do not coordinate to the metal [106]. Bis(cyclopentadienyl)- and bis(indenyl)-oxomolybdenum dichlorides react with thiocarboxylic acid potassium salts to give the corresponding thiocarboxylato complexes of oxomolybden $R'_2MoO(RCOS)_2$ (R'=cyclopentadienyl, indenyl) (for bond distances, see Table 2) [84].

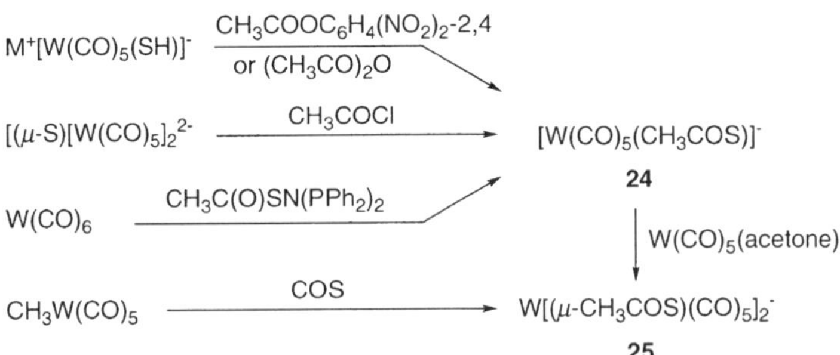

2.2.4.3
Tungsten

The mononuclear $W(SH)(CO)_5^-$ and bridged dinuclear μ-S$[W(CO)_5]_2^{2-}$ complexes which are prepared by the reaction of $W(CO)_6$ with NaSH react with acetic acid anhydride or 2,4-dinitrophenyl acetate to give the thioacetato complexes **24** in which the thioacetato ligand coordinates through the sulfur atom (Scheme 3) [78]. The mononuclear anion complex **25** can also be obtained in

Scheme 3 Syntheses of thiocarboxylato tungsten complexes

$$[CH_3C(=O\cdots H)S]W(CO)_5$$

26

26a ⇌ **26b**

good yields by the reaction of $W(CO)_6$ with the corresponding *N*-acetylsulfe-nate. On the orther hand, treatment of **24** with $W(CO)_5$(acetone) affords the anion complex **25** [78]. The direct reaction of $CH_3W(CO)_5^-$ with COS under 100–200 kPa pressure [46] affords the complex **25**. While synthesizing the thioketone-coordinated tungsten complex $(R_2CS)W(CO)_5(THF)$, a new type of tungsten complex **26**, with a protonated thioacetate group, was found, possessing resonance structures **26a** and **26b** [52].

2.2.5
Group 7 Element Compounds (Mn, Tc, Re)

2.2.5.1
Manganese

Whereas the reaction of $Mn(CO)_5Br$ with potassium thiobenzoate gives the dinuclear complex **27** in which the sulfur atoms of three thiobenzoato ligands triply bridge two manganese metals, the reaction with thiobenzoic acid forms the mononuclear $Mn(CO)_4(PhCOS)$ in which the thiocarboxylato ligand is chelated [98]. The reaction of $Cs_2(SOC-COS)$ with $Mn(CO)_5Br$ yields the μ-1,2-dithiooxalato complex **28** where the thiooxalato ligand has a *trans*-conformation and binds to the two Mn atoms through the sulfur atoms [326]. The homoleptic thiocarboxylato complex **29** of formula $(Ph_4P)[Mn(Ph_3CCOS)_3]$ in which the thiocarboxylato ligands act as a bidentate has been prepared by the reaction of triethylammonium thiobenzoate with $MnCl_2 \cdot 4H_2O$ in the presence of Ph_4PBr [53]. In the anion, the Mn metal is situated on the C-3 axis, and the three PhCOS ligands are related by three-fold rotational symmetry [53]. One phenyl ring of the Ph_4P group was disordered. The magnetic moment (5.99 μ_B) is higher

$$K^+[(CO)_3Mn(m\text{-PhCOS})_3Mn(CO)_3]^-$$

27

28

$$(Ph_4P)[Mn(Ph_3CCOS)_3]$$

29

than those of the corresponding structures [M(Ph$_3$CCOS)$_3$]$^-$ (M=Co, Ni) [53]. The reaction of Na$_2$[Mn(Se)(CO)$_4$]$_2$ with benzoyl chloride gives the dimeric selenocarboxylato manganese complex [Mn(PhCOSe)(CO)$_4$]$_2$, where each selenocarboxylato ligand bridges to the two Mn metals through the selenium atom [Mn-Se: 2482(1), 2.519(1) Å] and the selenocarboxyl oxygen atoms do not coordinate (for bond distances, see Table 2) [56].

2.2.5.2
Technetium

No technetium complexes with chalcogenocarboxylato ligands are known.

2.2.5.3
Rhenium

The reaction of Re(CO)$_5$Br with PhCOSH gives the dimeric complex **30**, where the thiocarboxyl sulfur atom is coordinated to the two rhenium atoms [99], whereas heating with cesium thiobenzoate in ethanol leads to the anion complexes **31** in which the *S*-thiobenzoato ligands are triply bridged to the two rhenium atoms [213]. In addition, reaction with Cs$_2$(SOC-COS) affords (OC)$_5$Re(SOC-COS)Re(CO)$_5$ [326], the structure of which resembles that of the corresponding manganese complex **28**.

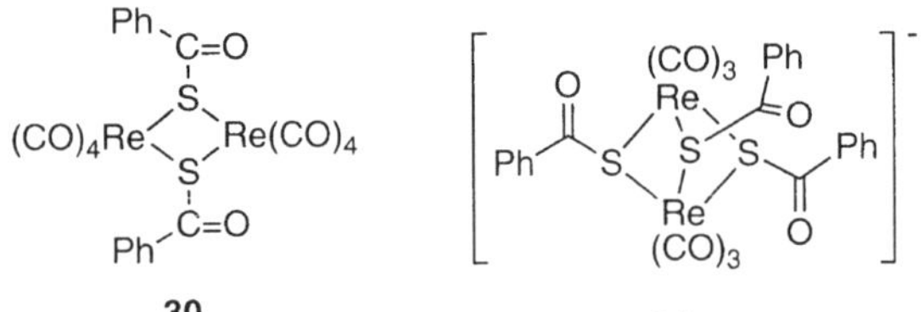

2.2.6
Group 8 Element Compounds (Fe, Ru, Os)

2.2.6.1
Iron

In 1966, King reported the first synthesis of thiocarboxylato iron complex CpFe(CO)$_2$(PhCOS), achieved by treating [CpFe(CO)$_2$]$_2$ with thiobenzoic acid [182]. The monoanions and dianions generated by the reaction of (μ-S$_2$)-[Fe(CO)$_3$]$_2$ with alkyl-lithium or Grignard reagents readily react with acyl chlorides to give clusters with thiocarboxylato ligands **32** [288]. The sulfur and selenium atoms in (μ-E$_x$)-[Fe(Cp′)(CO)$_2$]$_2$ (E=S, Se; Cp′=C$_5$H$_5$, *tert*-C$_4$H$_9$C$_5$H$_4$, 1,3-(*tert*-C$_4$H$_9$)$_2$C$_5$H$_3$, x=3, 4), which are prepared from the reaction of [CpFe-(CO)$_2$]$_2$ with elemental sulfur or selenium, react smoothly with acyl chlorides to give the corresponding thio- [57–60] and selenocarboxylato iron complexes **33** [CpFe(CO)$_2$(RCOE)]$_2$ (E=S, Se) [61, 125]. The thio- and seleno-carboxylato ligands in these complexes bind to the metal through the chalcogen atom [58]. The reactions of the iron thio- and seleno-terephthaloyl chloride complexes **34** with nucleophiles such as amines and alcoholates yield the corresponding amides, esters [125, 127], and bis(organoiron) complexes **35** [125, 126, 128]. Moreover, the irradiation of **33** in the presence of Ph$_3$E (E=P, As, Sb) leads to Fe(Cp)(CO)(Ph$_3$E)(RCOS) [122].

32

34 (E = S, Se; Y = Cl)
35 [E = S; Y = SFe(Cp')(CO)$_2$]

2.2.6.2
Ruthenium

Gould and Stephenson et al. [85] and Gilvert and Wilkinson [76] found that RuCl$_3$[Ph(CH$_3$)$_2$P]$_3$ or RuCl$_2$[Ph$_3$P]$_3$ readily react with ammonium or sodium thiocarboxylates to give thiocarboxylato ruthenium complexes **36**, which react further with CO or amines, giving the adducts **37** [85]. The bridged poly-chalcogen complexes (μ-E$_5$)[Ru(Cp')(CO)$_2$]$_2$ (E=S, Se; Cp'=C$_5$H$_5$, [(CH$_3$)$_3$C]C$_5$H$_4$, 1,3-[(CH$_3$)$_3$C]$_2$C$_5$H$_3$, x=3, 4) are acylated by acyl chlorides, giving the corresponding thio- **38** and seleno-carboxylato ruthenium complexes **39** [59, 123]. The chalcogenocarboxylato ligands of **36–39** are monodentate (for bond distances, see Table 2).

Ru(PhCOS)$_2$(R$_3$P)$_3$ Ru(PhCOS)$_2$(R$_3$P)$_2$(L)$_2$ Ru(Cp')(CO)(PhCOS)$_2$

36   **37**   **38** (E = S)
                                    39 (E = Se)

R$_3$P = Ph(CH$_3$)$_2$P, Ph$_3$P   L = CO or R'$_3$N

2.2.6.3
Osmium

Only one osmium complex **40** with a thiobenzoato ligand has been synthesized, by the reaction of Os$_3$(CO)$_{10}$(CH$_3$CN)$_2$ with thiobenzoic acid [2]. The thiobenzoato ligand is bound to the metal in the μ,η^1-mode through the

40

sulfur atom and it bridges the two osmium atoms of the triangle, but no intramolecular interactions between the carbonyl oxygen and the osmium atoms are observed (for bond distances, see Table 2) [2].

2.2.7
Group 9 Element Compounds (Co, Rh, Ir)

2.2.7.1
Cobalt

Using the stoichiometric reaction of cobalt sulfate with potassium thiobenzoate in aqueous solution, the complex $[Co(PhCOS)]_2$ can be obtained in low yields (<10%) [204]. Similar treatment of $Co_2(CO_3)_3$ or cobalt oxalate with excess thiobenzoic acid or its sodium salt leads to the Co(III) complex $Co(PhCOS)_3$ [204]. The reaction of $Co(NO_3)_2 \cdot 6H_2O$ with thiobenzoic acid in the presence of Ph_4PCl gives the anionic cobalt complex **41** $(Ph_4)P[Co(PhCOS)_3]$ in good yields [53]. The structure of **41** resembles that of the manganese complex **29** [321], where the sulfur and oxygen atoms from each thiocarboxylato ligand are bounded to the central metal atom, giving a distorted octahedral geometry (for bond distances, see Table 2) [53]. The anion $(C_2H_5)_4N[Co(pdtc)_2]$ **42** [pdtc= pyridine-2,6-bis(thiocarboxylate)(2–)], in which the thiocarboxylato ligands are monodentate through the sulfur atom, is formed from the reaction of $Co(CH_3COO)_2(4H_2O)$ with pyridine-2,6-bis(thiocarboxylate) in the presence of $(C_2H_5)_4NCl$ [193].

$(Ph_4P)^+[Co(PhCOS)_3]^-$

41

$Rh_2(CH_3COS)_4(Py)_2$

43

42

2.2.7.2
Rhodium

Dirhodium terakisformate $Rh(HCOO)_4(0.5H_2O)$ reacts with thioacetic acid to give rhodium tris(thioacetate), $Rh(CH_3COS)_3$. Treatment of the tris(thioacetato) complex with pyridine leads to the adduct **43** $Rh_2(CH_3COS)_4(Py)_2$ [9].

2.2.7.3
Iridium

No iridium complexes with monochalcogenocarboxylato ligands are known.

2.2.8
Group 10 Element Compounds (Ni, Pd, Pt)

2.2.8.1
Nickel

In 1952 Hieber and Brück obtained the dark violet crystals of a nickel bis-(thiobenzoate) complex by reacting $NiCl_2 \cdot 6H_2O$ with thiobenzoic acid [97]. Melson and his coworkers isolated the thiocarboxylato nickel complexes $Ni(RCOS)_2 \cdot 0.5C_2H_5OH$ ($R=CH_3, C_2H_5, C_6H_5$) (44) using $NiCO_3 \cdot 2Ni(OH)_3 \cdot 4H_2O$ as the source of nickel [222], and revealed that the molecular structure of the thiobenzoate derivative (44, R=Ph) is a dinuclear complex with formula $Ni_2(PhCOS)_4 \cdot C_2H_5OH$, in which the thiocarboxylate groups behave as bridging ligands between two nickel atoms to form a structure with a short $Ni \cdots Ni$ distance (2.49 Å), similar to those that have been found for several transition metal carboxylates [26, 221]. The complexes 44 readily react with amines such as pyridines (Py) to give the adduct $Ni(RCOS)_2(Py)_2$ 45 [26, 191]. X-Ray structural analysis showed that 45 (R=CH_3 [34], Ph [234]) has an octahedral geometry where the thiocarboxylato ligands are bidentate and the two nitrogen atoms are in a *cis*-configuration [34, 234].

$Ni(RCOS)_2 \cdot 0.5C_2H_5OH$ $Ni(RCOS)_2(Py)_2$

44   **45**

A series of thio- 46 [51, 168] and seleno-carboxylato complexes 47 [175] with *trans*-configurations have been synthesized from the reaction of $NiCl_2(R'_3P)_2$ with alkali metal or *O*-trimethylsilyl selenocarboxylates (Scheme 4). Telluro-carboxylato nickel complexes 48 $Ni(RCOTe)_2(R'_3P)_2$ are too labile to isolate, although their formation has been confirmed by their ^1H, ^{13}C, and ^{125}Te NMR and IR spectral data [175]. The anionic thiocarboxylato nickel complex 49 is

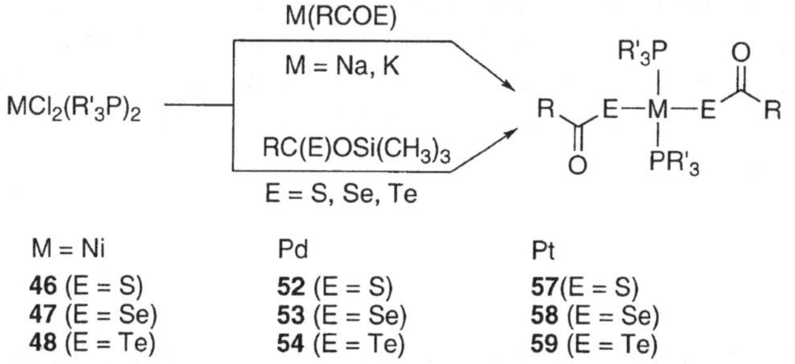

M = Ni	Pd	Pt
46 (E = S)	**52** (E = S)	**57** (E = S)
47 (E = Se)	**53** (E = Se)	**58** (E = Se)
48 (E = Te)	**54** (E = Te)	**59** (E = Te)

Scheme 4   Syntheses of bis(chalcogenocarboxylato)bis(phosphine)-nickel, -palladium and -platinum

synthesized in high yields from the reaction of $NiCl_2 \cdot 6H_2O$ with triethylammonium thiobenzoate. X-Ray structure analysis shows that the nickel atoms in **49** are situated on the C-3 axis, and the three PhCOS ligands are related by three-fold rotational symmetry (for bond distances, see Table 2) [53]. Holm and his coworker synthesized anionic (M)[Ni(pdtc)$_2$] **50** [(M=C$_2$H$_5$)$_4$N or Ph$_3$PCH$_2$Ph; pdtc=pyridine-2,6-bis(thiocarboxylate)] with a tetragonally distorted octahedral structure as a model compound of [NiFe]-hydrogenase from the reaction of Ni(CH$_3$COO)$_2$(4H$_2$O) with pyridine-2,6-bis(thiocarboxylate) in the presence of (C$_2$H$_5$)$_4$NCl or (Ph$_3$PCH$_2$Ph)Br [193].

$(Ph_4P)^+[Ni(PhCOS)_3]^-$

49

50

2.2.8.2
Palladium

It is well-known that the reaction of palladium(II) acetate with an excess of thiobenzoic acid in benzene gives the polymeric thiobenzoato palladium complex **51** [Pd(PhCOS)$_2$]$_x$ (x≥3) [83], although the complex formed from the reaction of PdCl$_2$ with sodium thiobenzoate in water was reported to be dimeric [Pd(PhCOS)$_2$]$_2$ in benzene [253]. The thiobenzoato complexes **51** obtained by the reaction of Na$_2$PdCl$_4$ with sodium thiobenzoate react with Lewis bases to give the neutral monomeric complexes *trans*-Pd(PhCOS)$_2$(L)$_2$ (L=R$_3'$P, R$_3'$As, R$_3'$Sb, pyridines, and so on), where the PhCOS ligands are monodentate [83]. ^{1}H and ^{31}P NMR studies suggest a *trans*-configuration for the thiocarboxylato ligands [83]. A series of thio- **52**, seleno- **53** [175] and telluro-carboxylato palladium complexes **54**, *trans*-Pd(RCOE)$_2$(R$_3'$P)$_2$ (E=S, Se, Te), have been synthesized by similar methods to those shown in Scheme 4 [168]. The monochalcogenocarboxylato ligands of these complexes connect to the central metal via the chalcogen atom [168, 175]. The chelating diphenylphosphinoferrocene (dppf) imposes a *cis*-configuration on Pd(RCOE)$_2$(dppf) (for bond distances, see Table 2) [241].

2.2.8.3
Platinum

Similar to the palladium complexes mentioned above, the complex Pt(PhCOS)$_2$]$_x$ (**55**), obtained by the reaction of Na$_2$PtCl$_4$ with sodium thiobenzoate, also reacts with Lewis bases such as phosphines and pyridines giving the neutral mono-

$$[Pd(PhCOS)_2]_x \ (x=>3)$$

51

$$Pt(PhCOS)_2]x \qquad\qquad\qquad trans\text{-}Pt(RCOS)_2L_2$$

55   **56**

L=R'$_3$P, R'$_3$As, R'$_3$Sb
pyridines etc.

meric complex **56** *trans*-Pt(RCOS)$_2$L$_2$ (L=R$_3'$P, R$_3'$As, R$_3'$Sb, and pyridines) **56**
[83]. The thiocarboxylato ligands were found to act as monodentate ligands on
the basis of the C=O stretching frequencies at ~1660 cm^{-1} [83]. Thio- **57**, seleno-
58, and telluro-carboxylato platinum complexes **59** *trans*-Pt(RCOE)$_2$[(C$_2$H$_5$)$_3$P]$_2$
(E=S, Se, Te) can be readily synthesized by the reaction of chalcogenocarboxylic
acid sodium salts or trimethylsilyl esters with *cis*- and *trans*-PtCl$_2$(R$_3'$P)$_2$ [168].
In addition, the complexes **57–59** can be obtained by the oxidative addition
of bis(acyl) dichalcogenides to Pt[(C$_2$H$_5$)$_3$P]$_4$ [168]. Note that whether *cis*- or
trans-PtCl$_2$[(C$_2$H$_5$)$_3$P]$_2$ is used as the starting material, only the *trans*-isomer is
obtained [168]. The chalcogenocarboxylato ligands in **57–59** bind to the central Pt metal through the chalcogen atoms and there are no interactions between the carbonyl oxygen and the central metal [168].

When one of the two oxygen atoms of the carboxyl group in RCOO-Z
(Z=all of the elements in the periodic table) is replaced by a sulfur, selenium
or tellurium atom, some high-lying π-orbitals and low-lying empty π*-or-
bitals that involve carbon and chalcogen atoms may make the compound more
reactive, in the order to S<Se<Te. In fact, the stability of the platinum complexes Pt(RCOE)$_2$(PR$_3'$)$_2$ (E=S, Se, Te) decrease in the order: E=S>Se≫Te. The
aliphatic complexes {M(RCOE)$_2$(PR$_3'$)$_2$, R=alkyl} are more labile than the
aromatic ones (R=aryl) [169, 175]. With Group 10 metals, M(RCOTe)$_2$(PR$_3'$)$_2$
becomes more labile in the order: M=Pt>Pd≫Ni. Upon exposure of the palladium **54** and platinum complexes **59** to air, no appreciable change is observed
for over week. The nickel complexes **48** with tellurocarboxylato ligands are too
sensitive toward oxygen to isolate.

2.2.9
Group 11 Element Complexes (Cu, Ag, Au)

2.2.9.1
Copper

The first preparation of a chalcogenocarboxylato copper complex was described in the article by Engelhardt et al. [62] in 1868, who prepared thiobenzoic acid copper from the reaction of potassium thiobenzoate with CuSO$_4$ in
aqueous solution. The chemistry of this class of compounds, however, is not

known. The reaction of CuCl or $Cu(CH_3COO)_2$ with thiobenzoic acid [239] and its sodium [271] or triethylammonium salts [268] results in yellow-green diamagnetic thiobenzoato copper(I) $Cu(C_6H_5COS)$. Its molecular structure analysis by X-ray diffraction remains to be done. 9,10-Phenanthrenesemiquinonatobis(triphenylphosphine)copper(I) readily reacts with dibenzoyl disulfide to give the dinuclear Cu(I) complex **60**, in which the thiobenzoato ligand is monodentate and bridges between the two copper ions [294].

$$O=CPh$$
$$|$$
$$S$$
$$(Ph_3P)Cu\diagdown\diagup Cu(PPh_3)_2$$
$$S$$
$$|$$
$$PhC=O$$

60

$$M^+[Cu(RCOS)_2]^-$$

61

$$M = (C_2H_5)_3NH, Ph_4P$$

Thiocarboxylato Cu(I) complexes with triphenylphosphine $[Cu_4(CH_3COS)_4$-$(Ph_3P)_4$, $Cu_4(CH_3COS)_4(Ph_3P)_3$, $Cu_2(CH_3COS)_2(Ph_3P)_4$, $Cu_2(PhCOS)_2(Ph_3P)_3$, $Cu(PhCOS)(Ph_3P)_2]$ are synthesized in high yields by the reactions of alkali metal thiocarboxylates with mixtures of CuCl and Ph_3P in stoichiometric ratios, and their structures have been determined by X-ray diffraction analysis [49]. The isolation of anionic thiocarboxylato copper **61** (R=Ph, M=Cu, M′=Ph_4P) has been reported from triethylammonium thiobenzoate with CuCl in the presence of Ph_4PCl, and structural analysis has indicated that the two thioacetato ligands are bound to the metal through the sulfur atom and no intramolecular interactions occur between the metal and the carbonyl oxygen atoms [268]. The selenocarboxyalto copper complexes Cu(RCOSe) and Li[Cu(RCOSe)(CN)], generated by bubbling COSe through PhCu or $R_2Cu(CN)Li$ solution, readily react with alkyl iodides to give the corresponding *Se*-alkyl selenoesters in good yields [72].

2.2.9.2
Silver

Thiocarboxylato complexes of silver(I) [148, 271] have been synthesized in good yields from the stoichiometric reaction of sodium or piperidinium thiocarboxylates with $AgNO_3$, respectively. Like the copper complexes, two anionic thiocarboxylato silver complexes $(Ph_4P)^+[Ag(CH_3COS)_2]^-$ and $[(C_2H_5)_3NH]^+$-$[Ag(PhCOS)_2]^-$ have been identified, where the two thiocarboxylato ligands are monodentate through the sulfur atom [268]. The reaction of Cp_2MoH_2 with sodium thiobenzoate and potassium thioacetate in the presence of $AgBF_4$ gives the dimeric $[(Cp_2MoH_2)Ag(PhCOS)]_2$ and a polymeric thiocarboxylato complex $[(Cp_2MoH_2)Ag(PhCOS)_3]_n$ of silver, respectively [38].

2.2.9.3
Gold

Recently, Schmidbaur and his coworkers found that thiocarboxylic acids are aurated by $\{[(R'_3P)Au]_3O\}^+BF_4^-$ to give novel thiocarboxylato gold(I) complexes $Au(RCOS)(R'_3P)$ (**62**) in which the thiocarboxylato ligand is exclusively bonded through the sulfur to the central gold atom [261]. In the solid state, the complex ($R=CCl_3$, $R'=CH_3$) exists in a dimer with weak intermolecular $Au\cdots S'$ (3.722 Å) and $Au\cdots Au'$ bonds (4.012 Å), but no intramolecular interaction between the carbonyl oxygen and Au atoms is observed. For the complex ($R=CH_3$, $R'=2$-$CH_3C_6H_4$), such $Au\cdots Au'$ bonds are not observed, owing to the steric hindrance arising from bulky $2\text{-}CH_3C_6H_4$ groups (for bond distances, see Table 2).

62

2.2.9.4
Reactions

Copper(I) thiocarboxylates Cu(RCOS) react with aryl iodides to give *S*-aryl thioesters in good yields [254]. Treatment with aqueous solution of pyridine affords CuS and $Cu(PhCOO)_2(Py)_2(H_2O)$ with evolution of H_2S [237]. Moreover, heating of **63** with 2-iodoaniline at 100–110 °C in hexamethylphosphorotriamide yields 2-arylbenzothiazoles [255]. Reacting silver thiocarboxylates with thioaroylsulfenyl bromides RC(S)SBr leads to the formation of the asymmetrical acy thioacyl disulfides RC(O)SS(S)CR [155], and additional treatment with Br_2 or *N*-bromo-succinimides (NBS) and *N*-iodo-succinimides (NIS) gives the acylsufenyl halides RCOSX (Br, I) (see Sect. 2.3.5) [149].

2.2.10
Group 12 Element Compounds (Zn, Cd, Hg)

2.2.10.1
Zinc

Sodium thiocarboxylates readily react with zinc salts such as $ZnCl_2$ or $Zn(NO_3)_2$ in aqueous solution to give the thiocarboxylato zinc complexes **64**, possessing two crystallizing H_2O molecules [271]. Reaction of thiocarboxylic acid with diethyl zinc in the presence of lutidine has been found to give the adduct $Zn(RCOS)_2(L)_2$ (L=3,5-lutidine) **65** in good yields [249]. Anionic thiocarboxylato complexes **66** (M=Zn) have also been isolated from the reaction of

$Zn(RCOS)_2 \cdot (H_2O)_2$ $Zn(RCOS)_2 L_2$

64   **65**

                                      L=3,5-lutidine

$M'^+[M(RCOS)_3]^-$ $Cd(PhCOS)_2 \cdot C_2H_5OH$

66   **67**

M=Zn, Cd, Hg
M'=$(C_2H_5)_3$NH,
    Ph_4P, Ph_4AS $Cd(PhCOS)_2 L_2$

68

L=pyridine
3,5-lutidine

triethylammonium thiobenzoate with $Zn(NO_3)_2$ in the presence of tetraphenyl-phosphonium and -arsonium chlorides [269, 321].

2.2.10.2
Cadmium

The preparation of $Cd(PhCOS)_2$ by reacting $CdCl_2$ with sodium thiobenzoate in aqueous solution was described in 1970 [272], although details of the reaction conditions were not given. The isolation and X-ray structural analysis of the ethanol adduct $Cd(PhCOS)_2 \cdot C_2H_5OH$ were reported by Russian chemists in 1977 [235]. The complex **67** reacts with pyridines to give the adducts $Cd(RCOS)_2(L)_2$ (L=pyridine or 3,5-lutidine) (**68**) [190], which are obtained from the direct reaction of RCOSH with cadmium carbonate in the presence of 3,5-lutidine [249]. Two types of anion complexes, **66** (M=Cd) and **69** (M=Na, M'=Cd), have been synthesized from the reaction of $Cd(NO_3)_2 \cdot 6H_2O$ with Na(PhCOS) in the presence of Ph_4ECl (E=P, As) [48, 269, 321], of $Cd(NO_3)_2 \cdot 4H_2O$ with $Et_3NH(PhCOS)$ [323], and of $Cd(NO_3)_2 \cdot 4H_2O$ with M(PhCOS) (M=Na, K) in the presence of $(CH_3)_4NCl$ [50, 319], respectively.

$(CH_3)_4N^+[MM'(PhCOS)_3]^-$ PhHg(RCOS)

69   **70**

M = Na, K
M' = Cd, Hg

Hg(PhCOS)(RCSS)

71

2.2.10.3
Mercury

Three preparation methods for aromatic mercury bis(thiocarboxylates) have been reported: (1) reaction of $HgCl_2 \cdot 4H_2O$ with Na(RCOS) [271]; (2) insertion of mercury into diacyl disulfides [300]; (3) reaction of Ph_2Hg with two molar amounts of RCOSH [145]. The phenylmercury thiocarboxylates PhHg(RCOS) (**70**) can be obtained by reacting Ph_2Hg with an equivalent RCOSH, or PhHgBr with potassium or piperidinium thiocarboxylates [147]. The compound **70** reacts with dithiocarboxylic acids affording the complex **71** with both thio- and dithio-carboxylato ligands [147]. The anionic complexes **66** (M=Hg) [269, 321, 322] and **69** (M=K, M′=Hg) [50] of mercury were prepared using triethylammonium, sodium or potassium thiocarboxylates, and mercury dichloride, and their structures are characterized by X-ray diffraction analysis [270, 322, 323]. The reaction of $Hg(PhCOS)_2$ with $CdCl_2$ or $HgCl_2$ in the presence of $Ph_4AsCl \cdot H_2O$ has been found to lead to the formation of a new type of ionic complex $(Ph_4As)_2[Hg(PhCOS)_2(MCl_2)]^{2-}$ (M=Cd, Hg) (**72**), in which the thiobenzoato ligands are monodentate through the sulfur atom and the central Hg metal is linked to MCl_4 through two chlorines [324].

72 (M = Cd, Hg)

2.2.10.4
Structures

The coordination environment around the zinc in **64** (R=Ph) is a distorted tetrahedron, consisting of two sulfur atoms from two equivalent ligand molecules and two oxygen atoms from two equivalent water molecules (bridged via hydrogen bonding) (for bond distances, see Table 2) [27]. The complexes $M(RCOS)_2(L)_2$ (M=Zn, Cd; L=pyridine, 3,5-lutidine) (**65 and 68**) are monomeric and approximately tetrahedral with approximate C_{2v} symmetry and with monodentate thiocarboxylato ligands binding to the metal through the sulfur atoms [236, 249]. It is worth noting that the oxygen atoms of two thiocarboxyl groups in **68** (M=Cd) do not participate in the coordination [236]. On the other hand, the ethanoladduct of bis(thiobenzoato) cadmium **68** exists as a dimer [Cd(PhCOS)_2-$(C_2H_5OH)]_2$ and is a distorted octahedron, where the two thiocarboxyl groups coordinate to the Cd metal by chelating and bridging, respectively, three sulfur atoms, two oxygen atoms (Cd-O: 2.68(3), 2.54(3) Å) from four thiocarboxyl groups, and one oxygen atom (Cd-O: 2.43(3) Å) from ethanol [235]. Structural analysis of the anionic complexes $[Zn(CH_3COS)_3(H_2O)]^-$ shows that the three thioacetato ligands bind to the metal through sulfur atoms, and in addition the two hydrogen atoms of an H_2O molecule are bound by hydrogen-bonding to the two oxygen atoms of two thiocarboxylate groups [269]. In contrast, the three thiocarboxylato ligands of the cadmium anion complexes **66** (M=Cd) are in a propeller-like arrangement and are bidentate [48,269,323]. The structures of the complexes of **66** all resemble each other; the metals are surrounded by three sulfur atoms in trigonal-planar fashion, and each metal atom is bound weakly to oxygen atoms (the coordination number of the Hg is six and is best described as arranged in a distorted octahedron) [321, 322]. It is noted that the CdS_3 kernel of **66** (M=Cd) has two crystallographically different structures: trigonal pyramidal (rhombohedral) and planar (monoclinic) [48]. The trinuclear anion in **69** (M=Cd) contains a distorted octahedral NaO_6 kernel [Na–O distances: 2.303(2)–2.470(5) Å], where an almost planar coordination of the three sulfur atoms and weak interactions of the three oxygen atoms are observed around each Cd [319]. In the new ionic complex **72**, the two thiobenzoato ligands are bound linearly to the central Hg metal through sulfur (S-Hg-S angle: 174.9°) [324].

2.2.10.5
Reactions

Heating **65**, **68** (M=Zn, Cd; R=CH$_3$, *tert*-C$_4$H$_9$; L=3,5-lutidine) gives metal sulfides in quantitative yields with the elimination of diacyl sulfides [249]. Treatment of mercury bis(thioacetate) with an aqueous solution of KOH gives mercury sulfide and potassium acetate [307]. Phenylmercury thiocarboxylates **70** react with *N*-bromo- [148] and *N*-iodo-succinimides [146] to give the corresponding acylsulfenyl halides (Scheme 5) (see Sect. 2.3.5).

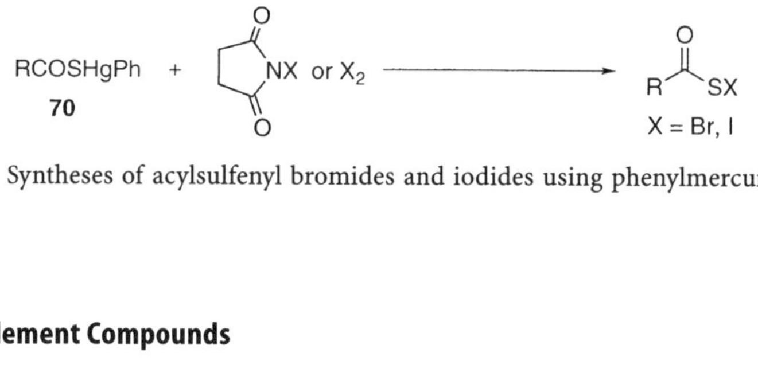

Scheme 5 Syntheses of acylsulfenyl bromides and iodides using phenylmercury thiocarboxylates

2.3
Typical Element Compounds

2.3.1
Group 13 Element Compounds (B, Al, Ga, In, Tl)

2.3.1.1
Boron

Thiocarboxylic acid borylesters **73** have been prepared by thioboration of the corresponding esters with bis(1,5-cyclooctanediylboryl) sulfide [185]. Treatment of [B$_3$H$_7$]$_2$ or B$_3$H$_7$/THF with thioacetic acid has been shown to lead to acylthiotriborane **74** [18].

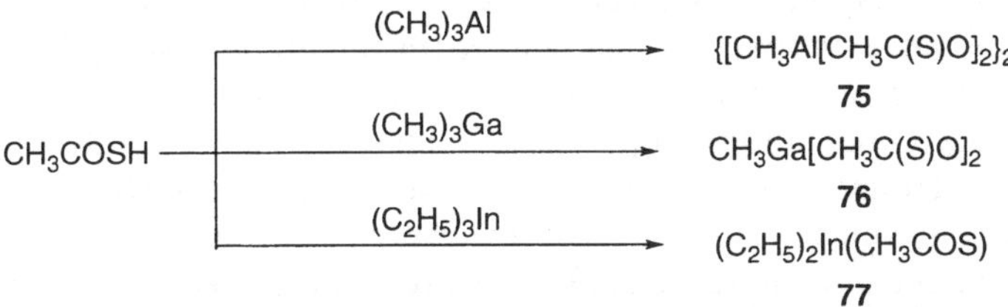

$$\textbf{73} \qquad\qquad \textbf{74}$$

2.3.1.2
Aluminum

The trialkyls $R_3'M$ (M=Al, Ga, In) undergo insertion reactions with COS to form alkylmetal thiocarboxylates $(RCOS)_xMR_{3-x}'$ (M=Al, Ga, In; x=1–3), which are difficult to separate [327]. Treatment of trimethylaluminum with thioacetic acid, however, leads to a good yield of dimethylaluminum thioacetate **75** as yellow crystals that sublime (Scheme 6) [327]. On the basis of infrared and ^1H NMR spectra, the structure of **78** has been deduced to be a dimer [327]. Syntheses of Group 13 metal derivatives of selenocarboxylic acids have not been reported, although the formation of $RCOSeAl(CH_3)_3$ has been described as a short-lived intermediate in the reaction of $RCOOR'$ with $[(CH_3)_2Al]_2Se$ to form diacyl diselenide [281].

$$CH_3COSH \xrightarrow{\quad (CH_3)_3Al \quad} \{[CH_3Al[CH_3C(S)O]_2\}_2 \quad \textbf{75}$$
$$\xrightarrow{\quad (CH_3)_3Ga \quad} CH_3Ga[CH_3C(S)O]_2 \quad \textbf{76}$$
$$\xrightarrow{\quad (C_2H_5)_3In \quad} (C_2H_5)_2In(CH_3COS) \quad \textbf{77}$$

Scheme 6   Syntheses of aluminum, gallium and indium thiocarboxylates

$$\textbf{78}$$

2.3.1.3
Gallium

In contrast to the above aluminum compound **75**, dimethylgallium thioacetate **76**, obtained from reacting trimethylgallium with thioacetic acid, is monomeric (Scheme 6) [327]. A similar reaction in the presence of 3,5-lutidine (Lut) forms the methylgallium bis(thioacetate) adduct **79** and the lutidine in which the central Ga metal is tetrahedrally coordinated by two sulfur atoms from monodentate thioacetato ligands, one methyl carbon, and one nitrogen from lutidine [327]. Gallium tris(thioacetates) obtained by the stoichiometric reaction of $(CH_3)_3Ga$ and thioacetic acid also form a 1:1 adduct $Ga(CH_3COS)_3(3,5$-lutidine), where the two thiocarboxyl groups and the gallium atom exist in the same plane, and the carbonyl oxygen atoms are oriented in the same direction (for bond distances, see Table 3) [289].

$$CH_3Ga[CH_3C(S)O]_2(Lut)$$

79

Lut = 3,5-lutidine

2.3.1.4
Indium

The ethylindium bis(thioacetates) **77** have been isolated from the reaction of the corresponding trialkylindium with thioacetic acid (Scheme 6) [91, 327]. X-Ray structural analysis of **77** shows that the In···O distance within the four-membered ring is nearly the same length as the distance between the oxygen and the indium atom of the adjacent ring, and that the coordination number of indium is five [90, 91]. The reaction of $InCl_3$ with potassium thiobenzoate in a 1:3 ratio gives a high yield of indium tris(thiobenzoate) which quickly decomposes to the sulfido complex **80** and dibenzoyl sulfide, while the reaction with triethylammonium thiobenzoate in a 1:4 molar ratio yields the anionic complex **81** in which the thiocarboxylato ligands are bound to the indium metal in a monodentate manner through the sulfur atoms [291]. The coordination environment of the indium is a distorted tetrahedral. No significant intramolecular interactions exist between the carbonyl oxygen and the central In metal (for bond distances, see Table 3) [291].

$$In(S)(PhCOS)_3 \qquad [(C_2H_5)_3NH]^+[In(PhCOS)_4]^-$$

80   **81**

In general, the thiocarboxylato Group 13 element derivatives **75–77** are labile and soluble in common aprotic solvents. It is worth noting that the indium compound **77** is even soluble in cold water.

2.3.1.5
Thallium

The reaction of $Tl(CH_3COO)_3$ with three equivalents of thiobenzoic acid produces thallium(III) tris(thiobenzoate) **82** in good yield, while the similar reac-

Table 3 Selected bond distances of main-group-element compounds of monochalcogenocarboxylic acids

M(RCOE)x	M	E	Distance [Å]				References
			$M \cdots O^a$	M–E	C=O	C–E	
$CH_3Ga(CH_3COS)_2(3,5\text{-}Lut)^b$	Ga	S	2.998	2.297(1) 2.288(2)	1.210(12)	1.716(12)	[289]
$(C_2H_5)_2In(Ch_3COS)$	In	S	2.56	2.63			[90]
$[(C_2H_5)_3NH][In(PhCOS)_2]_4$	In	S	2.557(2) −3.019(4)	2.478(1) −2.496(1)	1.222(5)	1.67	[291]
$CH_3C(S)OSiH_3$	Si	S	3.185(9)	1.699(5)	1.627(6) (C=S)	1.319(3) (C–O)	[10]
$4\text{-}CH_3C_6H_4COSGePh_3$	Ge	S	3.003(2))	2.2547(8)	1.209(3)	1.790(3)	[303]
$4\text{-}CH_3C_6H_4COSeGePh_3$	Ge	Se	3.131(2)	2.3760(4)	1.210(3)	1.953(3)	
$4\text{-}CH_3C_6H_4COTeGePh_3$	Ge	Te	3.131(2)	2.574(1)	1.210(3)	1.953(3)	[306]
$C_6H_5COSSnPh_3$	Sn	S	2.931(1)	2.435(5)	1.22(2)	1.76(2)	[308]
$4\text{-}CH_3C_6H_4COSSnPh_3$	Sn	S	2.907(2)	2.4453(9)	1.208(2)	1.775(3)	[303]
$4\text{-}CH_3C_6H_4COSeSnPh_3$	Sn	Se	3.068(2)	2.5515(7)	1.199(6)	1934(5)	[306]
$4\text{-}CH_3C_6H_4COTeSnPh_3$	Sn	Te	3.068(2)	2.745(1)	1.199(6)	1.934(5)	[306]

[a] Intramolecular $M \cdots O = C$.

Table 3 (continued)

M(RCOE)x	M	E	Distance [Å]				References
			M···O[a]	M–E	C=O	C–E	
$(C_6H_5COS)_2SnPh_2$	Sn	S		2.488(2)	1.211(12)	1.740(10)	[197]
				2.488(2)			
4-$CH_3C_6H_4COS)_3SnPh$	Sn	S	2.75(2)	2.75(2)	2.75(2)	2.75(2)	[291]
			2.95(3)	2.95(3)	2.95(3)	2.95(3)	
			2,752(6)	2,752(6)	2,752(6)	2.752(6)	
$(C_6H_5COS)_2SnCl_2$	Sn	S	2.267(2)	2.468(2)	1.262(3)	1.732(3)	[291]
4-$CH_3C_6H_4COSPbPh_3$	Pb	S	2.990(2)	2.539(2)	1.770(6)		[303]
4-$CH_3C_6H_4COSePbPh_3$	Pb	Se	3.130(2)	2.6365(5)	1.205(6)	1.934(5)	[306]
4-$CH_3C_6H_4COTePbPh_3$	Pb	Te	3.130(2)	2.815(1)	1.22(1)	2.18(1)	[306]
4-$ClC_6H_4COSAsPh_2$	As	S	2.943(3)	2.136(1)	1.207(4)	1.802(4)	[305]
(4-$CH_3C_6H_4COS)_2PPh$	P	S	2.747(3)	2.144(2)	1.211(5)	1.75(1)	[304]
			2.784(3)	2.146(3)	1.214(5)	1.74(1)	
(4-$CH_3OC_6H_4COS)_2AsPh$	As	S	2.708(3)	2.286(1)	1.223(4)	1.777(4)	[305]
			2.731(3)	2.280(1)	1.216(4)	1.797(4)	
$(C_6H_5COS)_2SbPh$	Sb	S	2.504(2)	2.487(1)	1.237(6)	1.761(1)	[292]
				2.504(2)	1.231(7)	1.766(6)	

Table 3 (continued)

M(RCOE)x	M	E	Distance [Å]				References
			M···O[a]	M–E	C=O	C–E	
$(4\text{-}CH_3OC_6H_4COS)_2BiPh$	Bi	S	2.71(1)[a]	2.644(4)	1.25(2)	1.76(2)	[181]
			2.63(1)[a]	2.614(4)	1.22(2)	1.77(2)	
			3.01(1)[c]				
			3.426(15)[d]				
$(C_6H_5COS)_2Se$	Se	S	3.11(5)	2.174(2)	1.187(7)	1.827(5)	[4]
$(C_6H_5COS)_2Te$	Te	S	3.242(5)	2.372(2)	1.199(9)	1.832(6)	[296]
$(C_6H_5COS)_2$	S	S			1.212(7)	1.811(6)	[267]
					1.198(7)	1.827(6)	
$(2\text{-}CH_3OC_6H_4COS)_2$	S	S	2.960(5)[e]	2.039(2)	1.216(7)	1.811(6)	[247]
			2666(4)[f]	(S–S)	1.199(6)	1.802(6)	
$(2\text{-}CH_3OC_6H_4COSe)_2$	Se	Se	3.133(3)	2.3132(6)	1.202(5)	1.957(4)	[247]
			2.686(3)[f]	(Se–Se)	1.202(5)	1.969(4)	
$(4\text{-}CH_3OC_6H_4COSe)_2$	Se	Se	3.162(5)[e]	2.2888(9)	1.191(7)	1.986(7))	[246]
			3.184(5)	(Se–Se)	1.202(7)	1.973(7)	
$(4\text{-}CH_3OC_6H_4COTe)_2$	Te	Te	3.11(1)[e]	2.739(2)	1.18(1)	2.15(1)	[247]
			2.765(9)[f]	(TeTe)			

[a] Intramolecular M···O=C; [b] 3,5-Lut=3,5-dimethylpyridine; [c] Intermolecular M···C=C; [d] Intermolecular M–O (4-methoxy oxygen); [e] Intermolecular M···O (2-methoxy oxygen); [f] Intermolecular M···O=C (the opposite carbonyl oxygen).

tion with thioacetic acid leads to a quantitative yield of Tl(I) thioacetate [313]. The compound **82** has been used as the starting compound for the preparation of thiocarboxylato niobium and tantalum complexes [110]. Heating of **82** produces dibenzoyl disulfide and Tl(I) thiobenzoate [313].

$$Tl(PhCOS)_3$$

82

2.3.2
Group 14 Element Compounds (M=Si, Ge, Sn, Pb)

2.3.2.1
Silicon

In 1966 Martel and Duffaut [210] and Gronowicz and Ryan [86] reported the first synthesis of *O*-triorganosilyl thiocarboxylates **83b** from reacting thiocarboxylic acids with triorganosilyl chlorides or triorganosilylamines, respectively. Since then, several preparation routes have been developed: (1) RCOSH or M(RCOS) (M=alkali metal) + R'_3SiCl [144]; (2) RCOCl+R'_3SiSLi [77] or $(R'_3Si)_2S$ [202]; (3) acyl thiocarbamoyl sulfides + R'_3SiCl in the presence of triethylamine [191]; (4) RCOCl with lithiated *N*-trimethylsilylthioacetamide [40] or $(R'_3Si)_2S$ [192]; (5) RCOSH+$(H_3Si)_2N=C=N(SiH_3)_2$ [54]; (6) $RC(O)SSnR'_3+R'_3SiBr$ [10]. Amongst these routes, methods (1) and (2) lead to good yields and are considered to be the methods of choice considering the availabilities of the starting compounds and the simple procedures (Scheme 7). Although *S*-organosilyl thiocarboxylates **83a** such as $CH_3C(O)SSiPh_3$ and $CF_3Cl(O)SSi(CH_3)_3$ [77, 228] have been reported to form, their spectroscopic detection appears to be difficult. *O*-Trimethylsilyl selenocarboxylates **84b** have been synthesized by reacting selenocarboxylic acid potassium salts with trimethylsilyl chloride (Scheme 8)

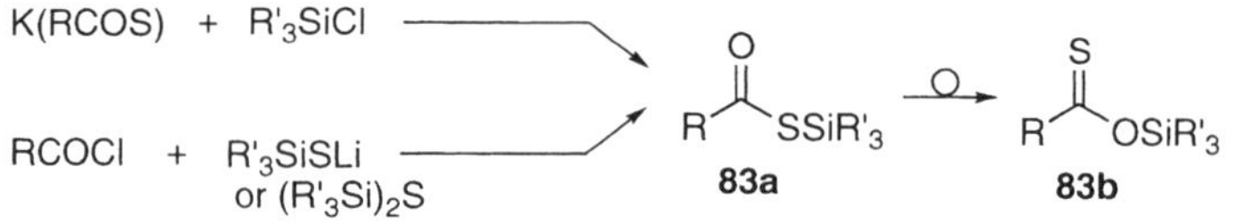

Scheme 7 Syntheses of *O*-triorganosilyl thiocarboxylates

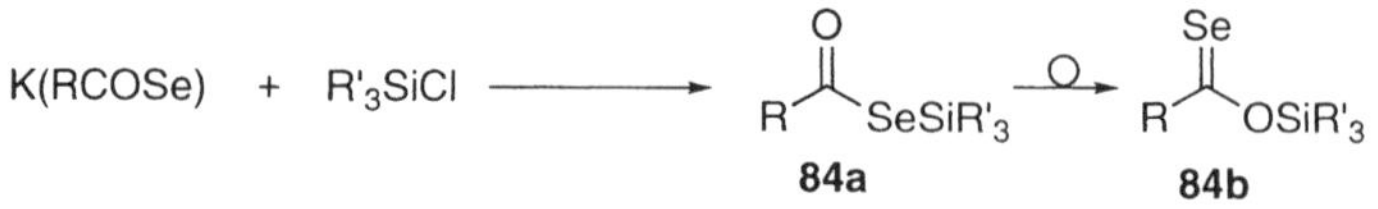

Scheme 8 Syntheses of *O*-triorganosilyl selenocarboxylates

[113, 166] and these have been shown to be useful starting compounds for the preparation of selenocarboxylic acid derivatives. For the tellurium isologue, both *Te*-triorganosilyl **85a** and *O*-triorganosilyl telluroesters **85b** have been observed spectroscopically (Scheme 9) [162, 286]. In the case of trimethylsilyl telluropivaroate (R=*tert*-C$_4$H$_9$; R′=CH$_3$), **85a** and **85b** exist in an equilibrium of 65:35 [286]. No silicon compounds bearing three or four chalcogenocarboxyl groups [(RCEO)$_x$SiR′$_{4-x}$; x=3, 4; E=S, Se, Te] are known.

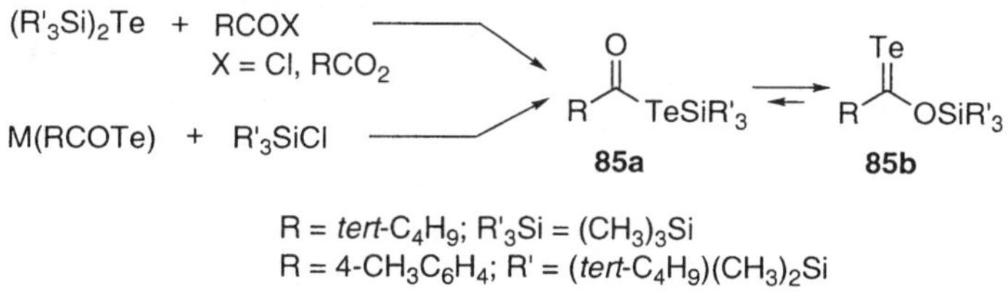

$$(R'_3Si)_2Te \;+\; RCOX$$
$$X = Cl,\; RCO_2$$

$$M(RCOTe) \;+\; R'_3SiCl$$

85a   **85b**

R = *tert*-C$_4$H$_9$; R′$_3$Si = (CH$_3$)$_3$Si
R = 4-CH$_3$C$_6$H$_4$; R′ = (*tert*-C$_4$H$_9$)(CH$_3$)$_2$Si

Scheme 9 Syntheses of *O*-triorganosilyl tellurocarboxylates

O-Orgnosilyl thiocarboxylates **83b** are stable thermally, but readily hydrolyze to thiocarboxylic acid and hexaorganylsiloxane. The selenium isologues **84b** RC(Se)OSiR′$_3$ are labile towards oxygen and moisture [112]. The tellurium isologues **85b** (E=Te, x=1) (dark blue to green) are extremely labile toward oxygen [286]. *O*-silyl telluroacetate CH$_3$C(Te)OSi(CH$_3$)$_3$ decomposes at room temperature to yield (*E*)- and (*Z*)-2-butene derivatives with the liberation of elemental tellurium [286, 287].

2.3.2.2
Germanium

The following three reactions have been used to prepare the *S*-triorganogermyl thiocarboxylate **86** (M=Ge, E=S): (1) RCOCl+R′$_3$GeSM (M=Li, Na) [47, 95, 97, 279]; (2) M(RCOS) (M=alkali metal) + organogermyl halides [143, 271, 303]; (3) RCOSH+H$_3$GeN=C=NGeH$_3$ [54]. Among these, method (2) appears to be the most convenient and versatile (Scheme 10).

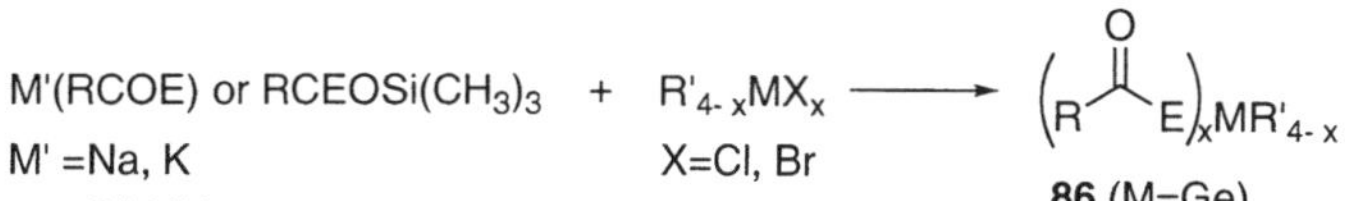

$$M'(RCOE) \text{ or } RCEOSi(CH_3)_3 \;+\; R'_{4-x}MX_x \longrightarrow \left(R\!-\!\!\!\underset{\;}{\overset{O}{C}}\!\!\!-E\right)_x MR'_{4-x}$$

M′ =Na, K   X=Cl, Br
R″$_4$NH
E=S, Se, Te

86 (M=Ge)
87 (M=Sn)
88 (M=Pb)

Scheme 10 Syntheses of organo–Group 14 metal chalcogenocarboxylates

Se-Organogermyl selenocarboxylates **86** (M=Ge, E=Se) can easily be synthesized by reacting alkali metals [165] or *O*-triorganosilyl selenocarboxylates

[132] with triorganogermyl chlorides. Recently, the isolation of *Te*-triorgano-germyl tellurocarboxylate **86** (M=Ge, E=Te) has been reported, obtained from reacting sodium or potassium tellurocarboxylates with triphenylgermyl chloride [306]. No *O*-organogermyl chalcogenocarboxylates are known.

2.3.2.3
Tin

A large number of organostannyl chalcogenocarboxylates with formulae $(RCOE)_xSnR'_{4-x}$ (E=S, Se, Te; x=1–4) and $(RCOS)_2SnX_2$ (X=halogen) are known. Five synthetic methods have been developed for the sulfur derivatives **87** (E=S),: (1) R'_3SnSLi [275–277] or $R'_3SnSNa+RCOCl$ [265]; (2) R_2SnO [37] or $RSn(O)OH$ [271] with RCOSH; (3) PhC(S)OAr with $(n$-$C_4H_9)_3SnH$ [11]; and (4) R_xSnX_{4-x} (X=halogen; x=0–3) with alkali metal or piperidinium thiocarboxylates [143, 271, 303]; (5) $CH_3COOSn(CH_3)_3$ [75] or $(CH_3COO)_4Sn$ [220]+CH_3COSH [75]. Amongst these methods, methods (1) (using lithium organotinthiolate) and (4) (using potassium thiocarboxylates) are the most practical from the point of view of good yields and simple procedures. A series of *Se*-organostannyl seleno- **87** (E=Se) [112, 117] and telluro-carboxylates **87** (E=Te) [307] have also been synthesized in moderate to good yields from the reaction of alkali metal seleno- and telluro-carboxylates and organotin chlorides. Tin compounds possessing both thiocarboxylato and xanthato ligands **89** are obtainable [311]. The anionic tin complexes **90** have been isolated from the reaction of tin dichloride with triethyl-ammonium thiobenzoate in the presence of $(CH_3)_4NCl$ [320]. Cyclic thio- **91** [180] and seleno-carboxylic acid tin esters **92** [212] have been obtained by the reaction of mercarptacids with trimeric dibutyltinsulfide [180], respectively.

$(RCOS)(R'_2NCSS)SnR''_2$ $(CH_3)_4N^+[Sn(PhCOS)_3]^-$

89   **90**

91 (E=S)
92 (=Se)
Tbt = 2,4,6-$[(Me_3Si)_2CH]_3C_6H_2$

$Ph_4As^+[Pb(PhCOS)_3]^-$

93

2.3.2.4
Lead

Attempts to prepare thiocarboxylic acid lead esters using thiobutyric acid were described by Ulrich in 1859 [314]. Lead bis(thiobenzoate) can be obtained in good yields by the reaction of $(CH_3COO)_2Pb$ or $(PhS)_2Pb$ with thiobenzoic acid [39]. Four routes to organoplumbyl thiocarboxylates $RCOE)_xPbR'_{4-x}$ (E=S; x=1, 2) have been developed: (1) $Ph_3PbSLi+RCOCl$ [77, 280]; (2) R'_3PbCl + alkali metal or piperidinium thiocarboxylates [304]; (3) RCOSH+$(C_2H_5)_4$Pb [94] (or Ph$_4$Pb [93, 189], Ph$_3$PbOH [189] and Ph$_6$Pb$_2$ [189]); (4) (RCOS)$_2$Pb+Ph$_2$Pb [189]. Anionic lead complexes **93** Ph$_4$As$^+$[Pb(PhCOS)$_3$]$^-$ with thiocarboxylato ligands also are obtained from two routes: (1) reaction of Na(PhCOS) with Pb(NO$_3$)$_2$ in the presence of Ph$_4$AsCl, and (2) reaction of (PhCOS)$_2$Pb with Ph$_4$As(PhCOS) [39]. A series of organoplumbyl selenocarboxylates **88** (E=Se; x=1, 2) have been synthesized in moderate to good yields by the stoichiometric reaction of the corresponding organolead chlorides with alkali metal or piperidinium selenocarboxylates [160]. Recently, triphenylplumbyl tellurocarboxylates **88** (E=Te; x=1) have been isolated from the reaction of Ph$_3$PbCl with the corresponding tellurocarboxylic acid sodium salts [306]. Lead complexes bearing two or more seleno- or telluro-carbxylato ligands are unknown.

The chalcogen-substituted Ge, Sn and Pb derivatives **86–88** are stable towards both oxygen and moisture. Their stability appears to decrease in the following order: E=S>Se≫Te; M=C≫Sn>Ge>Pb. The *Te*-substituted derivatives RCOTeMR$_3'$ (M=Ge, Sn, Pb) decompose in solutions (of CH$_2$Cl$_2$ or CHCl$_3$, for example) at –20 °C to give (R$_3'$M)$_2$Te (M=Ge, Sn, Pb) with the liberation of black tellurium [306].

2.3.2.5
Structures

The C–O, C–S, and C=O···M (Ge, Sn, Pb) distances for typical Group 14 element chalcogenocarboxylates are shown in Table 3. *O*-Silyl thioacetate [CH$_3$C(S)OSiH$_3$] shows no intramolecular interactions between the thiocarbonyl sulfur and the silicon atoms in both the solid and the gas state [10]. Instead, intermolecular interactions between C=S and Si atoms (3.19 Å) are observed; the resulting coordination number of the central silicon atom is five [10]. The structures of a series of Group 14 element derivatives of monochalcogenocarboxylic acids such as (PhCOS)$_2$Sn(CH$_3$)$_2$ [197], PhCOSSnPh$_3$ [308], ArC(O)EMR$_3$ (E=S, Se, Te; M=Ge, Sn, Pb) [303, 306], and (PhCOS)$_3$Sn(*n*-C$_4$H$_9$) [291] have been revealed. The structures of ArC(O)EMPh$_3$ (E=S, Se, Te; M=Ge, Sn, Pb) are isomorphous, where the complex containing the central Ge, Sn and Pb atoms forms a distorted tetrahedron and the carbonyl C–O, C–E and E–M distances are comparable to the C=O double and C–E and E–M single bonds, respectively. The distances between the carbonyl oxygen and the central Ge, Sn or Pb atom are all within the sum of the van der Waals radii of both the atoms, indicative of intramolecular attractions between the two atoms. It is worth noting that the C=O···Sn distance in 4-CH$_3$C$_6$H$_4$COSMPh$_3$ (M=Ge, Sn, Pb) is shorter than that of the corresponding C=O···Ge, despite the fact that the atomic radius of Sn is larger than that of Ge [303]. A similar shortening of the C=O···Sn distance is observed for the corresponding selenium and tellurium derivatives [306]. Their molecular orbital calculations using the model compounds CH$_3$C(O)EM(CH$_3$)$_3$ (E=S, Se, Te; M=Ge, Sn, Pb) support these results. The NBO (natural bond orbital) analysis of CH$_3$COSM(CH$_3$)$_3$ (M=Ge, Sn, Pb)

94 **95**

Table 4 NBO analysis of at B3LYP/LAND2DZ+p Level [303]

$$CH_3COSM(CH_3)_3$$

M	ΔE [kcal mol^{-1}][a]	
	$n_O \to \sigma^*_{MC7}$	$n_O \to \sigma^*_{MS}$
Ge	1.82	0.64
Sn	2.46	2.25
Pb	2.44	2.05

[a] ΔE=Stabilization energies associated with delocalization.

indicates that the orbital interactions between the nonbonding orbitals on the carbonyl oxygen (n_O) and the σ^*_{MS} orbitals **94** play a more important role than those between n_O and the σ^*_{MC} orbitals **95** (Table 4) [303], whereas in the case of the selenoacetate both interactions ($n_O \to \sigma^*_{MSe}$ and $n_O \to \sigma^*_{MC}$) are important, and the $n_O \to \sigma^*_{MC}$ interactions play a dominant role in telluroacetate [306]. The magnitude of the interactions (the affinity) between Group 14 metals and oxygen or chalcogen atoms such as sulfur, selenium, and tellurium is considered to be Si$\cdots$O>Sn$\cdots$O>Pb$\cdots$O>Ge$\cdots$O$\gg$C$\cdots$O [306]. In the triligated anion complexes [(CH$_3$)$_4$N]$^+$[Sn(PhCOS)$_3$]$^-$ [39] and (Ph$_4$As)$^+$[Pb(PhCOS)$_3$]$^-$ [320], the coordination geometries around the Sn and Pb atoms are trigonal pyramidal. Very weak interactions between the oxygen atoms of the carbonyl groups and the central metal ions are observed. The thiocarboxylate group of dichlorostannyl bis(thiobenzoate), which possesses electron withdrawing atoms on the tin atom, tends to become bidentate [291].

2.3.2.6
Reactions

The unimolecular gas-phase thermolysis of RR'CHC(S)OSi(CH$_3$)$_3$ at 380–760 °C gives the corresponding ketene as the major product [40]. The O-trimethylsilyl thiocarboxylates **83b** (R′=CH$_3$) react with ethanol and acetic acid to give thio-carboxylic acid [87]. Moreover, the compounds **83b** react with a variety of electrophiles, such as benzyl bromide [192] and acyl [192, 202, 226] and ethoxy- and thioethoxy-carbonyl chlorides [192], to afford the corresponding esters, *S*-alkyl thioesters [193], and diacyl sulfides [192, 226] in good yields, respectively. Treatment of **83b** with *tert*-butyl hypochlorite and NBS yields the corresponding acylsulfenyl halides [232]. Reacting ω,ω′-bis(*O*-trimethylsilyl) thio-esters [(CH$_3$)$_3$SiOC(S)-A-C(S)OSi(CH$_3$)$_3$, A=(CH$_2$)$_{2-8}$, -C$_6$H$_4$-] with ω,ω′-diacyl chlorides leads to polythioesters [193]. The selenium **84b** and tellurium iso-logues **85b** also undergo alkylation and acylation on the chalcogen atoms to give the corresponding *E*-alkyl chalcogenoesters (RCOER′, E=Se, Te) and diacyl chalcogenides [(RCO)$_2$E, E=Se, Te], respectively [171, 291], and in addition they react with arylchalcogenyl halides to give the corresponding *Se*-substituted selenides **96** (Scheme 11) [132]. The reactions of **84b** and **85b** with rubidium and cesium fluorides afford the corresponding heavy alkali metal seleno- **4** (M=Rb, Cs) [173] and telluro-carboxylates **5** (M=Rb, Cs) in good yields [174]. *S*-Organos-tannyl thiocarboxylates **87** (E=S; x=1, 2) are useful starting compounds for the synthesis of acylsulfenyl halides (RC(O)SX, X=Cl, Br, I) [150] (see Scheme 12).

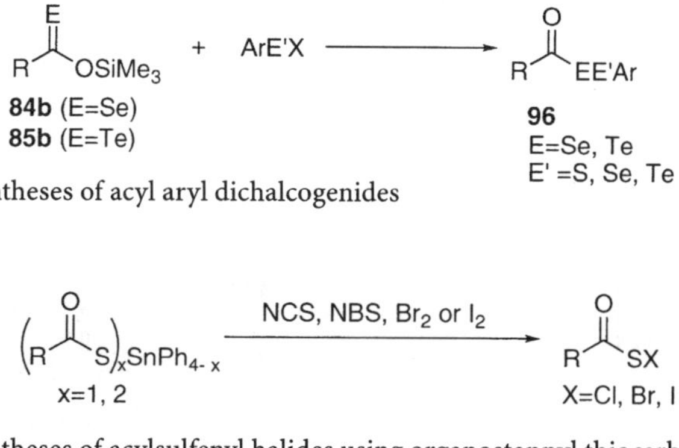

Scheme 11 Syntheses of acyl aryl dichalcogenides

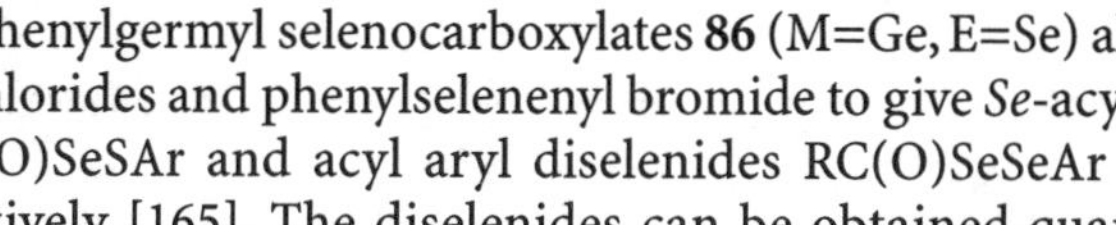

Scheme 12 Syntheses of acylsulfenyl halides using organostannyl thiocarboxylates

The *Se*-triphenylgermyl selenocarboxylates **86** (M=Ge, E=Se) also react with arylsulfenyl chlorides and phenylselenenyl bromide to give *Se*-acyl arylsulfenyl selenides RC(O)SeSAr and acyl aryl diselenides RC(O)SeSeAr in moderate yields, respectively [165]. The diselenides can be obtained quantitatively by reacting *Se*-triphenylstannyl selenocarboxylates with arylselenenyl bromides [117]. Treatment of the *Se*-triphenylstannyl **87** (M=Sn, E=Se, x=1) and *Se*-triphenylplumbyl selenocarboxylates **88** (M=Pb, E=Se; x=1) with *N*-bromo-succinimide produces dibenzoyl diselenide in quantitative yields, presumably via unstable intermediate PhCOSeBr [135]. Lead bis(thiocarboxylate) reacts with tetraphenylarsonium thiobenzoate to give anionic lead complexes **93** [39]. Organostannyl selenocarboxylates **87** (E=Se) do not react with alkyl halides such as benzyl bromide below 100 °C. In the presence of a catalytic amount of $AlCl_3$, however, the reaction proceeds to give the corresponding *Se*-benzyl selenoesters [117]. Although $(PhCOS)_2Pb$ does not dissolve in common solvents, $(cyclo\text{-}C_6H_{11})_3PCSS$ and $(cyclo\text{-}C_6H_{11})_3PO$ solubilize $(PhCOS)_2Pb$ to give a 1:1 adduct $[(cyclo\text{-}C_6H_{11})_3PCSS]\cdot(PhCOS)_2Pb$ [39].

2.3.3
Group 15 Element Compounds (N, P, As, Sb, Bi)

2.3.3.1
Nitrogen

There are four preparation methods for *S*-acyl sulfenamides **96**: (1) RC(O)SX (X=Cl, Br, I)+R'_2NH [20, 284]; (2) M(RCOS) (M=Na, K)+R'_2NCl [78, 229, 262]; (3) M(RCOS) (M=Na, K) + sodium hydroxylamine-*O*-sulfonate (see Scheme 13) [262]; (4) addition reaction of RCOSH to activated N=N double bond [67]. The cyclic compounds **97**, such as benzoisothiazolinones, have also been obtained by oxidizing the corresponding isothiazolthiones with $KMnO_4$ [198] or isopropylethylenoxide [82] and by heating bis(*o*-aminobenzoyl) disulfide in acetic acid [64].

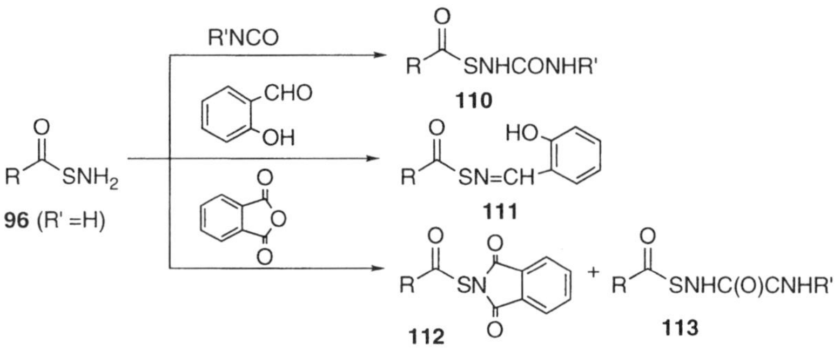

Scheme 13 Reactions of acylsulfenamides

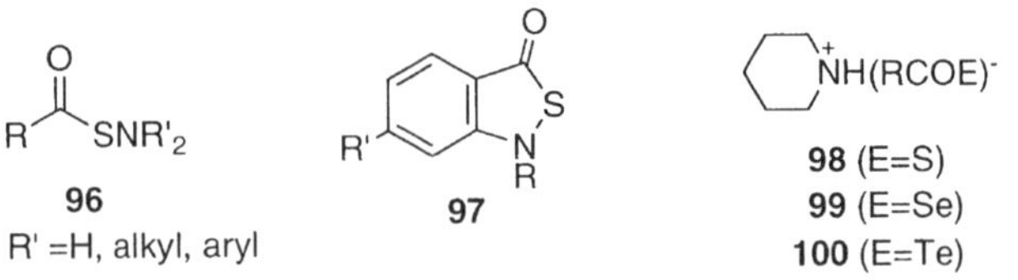

A variety of thiocarboxylic acid ammonium salts have been synthesized by the reaction of thiocarboxylic acids with amines such as piperidine and by salt-exchange reactions of alkali metal thiocarboxylate with tetraalkylammonium halides [141, 142]. Most of the lower aliphatic thiocarboxylic acid piperidinium salts (below C_{10}) are oils. The corresponding diethylammonium salts $(C_2H_5)_2NH_2(RCOS)$, however, are crystals [211]. Aromatic thio- [245], seleno- [245] and telluro-carboxylic acid tetraalkylammonium salts [174] can also be prepared in good yields by salt-exchange reactions between the corresponding alkali metal salts and $(CH_3)_4NCl$. Treatment of diacyl selenides [114], diselenides [115] or ditellurides [133] with primary and secondary amines such as cyclo-hexylamine and piperidine gives the corresponding ammonium seleno- **99** and telluro-carboxylates **100** [133]. The hydrogen selenide anion (HSe^-), generated by bubbling carbon monoxide into tetrahydrofuran solution containing Se powder, a stoichiometric amount of water and a base such as 1,8-diazabicy-clo[5,5,0]undec-7-ene (DBU), can be acylated with acyl chlorides to form selenocarboxylic acid salts $(DBU·H)^+(RCOSe)^-$ [243].

2.3.3.2
Phosphorus

In general, diorganylphosphorus monochalcogenocarboxylates $RCOEPR_2'$ (E=S, Se, Te) are oils and are difficult to purify due to their instabilities (Arbu-zov rearrangement readily occurs). The first example of chalcogenocarboxylic acid phosphorus compounds was reported in 1987 by Russian chemists, who synthesized tris(benzoylthio)phosphine $(PhCOS)_3P$ by reacting PCl_3 with thio-benzoic acid and revealed its structure by X-ray crystallographic analysis [3, 203]. A series of phosphorus compounds with thiocarboxyl groups **101** (E=S, x=1–3) were isolated from the reactions of Ph_2PCl, $PhPCl_2$ and PCl_3 with thiocarboxylic acid alkali metal salts [304] or O-trimethylsilyl esters [132], respectively. Some cyclic compounds **102–104** have been synthesized [45, 199]. The reaction of thiocarboxylic acid with $(RO)_2PSSH$ in the presence of tri-ethylamine produces acyl S-thiophosphoryl sulfides **105a**, which, when heated to 130 °C equilibrate with **105b** [335]. The reaction of the $PhRPSSe^-$ ion with acyl chlorides results in both the sulfide **106a** and the selenide **106b**, in the ratio 2:8 [233]. The reaction of $K[(R'O)_2PSSe]$ with RCOBr forms a selenide of type **106b** [337], whereas the reaction of $Na[(i\text{-}C_3H_7O)_2POSe]$ with benzoyl chloride affords only $PhCOOP(Se)(OC_3H_7\text{-}i)_2$ [81].

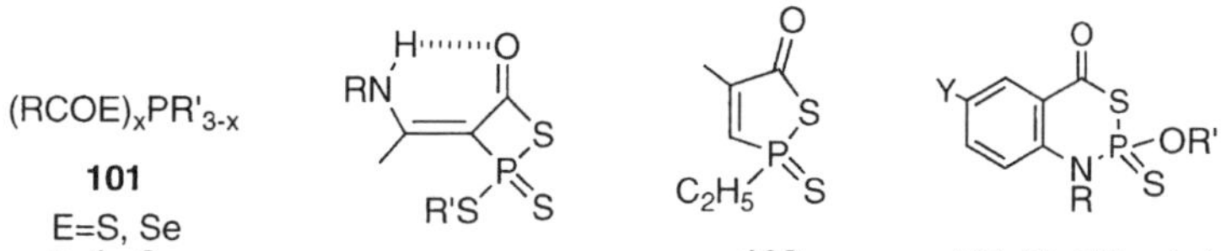

2.3.3.3
Arsenic

Thiocarboxylic acid arsenic derivatives **107** (E=S; x=1–3) have been synthesized by the following four methods: (1) $R'_2PhAsO+RCOSH$ [316]; (2) $R'_2PhAsS+RCOBr$ [317]; (3) $Ph_{3-x}AsCl_x+M(ArCOS)$ (M=K, $(CH_2)_4NH_2$) [305]; (4) $As_2O_3+RCOSH$ [to $(RCOS)_3As$] [292]. Among these, the methods (3) and (4) are preferable in terms of yield and simplicity of procedure. A series of selenium isologues **107** (E=Se) have been prepared in good yields using selenocarboxylic acid sodium salts or O-trimethylsilyl esters [132, 139]. Diphenylarsanyl selenocarboxylates **107** (E=Se, x=1) also are obtained by treating piperidinium diphenylselenoarsinate or -diselenoarsanate with acyl chlorides [137].

$$(RCOE)_x AsR'_{3-x}$$

107

E=S, Se
x=1– 3

2.3.3.4
Antimony

Although antimony tris(thiocarboxylate) $(RCOS)_3Sb$ (R=CH_3 [88, 89], Ph [292]) can be readily obtained in good yields from stoichiometric reactions of Sb_2O_3 with thiocarboxylic acids [88, 292] or $SbCl_3$ with potassium thiocarboxylates [181], the preparation and isolation of the organoantimony thiocarboxylates **108** $(RCOS)_{3-x}SbR'_x$ (x=1, 2) are difficult and have been limited to $PhCOSSbPh_2$ [292] and $(PhCOS)_2SbPh$ [292], most likely due to lack of availability of the starting compounds $R'_{3-x}SbX_x$ (X=Cl, Br, I; x=1, 2). The selenium isologues $(RCOSe)_2SbPh$ and $(RCOSe)_3Sb$ can be synthesized from the reactions of $PhSbCl_2$ and $SbCl_3$ with the selenocarboxylic acid alkali metal salts **4**, respectively [181]. Thiocarboxylato Sb(V) complexes $RCOSSbX(CH_3)_3$ (X=Cl, Br) and $(RCOS)_2Sb(CH_3)_3$ (x=0, 1) have been isolated [256].

$$(RCOE)_x SbR'_{3-x}$$

108

E=S, Se
x=1–3

2.3.3.5
Bismuth

Like the antimony derivatives, bismuth tris(arenecarbothioate) **109** (E=S; x=3) can be easily prepared using the reaction of $(CH_3COO)_3Bi$ [39], $(PhS)_3Bi$ [39], Bi_2O_3 [292] and BiX_3 [181] as the bismuth source. The organobismuth thiocarboxylates **109** are difficult to isolate due to the lack of availability of the corresponding organobismuth halogenides. Recently, a series of phenylbismuth bis(arenecarboselenoates) have been isolated from the reaction of a mixture of Ph_2BiI and $PhBiI_2$ with thiocarboxylic acids [181].

$$(RCOE)_xBiR'_{3-x}$$

109

R=aryl
E=S, Se, Te
x=2, 3

Thiocarboxylic acid Group 15 element derivatives $(RCOS)_xMPh_{3-x}$ (M=P, As, Sb, Bi; x=1–3) are very stable thermally and towards oxygen and moisture whereas the selenium isologues are sensitive toward oxygen and decompose in air with the liberation of red selenium [181]. Bismuth tris(arenecarbothioate) compounds show very poor solubilities in common organic solvents [39].

2.3.3.6
Structures

The selected bond distances of chalcogenocarboxylic acid Group 15 element derivatives are listed in Table 3. The C–O (1.217–1.237 Å) and C–S (1.697–1.723 Å) distances of 2,6-dimethylpiperidinium thiobenzoate [259], $(CH_3)_4N^+$-$(2\text{-}CH_3OC_6H_4COS)^-$ [245], $(CH_3)_4N^+(2\text{-}CF_3C_6H_4COS)^-$ [245], and $(C_2H_5)_4N^+(2\text{-}HOC_6H_4COS)^-$ [223] are long and short, respectively, compared to those of S-aryl thioesters, which indicates that the anion charge is localized on the chalcogenocarboxyl group [245]. A similar trend is observed for the C=O and C–Te distances of the tellurium isologue $(CH_3)_4N^+(2\text{-}CH_3OC_6H_4COTe)^-$ [174, 245]. The molecular structures of a series of thio- and seleno-carboxylic acid pnictogen derivatives $(RCOE)_{3-x}MPh_x$ (M=P, As, Sb, Bi; E=S, Se; x=0–2) have been analyzed by X-ray crystallographic analysis. The diphenyl derivatives $RCOSMPh_2$ (M=P, As, Sb, Bi) show analogous structure regardless of the nature of the central Group 15 atom, where the geometry around the pnictogen atoms is a distorted tetrahedral with a lone pair at the apex, and the thiocarboxyl group exists in the same plane as one of the two phenyl groups. In addition, weak interactions between the carbonyl oxygen and the central pnictogen atom are observed [174, 292]. This similarity in structure can also be observed for the bis- and tris-derivatives $(RCOS)_{3-x}MR'_x$ (M=P, Sb, Bi; x=0, 1) which exist in in C_3 symmetry [181, 292]. In addition, the two chalcogenocarboxylato ligands in the bis-derivatives exist almost in the same plane and the same direction [174]. Molecular orbital calculations, using the simplified models $(CH_3COE)_{3-x}M(CH_3)_x$ (M=N, P, As, Sb, Bi; E=S, Se, Te; x=0–2), indicate interactions between the unshared electron pair on the carbonyl oxygen and the σ^* orbitals of the M–S and/or M–C*ipso* bonds [174, 181, 292], and that the interactions in the bis-derivatives are stronger than those in the mono- (x=2) and tris-ones (x=0) [174]. It is noted that the C=O··· As distances in (RCOS)$_2$MPh (M=P, As, Sb, Bi) are shorter than that of the corresponding C=O···P, despite the fact that the atomic radius of As is larger than that of P [304, 305]. The tendency for this shortening of C=O··· As is observed in the tris-derivatives (RCOS)$_3$M (M=P, As, Sb) [181, 304, 305], reflecting the magnitude of the interactions (the affinity) of the P or As atoms with the oxygen atom.

2.3.3.7
Reactions

S-Acylhydrosulfenamides **96** (R′=H) readily react with phenyl isocyanate, aldehyde, and phthalic anhydride at room temperature to give the corresponding *S*-acylsulfenamides **110–113** (Scheme 13) [262]. Treatment of the cyclic 2,1-benzisothiazoline-3-ones **97** with oxalyl chloride gives the isothiazolium chlorides **114** (Scheme 14) [64]. The reaction of ethylphenylarsinic thioacetate with acetyl bromide gives diacetyl sulfide [317]. *N*-Acylthiosuccinimides **115** have been found to react readily with thio- and dithio-carboxylic acids and dithiophosphoric acids at room temperature to give the corresponding asymmetrical disulfides **125, 128, 130** in good yields (Scheme 15) [144, 229]. The reaction of diphenylarsanyl thio- and seleno-carboxylates **107** (E=S, Se; x=1) with piperidine affords the corresponding piperidinium diphenyldithioarsinate **116** and diphenyldiselenoarsinate **117** along with tetraphenyl diarsane and *N*-acylpiperidine (Scheme 16) [305]. On the other hand, the bis-derivative (ArCOS)$_2$AsPh forms the phenyltrithioarsonate dianion **118** in which two anion charges are delocalized on the AsS$_3$ moiety along with the

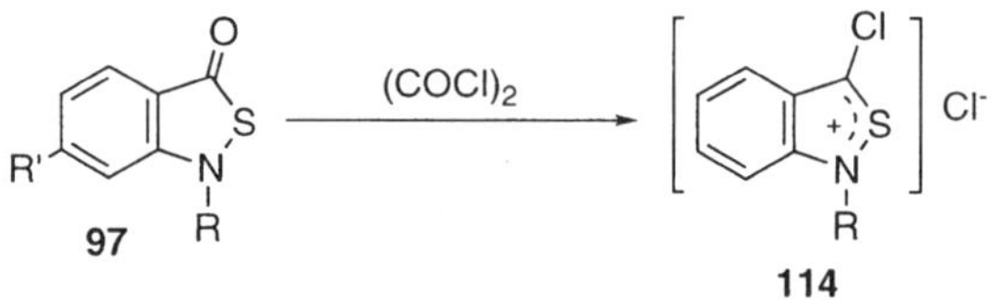

Scheme 14 Reactions of cyclic sulfenamides **97** with oxalyl chloride

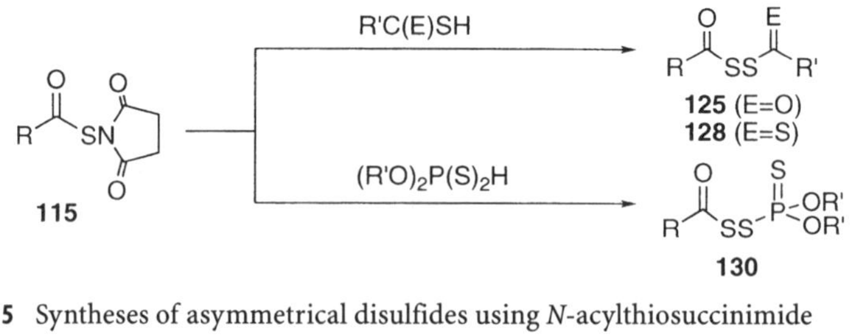

Scheme 15 Syntheses of asymmetrical disulfides using *N*-acylthiosuccinimide

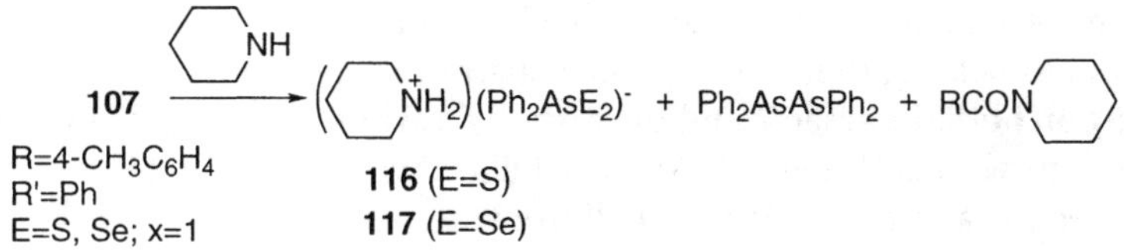

Scheme 16 Syntheses of acyl aryl dichalcogenides

cyclic tetramer **119** (Scheme 17) [305]. Treatment of the arsanyl selenoester **107** (E=Se; x=1) with *N*-bromosuccinimide generates the acylselenenyl bromide **120** which adds to alkenes to give the *cis*-adducts **121** (Scheme 18) [135]. Moreover, **107** reacts with activated alkynes such as phenylacetylene in benzene to give the corresponding (*E*)-addition products **122** with regio- and stereo-selectivity [138] (Scheme 18), whereas the reaction with the sodium alcoholate gives

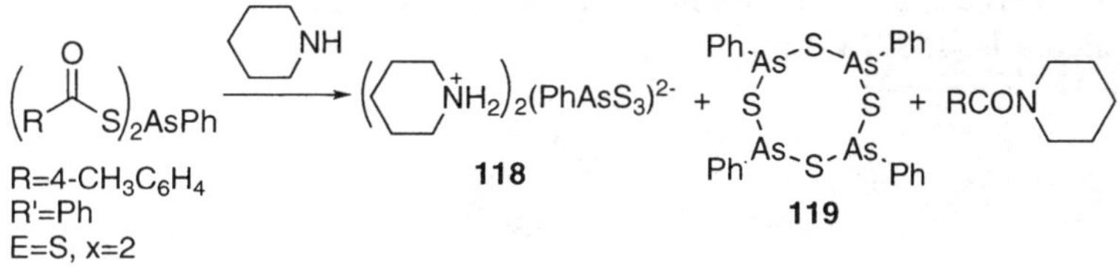

Scheme 17 Reactions of diphenylarsanyl thio- and seleno-carboxylates with piperidine

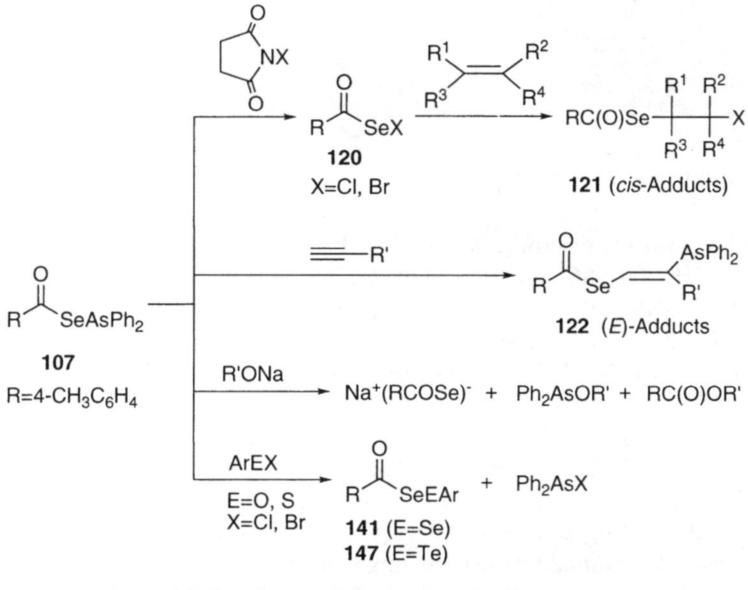

Scheme 18 Reactions of diphenylarsanylselenocarboxylates

the selenocarboxylic acid sodium salt along with the corresponding esters Ph$_2$AsOR′, indicating that nucleophilic substitution reactions take place competitively on both the arsenic and the carbonyl carbon atoms [139]. In contrast, the reactions with arylselenenyl and aryltellurenyl halides lead to the acyl aryl diselenides **141** and the *Se*-acyl aryltellurenyl selenides **147** along with the corresponding Ph$_2$AsX, respectively (Scheme 18) [139].

2.3.4
Group 16 Element Compounds (O, S, Se, Te)

2.3.4.1
Oxygen [RC(O)E-O-, E=S, Se, Te]

Regarding the *S*-acyl alkoxy chalcogenides [RC(O)EOR′; E=S, Se, Te], a few sulfur derivatives **123** RC(O)SOR′ have been prepared from acylsulfenyl chlorides with alcohols in the presence of pyridine [100, 102]. The aromatic derivatives **123** (R=aryl) can also be obtained in good yields using aroylsulfenyl bromides with alcohols [189, 253].

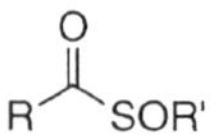

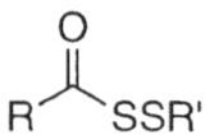

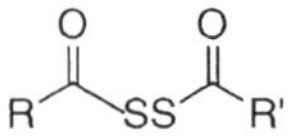

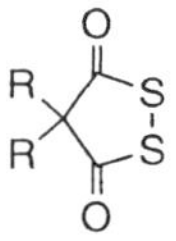

2.3.4.2
Sulfur [RC(O)E-S-, E=S, Se, Te]

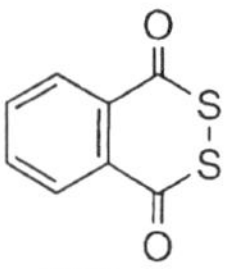

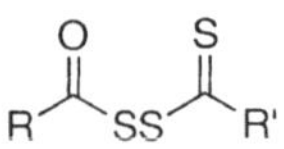

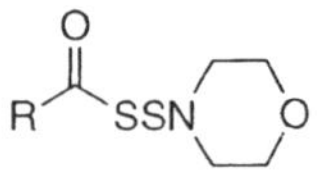

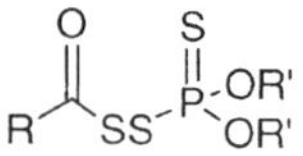

There are a variety of synthetic routes to the acyl organyl disulfides RC(O)SSR′ **124**: (1) CH$_3$C(O)SSCl + alkenes [20, 24, 283]; (2) RC(O)SCl+R′SH [176]; (3) RCOSH+R′SNR″$_2$ [176, 177, 282]; (4) RCOSH+R′SCl [66, 67, 186]; (5) RCOSH+ R′SSO$_3$Na [227]; (6) RCOSH + activated arenes [71]; (7) RCOSH+ thiasulfonium [R′SS$^+$(CH$_3$)$_2$] [55], R′OC(O)SSR″ [36], or organyldithiopyridine *N*-oxide [12]; (8) RC(O)SCN+R′SH [67]; (9) (RCOS)$_2$+R′SH [67]; (10) RC(O)SSO$_2$R+ R″SH [67]; (11) RC(O)SS(O)COR′+R″SH [95]; (12) ArCSSR′ [or ArC(=S= O)SR′]+*m*-ClC$_6$H$_4$C(O)OOH (*m*-CPBA) [225]. Fields and his coworkers have examined most of these routes and recommend method (4) in terms of yield and simplicity of procedure [65]. The symmetrical diacyl disulfides **125** (R=R′) can be readily obtained by the following methods: (1) air-oxidation of thiobenzoic acids [70]; (2) oxidation of alkali and alkali earth metal thiocarboxylates with oxygen [62], H_2O_2 [1], I_2/KI [1, 248], CH_3SO_2Cl [68], $ArSO_2Cl$ [68, 141], $(RO)_4Sn/FeCl_3$ [270], $Tl(CH_3COO)$ [313], K_2FeCN_6 [68], NiO_2 [239] or XeF_2 [244]; (3) heating $Tl(RCOS)_3$ (R=CH_3, Ph) [266, 313] or acylsulfenyl halides [266]; and (4) the reaction of RCOCl with Li_2S_2 [79, 80]. Amongst these, oxidizing potassium thiocarboxylates with an ethanolic solution of I_2/KI [248] or an acetonitrile solution of XeF_2 [244] in method (1) is the method preferred for the preparation of common diacyl disulfides. Three five-membered ring diacyl disulfides **126** have been synthesized by the iodo-oxidation of bis(thiomalonic acids) in the presence of pyridine [273], but the isolation of six-membered compounds like **127** which are more sterically hindered is still to be solved. The asymmetrical diacyl disufides **125** (R≠R′) can also be prepared by reacting thiocarboxylic acid with acetylsulfenyl chloride [22] or *N*-acylthiosuccinimides [229]. The acyl thioacyl disulfides **128** have been prepared by the following three reactions: (1) RCSSH+*N*-acylthiosuccinimides [144]; (2) RCOSH+ RC(S)SBr [155] and (3) RCSSH+RC(O)SBr [232]. The reactions of $CH_3C(O)SSCl$ with morphorine, of dithiophosphoric acids with *N*-acylthiosuccinimides, and of diacyl disulfides with chlorine yield the asymmetrical disulfides **129** [284], **130** [157] and **131** [20, 22], respectively, in which the nitrogen, phosphorus or chlorine atom is connected to the opposite sulfur atom of the disulfide sulfur moiety, respectively.

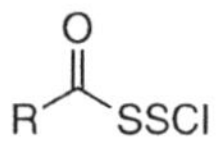

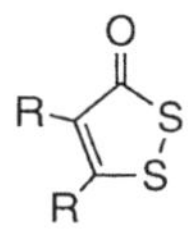

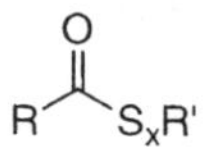

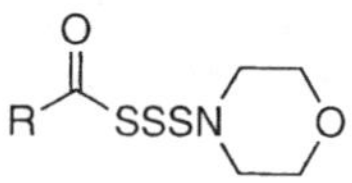

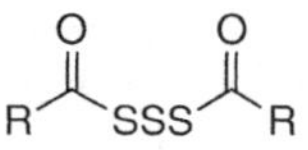

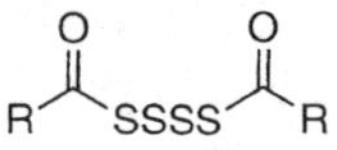

A variety of cyclic compounds **132**, including a -C(O)SS- moiety, have been synthesized [63, 209]. Acetyl alkyl trisulfides **133** have been obtained from the reactions of thiocarboxylic acid with alkylthiosulfenyl chlorides (RSSCl) [285] and acetylthiosulfenyl chloride $CH_3C(S)SCl$ with thiophosgen [219]. Acetyl trichloromethyl trisulfide generated from acetylthiosulfenyl chloride and Cl_2CS reacts with morphorine to give acetyl morphorinecarbonyl trisulfide **134** in good yields [219]. The acetyl methyl **133** and the diacetyl polysulfides **135** (R=CH_3, x=3, 4) can be obtained by the reaction of R′SH or CH_3COSH with $CH_3C(O)S_xCl$ (x=2, 3) or by oxidation of acetylthiosulfenyl chloride with an aqueous I_2/KI solution, respectively [23]. Aromatic diacyl tri- **135** (R=aryl; x=3) [14, 15, 19] and tetra-sulfides **136** (R=aryl; x=4) [14, 15, 19] can be easily produced by the reactions of thiobenzoic acid or its potassium salts with SCl_2 and S_2Cl_2, respectively. *Se*-Acyl arylsulfenyl selenides **137** are prepared by the reactions of selenocarboxylic acid alkali metal salts [116] or *O*-trimethylsilyl thioesters [132] with arylsulfenyl halides and of *Se*-triphenylgermyl selenocarboxylates with arylsulfenyl halides [117, 165]. *Te*-Acyl arylsulfenyl tellurides **138** are isolated from the reaction of potassium tellurocarboxylates with arylsufenyl chlorides [133].

$$\underset{\textbf{137}}{R\overset{O}{\overset{\|}{-}}SeSAr} \qquad \underset{\textbf{138}}{R\overset{O}{\overset{\|}{-}}TeSAr}$$

2.3.4.3
Selenium [RC(O)E-Se-, E=S, Se, Te]

For the synthesis of the acylsulfenyl aryl selenides **139**, the following five routes have been developed: (1) RCOSH or M(RCOS) (M=Na, K)+ArSeBr [152]; (2) (RCOS)$_2$ [152] or (RCOS)$_2$SnBr$_2$ [151]+PhSeBr [151, 152]; (3) RCOSH+ phenylselenylsuccinimides [309]; (4) (RCOSe)$_2$+ArSBr [153] and (5) ArC(S)SePh+ *m*-CPBA [159]. Among these, methods (1) and (2) are the most practical, with easy procedures and high yields. Thiocarboxylic acids and their piperidinium salts react with SeCl$_4$ or Na$_2$SeS$_4$O$_6$ to give the selenium bis(thiocarboxylates) **140** (E=Se) [4, 156]. The acyl aryl diselenides **141** are obtained in moderate to good yields by the reactions of selenocarboxylic acids [113, 116] and their alkali metal or piperidinium salts [113, 116] with arylselenyl halides. Instead of these acids and salts, *O*-trimethylsilyl [132] and *Se*-organo-germyl [165], -stannyl [117], -plumbyl [160] and -arsenic selenoesters [139] have also been used as starting compounds for the synthesis of **141**. The aliphatic diacyl diselenides **142** (R=alkyl) (except for di(stearoyl) diselenide) have not been isolated due to difficulties with purification. In 1932 Szperl and Wiorogorsky reported the isolation of dibenzoyl diselenide from the reaction of acyl chloride in the presence of AlCl$_3$ [301]. Since then, several routes to the diaroyl diselenides **142** (R=aryl) have been found: (1) oxidation of RCOSeH [104, 114, 120] or M(RCOSe) (M= alkali metals, piperidinium) with air, iodine [113] or XeF$_2$ [246]; (2) reaction of RCOCl with elemetal selenium/NaOH under phase-transfer conditions [318];

$$\underset{\textbf{139}}{R\overset{O}{\overset{\|}{-}}SSeR'} \qquad \underset{\textbf{140}}{R\overset{O}{\overset{\|}{-}}SSeS\overset{O}{\overset{\|}{-}}R} \qquad \underset{\textbf{141}}{R\overset{O}{\overset{\|}{-}}SeSeAr}$$

$$\underset{\textbf{142}}{R\overset{O}{\overset{\|}{-}}SeSe\overset{O}{\overset{\|}{-}}R} \qquad \underset{\substack{\textbf{143} (E=S)\\ \textbf{144} (E=Se)}}{\overset{O}{\overset{\|}{-}}\overset{E}{\underset{Se}{<}}} \qquad \underset{\textbf{145}}{R\overset{O}{\overset{\|}{-}}STeS\overset{O}{\overset{\|}{-}}R} \qquad \underset{\textbf{146}}{R\overset{O}{\overset{\|}{-}}STeR'}$$

(3) reaction of acyl chlorides with HSe$^-$ generated by bubbling CO into a suspension of elemental selenium in tetrahydrofuran solution containing stoichiometric amounts of H_2O and base, followed by air-oxidation [243]. Among these, method (1), using Na(RCOSe) and XeF_2, and method (2) are preferable. Cyclic compounds that include a -C(O)SE- (E=S, Se) [201] or -C(O)SeSe- moiety [201] in the molecules, as in **143** and **144**, have also been prepared.

2.3.4.4
Tellurium [RC(O)E-Te-, E=S, Se, Te]

The *S*-acyl aryltellurenyl sulfides **146** can be synthesized from the reactions: (1) RCOSH+PhTeOH [105]; (2) RC(O)SBr+ArTeTeAr [152]; (3) M(RCOS) (M=Na, K, piperidinium)+ArTeBr [151]. The crystalline *Se*-acyl phenyltellurenyl selenides **147** have been isolated in good yields from the reaction of *O*-trimethylsilyl selenocarboxylate with PhTeI [132]. Thiocarboxylic acids and their piperidinium salts react with $TeCl_4$ or $Na_2TeS_4O_6$ to give the tellurium bis(thiocarboxylates) **145** (E=Te) in good yields [156, 296]. In contrast to the sulfur and selenium compounds such as **125, 137, 139, 140, 141,** and **142**, the acyl aryl ditellurides**148** and the diacyl ditellurides **149** are thermally very labile and are air-sensitive, so their isolation has been limited to just the 2-methoxy derivative **149** (R=2-$CH_3OC_6H_4$) from the oxidation of the corresponding alkali metal or piperidinium salts with I_2 [133], $C_6H_5SO_2Cl$ [133], XeF_2 [247] or air [298].

$$
\begin{array}{ccc}
\underset{\textstyle R}{\overset{\textstyle O}{\|}}\!\!\diagdown\!\!SeTeR' & \underset{\textstyle R}{\overset{\textstyle O}{\|}}\!\!\diagdown\!\!TeTeR' & \underset{\textstyle R}{\overset{\textstyle O}{\|}}\!\!\diagdown\!\!TeTe\diagup\!\!\underset{\textstyle R}{\overset{\textstyle O}{\|}} \\
\mathbf{147} & \mathbf{148} & \mathbf{149}
\end{array}
$$

2.3.4.5
Structures

The molecular structures of selenium(II) [4] and tellurium(II) bis(thiobenzoate) **145** (R=C_6H_5) [296] show that the two thiocarboxyl groups are attached to the central selenium or tellurium atoms through the sulfur atom (for bond distances, see Table 3). The geometry around the selenium or tellurium atom appears to be a distorted tetrahedral, wherein the two carbonyl oxygen atoms associate with the central Se or Te atoms of other molecules. In the cases of diacyl dichalcogenides such as dibenzoyl disulfide [267], bis(4-chlorobenzoyl) disulfide [244] and bis(4-methoxybenzoyl) diselenide [246], the carbonyl oxygen atoms associate with the opposite chalcogen atom [C=O···S distance: <3.0 Å; C=O···Se: 3.0 Å] and the dihedral angles of -C(O)-E-E-C(O)- (E=S, Se) are 79–90° [244, 246, 267]. Interestingly, the ditelluride **149** (R=2-$CH_3OC_6H_4$) is rendered almost planar by the four non-bonding intramolecular interactions between the tellurium and the methoxy oxygen atoms and between the Te and the opposite carbonyl oxygen atoms (**150**), whereas the corresponding sulfur **125** and selenium isologues **142** exhibit dihedral angles of ~70° [247].

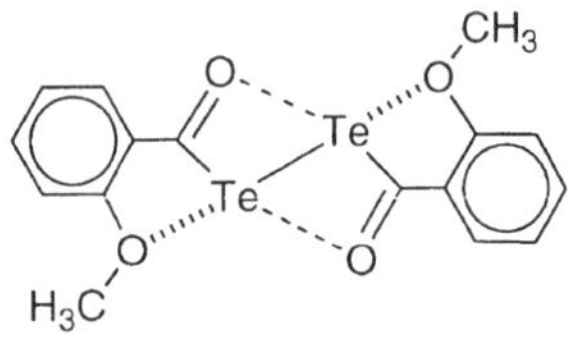

150

Molecular orbital calculations show that orbital interactions between the loan pair of the carbonyl oxygen and σ^*_{EC} (1,4-interaction) contribute to the stabilization of the planar conformer of the ditelluride, and also result in a shortening of the Te···O=C distance [247]. The orbital energies of the lone pair of the carbonyl oxygen in the dichalcogenides $RC(O)EEC(O)R$ ($R=2\text{-}CH_3OC_6H_4$, E=S, Se, Te) are almost equal, and the energy gap between n_O (donor) and σ^*_{TeC} (acceptor) is the smallest among the three compounds. The interactions between the n_O orbitals of the methoxy oxygen and the σ^*_{TeTe} orbitals play a more important role in the planarity of $RC(O)TeTeC(O)R$ ($R=2\text{-}CH_3OC_6H_4$) than those between the carbonyl oxygen and the σ^*_{TeTe} or σ^*_{TeC} orbitals [247].

Disulfides such as **124**, **125**, and **132** are highly stable. Diacyl diselenides **142** gradually decompose in air with liberation of elemental selenium to give diacyl selenides. Asymmetrical diacyl disulfides ($R\neq R'$) gradually undergo disproportionation reactions at room temperature to give the corresponding symmetrical diacyl disulfides [22]. The acyl aryl disulfides **124** (R'=aryl) also undergo disproportionation under heating to give the corresponding symmetrical diacyl and diaryl disulfides [65]. The structures of selenium [4] and tellurium bis(thiobenzoates) [296] have the carbonyl oxygen atoms associated with the central selenium or tellurium atom on another molecule.

Aromatic diacyl trisulfide **135** (R=aryl), selenium **140** (E=Se) and tellurium bis(thiocarboxylates) **145** (E=Te) are stable both thermally and towards oxygen. The introduction of functional groups such as CH_3O and $(CH_3)_2N$ that possess an electron pair on the *ortho*-position of the phenyl ring significantly enhances the stability of these aromatic chalcogenocarboxylic acid derivatives with the intramolecular coordination of the loan pair on the oxygen or chalcogen atom of the carbonyl or the chalcogenocarbonyl group to the central metal.

2.3.4.6
Reactions

Alcoholysis of the acyl organyl sulfides $RC(O)S_xR'$ (R, R'=alkyl, aryl; x=2, 3) in the presence of hydrogen chloride gives the organyl hydrodi- and hydrotrisulfides RS_xH (x=2, 3) in good yield [20, 21, 65, 176, 282]. Acyl alkyl disulfides react with thiols to yield asymmetrical disulfides [200]. Selenium bis(thiocarboxylates) **140** (E=Se) react with halogen to give acylsulfenyl halides which gradually decompose to the corresponding acyl chloride and diacyl disulfides [157]. Similar halogenation of tellurium bis(thiocarboxylate) **145** (E=Te) at 0 °C gives diacyl disulfides with the liberation of black tellurium, presumably via halotellurium bis(thiocarboxylate) **151** [156]. Treatment of the telluride with bromine at –30 °C affords dibromotellurium bis(thiocarboxylate) **151** (X=Br) [156].

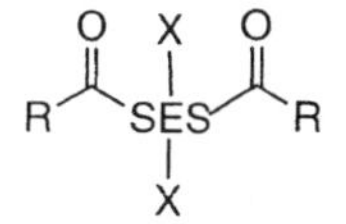

151

E=Se, Te; X=Cl, Br

Although aroyl aryl disulfides **124** (R, R′=aryl) do not react with methanol (even under reflux conditions), acetyl benzyl disulfides **124** (R=CH$_3$; R′=PhCH$_2$) are ethanolyzed to give the corresponding benzyl hydrodisulfides in good yields [176]. The reaction of **124** with thiols proceeds at room temperature to yield asymmetrical disulfides [200]. Refluxing 4-chlorobenzoyl phenyl diselenide in a mixed solvent of methanol/benzene produces the deselenized *Se*-phenyl selenoester in good yields [116], while distearoyl diselenide **142** (R= *n*-C$_{17}$H$_{35}$) in methanol is decomposed to methyl stearate with the evolution of hydrogen selenide and the liberation of elemental selenium [114]. Treatment of *S*-acyl organoselenenyl sulfides **139** with primary and secondary amines in the presence of Hg(CH$_3$COO)$_2$ results in good yields of the corresponding amides [310]. The compound **139** has also been found to react with cyclohexene to give the adduct **152** (Scheme 19) [310]. The acyl aryl **124** and diacyl disulfides **125** [151] are readily desulfurized by triorganophosphines to give *S*-aryl thioesters and diacyl sulfides in good yields, respectively. Reduction of diaroyl disulfides with LiAlH$_4$ affords the corresponding thiocarboxylic acid and benzylmercaptane [196]. Similarly, aroyl phenyl diselenides **141** (R=aryl; R′=Ph) [116] are deselenized with triphenylphosphine giving *Se*-aryl selenoesters [151]. In contrast, *S*-acyl phenyltellurenyl sulfides **146** (R′=Ph) are not desulfurized with trialkylphosphines even under reflux in toluene [151]. Refluxing diaroyl diselenides **142** (R=aryl) in a mixed solvent of methanol and benzene (1:1) in the presence of KOH yields potassium selenocarboxylates containing one selenium

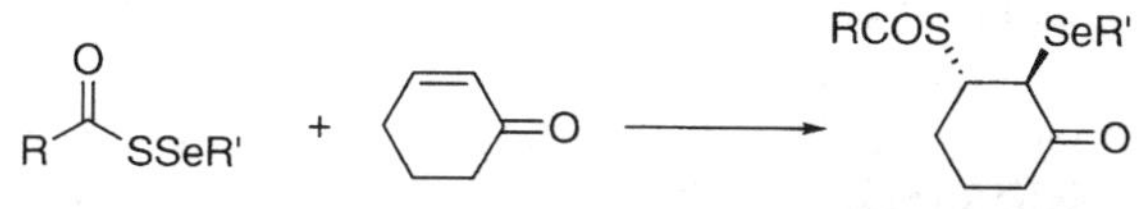

139 + → **152**

Scheme 19 Reactions of *Se*-aryl acylsulfenyl selenides with cyclohexenone atom as green crystals [104, 116]. The reaction of acetyl benzyl disulfide **124** (R=CH$_3$, R′=PhCH$_2$) with sodium benzylthiosulfate in water gives dibenzyl trisulfide in moderate yields, and the reaction with cyclohexylamine affords *N*-cyclo-hexylacetoamide [227]. Acetylthiosulfenyl chloride **131** (R=CH$_3$) reacts with alkenes to give both Markovnikov and anti-Markovnikov adducts (except for styrene, which gives only the Markovnikov oriented product) [230]. Treatment of the diaroyl di- and tri-sulfides (RCO)$_2$S$_x$ (x=2, 3) with aniline quantitatively produces benzanilide with the liberation of elemental sulfur or hydrogen sulfide [19]. The diaroyl disulfides **125** (R, R′=aryl) and diselenides **142** (R= aryl) react with piperidine to give the corresponding chacogenocarboxylic acid piperidinium salts [(CH$_2$)$_5$NH$_2$]$^+$(RCOE)$^-$ (E=S [170], Se [115], Te [133]). Dibenzoyl disulfide, and Lewis acids such as BX$_3$, AlX$_3$, SnCl$_4$, TiCl$_4$, ZrCl$_4$, and SbCl$_5$ form the 1:1 adducts MX$_x$·(PhCOS)$_2$ (M=metal; X=Cl, Br) [206]. Nakabayashi and his coworkers synthesized acetyl benzyl disulfide **124** (R′=PhCH$_2$) with ^{35}S at the acetylsulfenyl sulfur, and revealed that the acetylsulfenyl sulfur atom is swapped for acetyl benzyl disulfide with triphenylphosphine [177]. The diacyl diselenides **142** serve as useful acylating agents for a variety of nucleophiles such as amines, sodium alcoholates and thiolates, and organocopper compounds [242]. In 1907, Fromn and Schmoldt reported that the dry-distillation of dibenzoyl disulfide produces a crystalline compound with formula C$_{14}$H$_{10}$S$_4$: "tolanetetrasulfide" [68]. The structural analysis of this compound remains unresolved.

2.3.5
Group 17 Element Compounds (F, Cl, Br, I)

2.3.5.1
Fluorine

No acylsulfenyl fluorides have been known.

2.3.5.2
Chlorine

The first acylsulfenyl halides **153–154** RC(O)SX (X=Cl, Br, I) were reported in 1952 by Böhme and Clement [21] who isolated acetylsulfenyl chloride **153** (R=CH$_3$) using the reactions of thioacetic acid, diacetyl sulfide or acetyl benzoyl sulfide with chlorine below –10 °C [20]. Trifluoroacetyl derivatives also have been prepared by a similar method [228, 266]. This methodology, however, cannot be applied to the synthesis of other common aliphatic and aromatic acylsulfenyl halides due to their thermal instabilities and purification difficulties. In 1982, a general preparation method for acylsulfenyl halides was developed by reacting diphenylstannyl bis(thiocarboxylates) with a halogen or *N*-halorosuccinimide (NXS) (Scheme 20) [149]. Pivaloylsulfenyl chloride has been synthesized without an α-hydrogen atom via the reaction of *O*-trimethylsilyl thiopivaloate with *tert*-butyl hypochlorite [232].

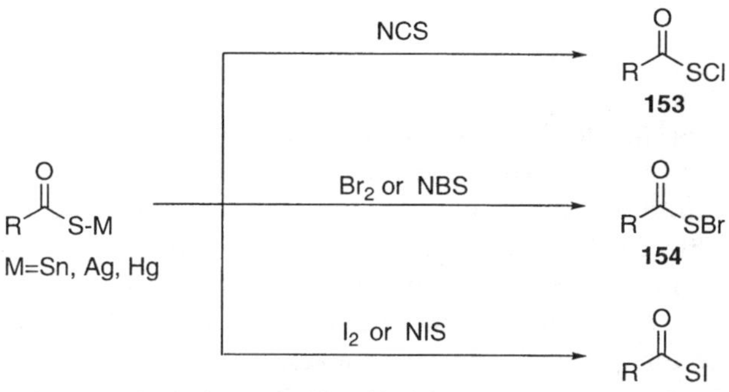

Scheme 20 Synthetic methods for acylsulfenyl halides

2.3.5.3
Bromine

Aromatic acylsulfenyl bromides **154** have been synthesized by reacting thio-carboxylic acid phenylmercury [148], diphenyltin [149], silver [149], or *O*-tri-methylsilyl derivatives [232] with bromine or *N*-bromosuccinimide (NBS) (Scheme 20). In contrast, the selenium isologues are very sensitive toward oxy-gen and are also thermally labile. In the reaction of the diphenylarsanyl seleno-carboxylates **107** with NBS, only the formation of the 4-methylbenzoylselenenyl bromide **120** (R=4-CH$_3$C$_6$H$_4$; X=Br) has been proved, by converting it into the β-bromoethyl selenocarboxylates **121** (X=Br) (see Scheme 18) [135].

2.3.5.4
Iodine

Like the chlorides **153** and the bromides **154**, treating phenylmercury thiocar-boxylates with iodine yields the acylsulfenyl iodides **155** in moderate yields (Scheme 20) [146]. Moreover, silver thiocarboxylates undergo *S*-iodination with *N*-iodosuccinimide (NIS) to give **155** [148]. Trifluoroacetylsulfenyl iodide has been isolated [228]. The use of diphenylstannyl bis(thiocarboxylates) instead of PhHg(RCOS) is more effective [149].

The acylsulfenyl halides **153–155** are labile thermally and toward oxygen. The stabilities toward oxygen and heat decrease in the order X=Cl>Br>I, while the stability toward moisture decreases in the order X=I>Br>Cl.

2.3.5.5
Structures

2-Methoxybenzoylsulfenyl bromide and benzoylsulfenyl iodide have mono-meric and nearly planar structures, where the carbonyl oxygen intramolecu-larly coordinates to the halogen atoms, but no intermolecular interactions between the oxygen and halogen or between halogens are observed [170].

2.3.5.6
Reactions

The acylsulfenyl halides **153–155** have been known to act as electrophiles. Some of their reactions are shown in Scheme 21. For example, acetylsulfenyl chloride reacts with amines such as aniline to give the corresponding *S*-acetylsulfenamides **96** [20, 262, 283], and it undergoes the acylthiolation reaction on the α-carbon with carbonyl compounds such as alkyl acetates, benzoylacetone, acetone and *cyclo*-hexanone, leading to the thioesters **156** [23]. Moreover, the reaction with unequivocally polarized alkenes yields the acetyl β-chloroalkyl sulfides **157** (anti-Markovnikov addition) [20, 24] along with the acyl β-halogenoalkyl disulfides **158** [20, 24, 286], whereas the reaction with a variety of activated aromatic compounds such as phenol and toluene gives the acetyl aryl disulfides **159** as a major product [25, 71]. Aroylsulfenyl bromides **154** (R=aryl) also add to alkenes to yield similar anti-Markovnikov adducts [147]. A formation mechanism for the disulfides **158** and **159** has been proposed, and is illustrated for the β-chloroalkyl disulfide **162** in Scheme 22. Here the episulfide intermediate **161** formed by splitting CH_3COCl from the episulfonium chloride **160** reacts further with another mole of acetylsulfenyl chloride [24]. The aroylsulfenyl bromides **154** and iodides **155** readily react with thio- and dithio-benzoic acids and their piperidinium salts to give asymmetrical acyl benzoyl **125** [22, 147] and acyl thiobenzoyl disulfides **128** in good yields, respectively [147, 232]. The trifluoroacetylsulfenyl halides $CF_3C(O)SX$ (X=Cl, Br) react with mercury at room temperature, giving $Hg(CF_3COS)_2$ and $(CF_3COS)_2$ [266]. The acylselenenyl halides **120** also react with alkenes such as cyclohexene

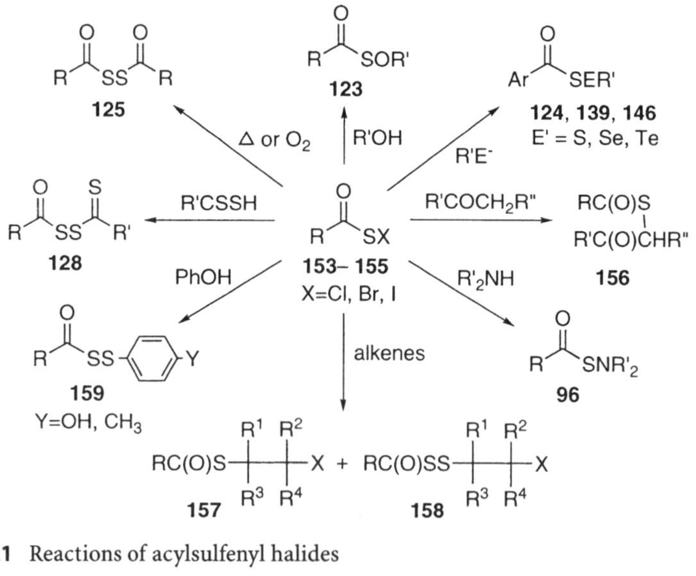

Scheme 21 Reactions of acylsulfenyl halides

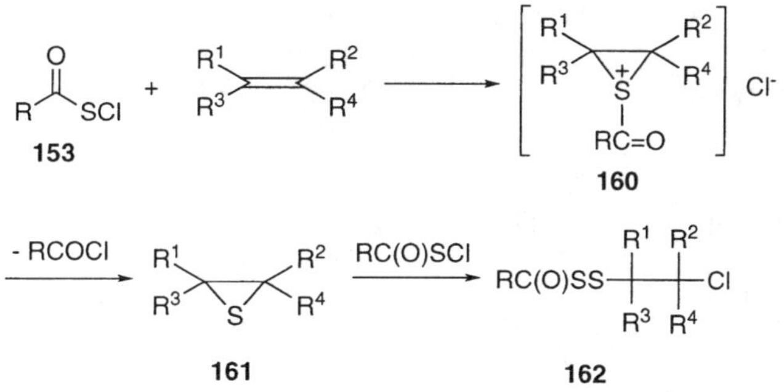

Scheme 22   Formation mechanism for chloroethyl acyl disulfides **162**

and enol silyl ether at room temperature to give *cis*-adducts **121** as major products (see Scheme 18) [135].

3
Infrared, NMR and Electron Spectra

3.1
Alkali Metal and Alkali Earth Metal Salts

C=O stretching frequencies and some ^{13}C=O, ^{77}Se, and ^{125}Te spectra of representative monochalcogenocarboxylic acid alkali and alkali earth metal salts are collected in Table 5. The asymmetric νCOE bands (E=S, Se, Te) of these ionic alkali metal salts can be observed in the range of 1510–1585 cm^{-1}. The ^{13}C=O, ^{77}Se, and ^{125}Te NMR spectra appear at δ=205–225, δ=210–230, and δ=50–470, respectively. Specific trends in the carbonyl stretching and ^{13}C=O, ^{77}Se, and ^{125}Te signals are not observed as we progress through the metal cations. Little is known about the IR and NMR spectra of the alkali earth metal monochalcogenocarboxylates.

3.2
Transition Metal Complexes

In Table 6, the C=O stretching frequencies and some ^{13}C=O, ^{77}Se, ^{125}Te and ^{31}P NMR spectra of transition metal complexes with monochalcogenocarboxylato ligands are listed. The C=O bands for the thiocarboxylato ligands bridged between the two molybdenum metals are observed at 1400–1450 cm^{-1} [107]. The bidentate thiocarboxylato ligands of homoleptic transition metal complexes such as (Ph$_4$P)M(PhCOS)$_3$ (M=Mn, Co, Ni) show the asymmetrical νC=O bands at 1540–1630 cm^{-1} [53]. The thio-, seleno- and telluro-carboxylato ligands of M(RCOE)$_2$(R$_3'$P) [M=Ni (**48**), Pd (**54**), Pt (**59**); E=S, Se, Te; R, R$'$=alkyl, aryl] are typical monodentate ligands and their asymmetrical carbonyl stretch-

Table 5 Spectral data for alkali and alkaline earth metal monochalcogenocarboxylates

Compounds, M(RCOE)	M	E	IR[a]/cm^{-1} ν (C=O)	NMR [δ]		References
				^{13}C=O[b]	^{77}Se or Te[b]	
Li(C$_6$H$_5$COS)((C$_2$H$_5$)$_2$O)	Li	S	1509			[150]
Li(CH$_3$COSe)(LiCl)	Li	Se	1557	222.5	409.0	[187]
Li(C$_6$H$_5$COSe)(LiCl)	Li	Se	1556	217.5	371,1	[187]
Li(CH$_3$COTe)(THF)	Li	Te	1584 1505	210.2	186	[164]
Li(C$_6$H$_5$COTe)/THF	Li	Te	1558	211.8	234.8	[164]
Na(CH$_3$COS)	Na	S	1517 1551 1562			[150]
Na(C$_6$H$_5$COS)	Na	S	1521	216.4		[150, 245]
Na(CH$_3$COSe)	Na	Se	1575	219.7		[150, 168]
Na(C$_6$H$_5$COSe)	Na	Se	1525	215.9	224.8	[162, 168]
Na(CH$_3$COTe)	Na	Te	1552	207.8	251.9	[167]
Na(C$_6$H$_5$COTe)	Na	Te		209.0	224.8	[167]
K(CH$_3$COS)	K	S	1528			[150]
K(C$_6$H$_5$COS)	K	S	1523	214.5		[150]
K(CH$_3$COSe)	K	Se		219.7	392.8	[246]
K(C$_6$H$_5$COSe)	K	Se	1538	216.0	364.0	[246]
K(CH$_3$COTe)	K	Te		207.1		[167]
K(C$_6$H$_5$COTe)	K	Te	1548	208.9	220.0	[167]
Rb(CH$_3$COS)	Rb	S	1525	220.2		[150, 245]
Rb(C$_6$H$_5$COS)	Rb	S	1520	214.5		[150, 245]
Rb(CH$_3$COSe)	Rb	Se	1538	219.7		[172]
Rb(C$_6$H$_5$COSe)	Rb	Se	1545	215.8	370.8	[171, 172]
Rb(CH$_3$COTe)	Rb	Te	1570	207.3	262,1[c]	[174]
Rb(C$_6$H$_5$COTe)	Rb	Te	1543	208.7	228.8[c]	[174]
Cs(CH$_3$COS)	Cs	S	1520			[150]
Cs(C$_6$H$_5$COS)	Cs	S	1526	214.2		[150, 245]
Cs(CH$_3$COSe)	Cs	Se	1539d	219.9		[172]
Cs(C$_6$H$_5$COSe)	Cs	Se	1547	215.7	388.7	[171, 172]
Cs(CH$_3$COTe)	Cs	Te	1574	207.2	286.9[c]	[174]
Cs(C$_6$H$_5$COTe)	Cs	Te	1555	208.6	254.4[c]	[174]
Mg(4-CH$_3$C$_6$H$_4$COS)$_2$(15-C-5)[e]	Mg	S	1487	208.0		[170]
Ca((4-CH$_3$C$_6$H$_4$COS)$_2$(18-C-6)[f]	Ca	S	1459	209.2		[170]
Sr(4-CH$_3$C$_6$H$_4$COS)$_2$(18-C-6)	Sr	S	1492	213.5		[170]
Ba(4-CH$_3$C$_6$H$_4$COS)$_2$(18-C-6)	Ba	S	1507	208.6		[170]

[a] KBr disc; [b] In CDCl$_3$; [c] In CH$_3$OD; [d] Nujol; [e] 5–C–5=15-crown ether-5; [f] 18–C–6=18-crown ether-6.

Table 6 Spectral data for monochalcogenocarboxylatato transition metal complexes

Complexes, RCOE–M	M	E	IR^a/cm^{-1} ν (C=O)	NMR [δ] ^{13}C=O^b	^{77}Se or Te^b	Hz ^{31}P^b	References
Sm[4-CH$_3$C$_6$H$_4$COS]$_3$·(THF)$_2$	Sm	S	1458^c	228.2^d			[140]
U([CH$_3$COS]$_2$(Cp)$_2$	U	S	1487				[5]
U([C$_6$H$_5$COS]$_2$(Cp)$_2$	U	S	1596				[5]
Nb[C$_6$H$_5$COS]$_2$(Cp)$_2$	Nb	S	1610				[110]
Ta[C$_6$H$_5$COS]$_2$(Cp)$_2$	Ta	S	1610				[110]
Cr(CH$_3$COS)$_3$	Cr	S	1635				[271]
Mo$_2$(CH$_3$COS)$_4$	Mo	S	1450				[107]
Mo$_2$(PhCOS)$_4$	Mo	S	1450, 1402				[107]
Cp$_2$MoO(CH$_3$COS)$_2$	Mo	S	–1520				[84]
W(CH$_3$COS)(CO)$_5$	W	S		197.19			[52]
Fe(Cp)(CO)$_2$(C$_6$H$_4$COS)	Fe	S	1596				[57]
Fe(Cp)(CO)$_2$(4-NO$_2$C$_6$H$_4$COS)	Fe	S	1590				[57]
cis-Ru[Ph(CH$_3$)$_2$P](Et$_2$NH)$_2$(C$_6$H$_4$COS)$_2$	Ru	S	1525				[85]
Ru[(CO)$_2$(*tert*-BuCp)$_2$(3-NO$_2$C$_6$H$_4$COS)	Ru	S	1590	195.37			[123]
Os$_3$H(CO)$_{10}$(C$_6$H$_5$COS)	Os	S	1918–2065	193.0			[2]
[Mn(CO)$_4$(C$_6$H$_5$COSe)]$_2$	Mn	Se	1780				[10]
[Re(CO)$_4$(C$_6$H$_5$COS)]$_2$	Re	S	1680				[99]
Co$_2$(CH$_3$COS)$_4$	Co	S	1565				[204]
[Ni(CH$_3$COS)$_2$]$_2$·0.5C$_2$H$_5$OH	Ni	S	1540				[221]

Table 6 (continued)

Complexes, RCOE–M	M	E	IR^a/cm^{-1} ν (C=O)	NMR [δ] ^{13}C=O^b	^{77}Se or Te^b	Hz ^{31}p^b	References
Ni(C$_6$H$_5$COS)$_2$(α-Pic)$_2$	Ni	S	1610				[222]
Ni(C$_6$H$_5$COS)$_2$(Py)$_2$	Ni	S	1600				[222]
trans-Ni(4-CH$_3$C$_6$H$_4$COSe)$_2$[(C$_2$H$_5$)$_3$P]$_2$	Ni	Se	1612	201.7	420.1(Se)		[175]
trans-Pd($_4$-CH$_3$C$_6$H$_4$COSe)$_2$[(C$_2$H$_5$)$_3$P]$_2$	Ni	Se	1616	200.9	361.2(Se)		[175]
trans-Pd(PhCOS)$_2$(Ph$_3$P)$_2$	Pd	S	1595, 1570			14.8	[83]
trans-Pt(PhCOS)$_2$(Ph$_3$P)$_2$	Pt	S	1600, 1575			19.7	[83]
trans-Pt(4-CH$_3$C$_6$H$_4$COS)$_2$[(C$_2$H$_5$)$_3$P]$_2$	Pt	S	1609	197.8		12.0	[168]
trans-Pt(4-CH$_3$C$_6$H$_4$COSe)$_2$[(C$_2$H$_5$)$_3$P]$_2$	Pt	Se	1619	199.6	347.3(Se)	7.6	[175]
trans-Pt(4-CH$_3$C$_6$H$_4$COTe)$_2$[(C$_2$H$_5$)$_3$P]$_2$	Pt	Te	1621	196.9	392.6(Te)	6.7	[168]
Cu(PhCOS)	Cu	S	1670				[238]
(Ph$_3$P)Cu(m-PhCOS)	Cu	S	1612	205.3		−1.83tms	[49]
(Ph$_3$P)$_3$Cu$_2$[(m-PhCOS)$_2$]	Cu	S	1609	203		−1.22tms	[49]
(Ph$_3$P)$_4$Cu$_2$(m-CH$_3$COS)$_2$	Cu	S	1618	208.9		−2.03	[49]
[(C$_2$H$_5$)$_3$NH][Cu(PhCOS)$_2$]	Cu	S	1636	208.24			[268]
(Ph$_3$P)Cu(μ-PhCOS)$_2$Cu(PPh$_3$)$_2$	Cu	S	1597, 1553	209.54			[294]
[Ph$_4$P][Cu(CH$_3$COS)$_2$]	Cu	S	1616				[268]
Ag(CH$_3$COS)	Ag	S	1605	208.13			[148]
[(C$_2$H$_5$)$_3$NH][Ag(PhCOS)$_2$]	Ag	S		209.97			[268]
[Ph$_4$P][Ag(CH$_3$COS)$_2$]	Ag	S		192.9			[268]
(CH$_3$)$_3$PAu[CCl$_3$COS]	Au	S		193.8		−2.0	[261]
Ph$_3$PAu[CCl$_3$COS]	Au	S		193.5		37.9	[261]
(2-CH$_3$C$_6$H$_4$)$_3$PAu[CCl$_3$COS]	Au	S		186.8		16.1	[261]

Table 6 (continued)

Complexes, RCOE–M	M	E	IR^a/cm^{-1} ν (C=O)	NMR [δ]		Hz	References
				^{13}C=O^b	^{77}Se or Te^b	^{31}P^b	
Ph$_3$PAu[CH$_3$COS]	Au	S				37.9	[261]
Zn(PhCOS)$_2$	Zn	S	1545				[271]
[Ph$_4$P][Zn(PhCOS)$_3$](H$_2$O)	Zn	S		212.2 2068			[289, 319]
[Ph$_4$P][Cd(PhCOS)$_3$]	Cd	S		213.2		287 (Cd)^e	[269, 321]
Cd(PhCOS)$_2$	Cd	S	1580				[271]
[(CH$_3$)$_4$N]{Na[Cd(PhCOS)$_3$]$_2$}	Cd	S		208.1		353.5 (111Cd)^e 0.02–(23Na)^f	[50, 319]
[(CH$_3$)$_4$N]{K[Cd(PhCOS)$_3$]$_2$}	Cd	S	1580	208.27			[50]
PhHg(CH$_3$COS)	Hg	S		197.3 [326]			[145, 325]
PhHg(PhCOS)	Hg	S					[145]
Hg(PhCOS)$_2$	Hg	S	1629				[145, 271, 324]
Hg(4-CH$_3$C$_6$H$_4$COS)$_2$	Hg	S	1625				[145]
Hg(4-CH$_3$C$_6$H$_4$COS)(4-CH$_3$C$_6$H$_4$CSS)	Hg	S	1629				[145]
[Ph$_4$P][Hg(PhCOS)$_3$]	Hg	S	1629	205.9, 199.7			[269, 321]
[Ph$_4$As][Hg(PhCOS)$_3$]	Hg	S	1643, 1630	200.0		−749 (199Hg)^g	[321]
(Ph$_4$As)$_2$[Cl$_4$Hg$_2$(PhCOS)$_2$]	Hg	S	1619	196.9			[324]
(Ph$_4$As)$_2$[CdCl$_4$Hg(PhCOS)$_2$]	Hg	S	1628	197.1			[324]
[(CH$_3$)$_4$N]{Na[Hg(PhCOS)$_3$]$_2$}	Hg	S		201.55			[50]
[(CH$_3$)$_4$N]{K[Hg(PhCOS)$_3$]$_2$}	Hg	S		201.16			[50]

^{a} KBr disc; ^{b} In CDCl$_3$; ^{c} Nujol; ^{d} In (CD$_3$)$_2$CO; ^{d} In CH$_3$CN; ^{e} 0.1Cd(ClO$_4$)/H$_2$O; ^{f} 0.1NaBPh$_4$/acetone; ^{g} (CH$_3$)$_2$Hg/CH$_3$CN.

ing frequencies are relatively high, at >1600 cm^{-1} [83, 168, 175, 241]. The νC=O bands for Pt(4-CH$_3$C$_6$H$_4$COE)$_2$[(C$_2$H$_5$)$_3$P]$_2$ (M=Pd or Pt; E=S, Se, Te) show low frequency shifts in the order: E=S, Se, and Te, although the C=O distances decrease in this order [168, 170].

The ^{13}C NMR spectra of the tellurocarboxylato palladium **54** and platinum complexes **59** are observed at δ=213.9±0.2 (R=*tert*-C$_4$H$_9$) and δ=196.8±1.6 ppm (R=aryl). The ^{125}Te NMR spectra of **54** and **59** show one signal at δ=329–532 ppm, which indicates lower field shifts than those of the corresponding alkali metal salts [M(RCOTe), M=Li, Na, K, Rb or Cs] [164, 167, 174]. For the platinum complexes **59**, the ^{125}Te resonances [with ^1J(PPt)=681–812 Hz] show greater field shifts (by 33.6±0.3 ppm) than those of the corresponding palladium complexes **54**. In the ^{31}P NMR spectra, the signals of the platinum complexes **59** (δ=−1.1–0.0 ppm) are shifted to somewhat higher fields than those of the corresponding palladium complexes **54** (δ=5.7–7.1 ppm) [168]. The coupling constants [^1J(PPt)] of the *trans*-isomers **59** are ~2400 Hz. ^{113}Cd(II) and ^{23}Na NMR signals of the anionic complexes [(CH$_3$)$_4$N]{Na[(Cd(PhCOS)$_3$)]$_2$} have been reported [319]. The ^{113}Cd(II) and ^{199}Hg NMR signals of (Ph$_4$E)-[M(PhCOS)$_3$](H$_2$O) (M=Cd, Hg; E=P, As) also are observed at δ=280–295 and δ=−740 to -760, respectively [269].

3.3
Main-Group-Element Compounds

The carbonyl stretching frequencies and ^{13}C=O, ^{77}Se and ^{125}Te NMR chemical shifts of representative monochalcogenocarboxylic acid main-group-element compounds are shown in Table 7. The structures of tin thiocarboxylates RCOSSnR$_3'$ and (RCOS)$_2$SnR$_2'$ [108, 163, 220, 311] have been discussed in the light of infrared spectra. The asymmetrical νC=O bands appear at 1550–1650 cm^{-1}. The ^{13}C=O, ^{77}Se and ^{125}Te chemical shifts are observed at δ=180–200, δ=300–600 and δ=300–450, respectively. The C=O stretching bands of ArC(O)-EMPh$_3$ (M=Sn, Pb; E=S, Se, Te) shift to higher frequencies upon going from S, to Se, Te. The ^{13}C=O signals shift upfield by 1–5 ppm in the same order, except in the case of Ge derivatives.

The infrared and Raman spectra of benzoyl- [154] and trifluoroacetyl-sulfenyl halides [228] have been measured and the bands at 522, 446 and 396 cm^{-1} have been assigned to the S–Cl, S–Br and S–I stretching frequencies, respectively, whereas the bands at 190, 152 and 138 cm^{-1} were assigned to the C–S–Cl, C–S–Br, and the C–S–I bend frequencies, respectively. On the basis of the Mössbauer spectra, diorganotin bis(thioacetate) is believed to have octahedral structure [260].

O-Triorganosilyl thio-, seleno- and telluro-carboxylates are pale yellow to yellow, orange to purple, and blue to green, respectively. The n-π^* transitions of the thio-, seleno-, and telluro-carbonyl groups in ArC(E)OSi(CH$_3$)$_3$ (Ar=aryl; E=S, Se, Te) appear at 425–445 (*cyclo*-hexane) [143], 495–545 (hexane) [166], and 670–730 nm (in acetonitrile) [174], respectively.

Table 7 Spectral data for main-group-element compounds of monochalcogenocarboxylic acids

Compounds RCOE-M	M	E	IR[a]/cm^{-1} v(C=O)	NMR [δ]		UV/Vis [nm] λ_{max}[c]	References
				^{13}C=O[b]	^{77}Se or Te[b]		
Al(CH$_3$)$_2$(CH$_3$COS)	Al	S	1461				[327]
Al(C$_2$H$_5$)$_2$(CH$_3$COS)	Al	S	1460				[327]
(CH$_3$)$_2$CH$_3$COS(Gal)	Ga	S	1488				[327]
CH$_3$Ga(CH$_3$COS)$_2$(3,5-Lut)[d]	Ga	S		201.95			[289]
In(CH$_3$)$_2$(CH$_3$COS)	In	S	1475				[91]
In(C$_2$H$_5$)$_2$(CH$_3$COS)	In	S	1438				[327]
In(S)(CH$_3$COS)	In	S	1626				[291]
In(PhCOS)$_3$	In	S	1618				[291]
[(C$_2$H$_5$)$_3$NH][In(PhCOS)$_2$]$_4$	In	S	1615				[291]
4-CH$_3$C$_6$H$_4$COSCH$_3$	C	S	1670	185.0	445.6		[167]
4-CH$_3$C$_6$H$_4$COSGePh$_3$	Ge	S	1651	191.7	356.5, 351.2[e]		[165, 303]
4-CH$_3$C$_6$H$_4$COSSnPh$_3$	Sn	S	1621	196.0	597.6		[303]
4-CH$_3$C$_6$H$_4$COSPbPh$_3$	Pb	S	1618	196.4	488.6		[303]
4-CH$_3$C$_6$H$_4$COSeCH$_3$	C	Se	1665	193.2	611.5		[170]
4-CH$_3$C$_6$H$_4$COSeGePh$_3$	Ge	Se	1665	192.8			[165]
4-CH$_3$C$_6$H$_4$COSeSnPh$_3$	Sn	Se	1644	194.9	350.2		[117]
4-CH$_3$C$_6$H$_4$COSePbPh$_3$	Pb	Se	1641 409.4	195.4	288.2		[160]
4-CH$_3$C$_6$H$_4$COTeCH$_3$	C	Te	1655	196.8			[167]
2-CH$_3$C$_6$H$_4$C(O)TeSi(CH$_3$)$_2$(n-C$_4$H$_9$)	Si	Te		200.0			[162]

Table 7 (continued)

Compounds RCOE-M	M	E	IR^a/cm^{-1} $\nu(C=O)$	NMR [δ]		UV/Vis [nm] $\lambda_{max}{}^c$	References
				$^{13}C=O^b$	^{77}Se or Te^b		
4-$CH_3C_6H_4COTeGePh_3$	Ge	Te	1676	190.0			[306]
4-$CH_3C_6H_4COTeSnPh_3$	Sn	Te	1654	189.8			[306]
4-$CH_3C_6H_4COTePbPh_3$	Pb	Te	1655	189.8			[306]
$(C_6H_5COS)_2Sn(CH_3)_2$	Sn	S	1568				[197]
$(C_6H_5COS)_3SnPh$	Sn	S	1620		1048		[291]
$(C_6H_5COS)_4Sn$	Sn	S	1597				[291]
4-$CH_3C_6H_4C(S)OSi(CH_3)_3$	Si	S	1238	210.9^f			[143, 163]
4-$CH_3C_6H_4C(Se)OSi(CH_3)_3$	Si	Se		222.0			[167]
$tert$-$C_4H_9C(Te)OSi(CH_3)_3$	Si	Te				350 624	[286]
$C_6H_5C(Te)OSi(CH_3)_3$	Si	Te				712^g	[174]
2-$CH_3C_6H_4C(Te)OSi(CH_3)_2(n$-$C_4H_9)$	Si	Te		232.1			[162]
4-$CH_3C_6H_4C(Te)OSi(CH_3)_3$	Si	Te				700^g	[174]
4-$CH_3C_6H_4COSNH_2$	N	S	1658				[262]
4-$CH_3C_6H_4COSPPh_2$	P	S	1652	189.8			[304]
4-$CH_3C_6H_4COSAsPh_2$	As	S	1644	192.1			[305]
$C_6H_5COSSbPh_2$	Sb	S	1614				[202]
4-$CH_3C_6H_4COSSbPh_2$	Sb	S	1624	194.2			[181]
4-$CH_3C_6H_4COSBiPh_2$	Bi	S	1657	196.5			[181]

Table 7 (continued)

Compounds RCOE-M	M	E	IR^a/cm^{-1} ν(C=O)	Raman νM–S	NMR [δ] ^{13}C=O^b	^{77}Se or ^{125}Te	References
4-CH$_3$C$_6$H$_4$COSePPh$_2$	P	Se	1682		192.2	582.7	[181]
4-CH$_3$C$_6$H$_4$COE)$_n$MPh$_{3-n}$							
4-CH$_3$C$_6$H$_4$COSeAsPh$_2$	As	Se	1665		192.5	597.6	[139]
4-CH$_3$C$_6$H$_4$COSeSbPh$_2$	Sb	Se	1643		194.2	488.6 (490.6)	[181]
(4-CH$_3$C$_6$H$_4$COS)$_2$PPh	P	S	1643		189.8		[304]
(4-CH$_3$C$_6$H$_4$COS)$_2$AsPh	As	S	1626		192.1		[305]
(CH$_3$COS)$_2$SbPh	Sb	S	1625				[89]
(C$_6$H$_5$COS)$_2$SbPh	Sb	S	1614		194.2		[202]
(4-CH$_3$C$_6$H$_4$COS)$_2$SbPh	Sb	S	1590		196.5S		[181]
(4-CH$_3$C$_6$H$_4$COS)$_2$BiPh	Bi	S	1550				[181]
(4-CH$_3$C$_6$H$_4$COSe)$_2$PPh	P	Se	1667		190.1		[170]
(4-CH$_3$C$_6$H$_4$COSe)$_2$AsPh	As	Se	1665		194.2	–	[181]
(4-CH$_3$C$_6$H$_4$COSe)$_2$SbPh	Sb	Se	1621	384 (νSb-S)	196.6	561.7 465.0	[181]
(4-CH$_3$C$_6$H$_4$COS)$_3$P	P	S	1643		189.8		[304]
(C$_6$H$_5$COS)$_3$As	As	S	1630				[202]
(4-CH$_3$C$_6$H$_4$COS)$_3$As	As	S	1626		192.1		[305]
(CH$_3$COS)$_3$Sb	Sb	S	1640, 1627	1652, 1642, 1632			[88]
(C$_6$H$_5$COS)$_3$Sb	Sb	S	1589		194.2		[181, 202]
(C$_6$H$_5$COS)$_3$Bi	Bi	S	1589				[202]
4-CH$_3$C$_6$H$_4$COS)$_3$Bi	Bi	S	1550		196.5		[181]
(CH$_3$COS)$_2$Sb(CH$_3$)$_3$	Sb	S	1639				[256]
(C$_6$H$_5$COS)$_2$Sb(CH$_3$)$_3$	Sb	S	1621	384 (νSb-S)			[256]

Table 7 (continued)

Compounds RCOE-M	M	E	IR[a]/cm^{-1} ν(C=O)	Raman νM–S	NMR [δ] ^{13}C=O[b]	^{77}Se or ^{125}Te	References
(C$_6$H$_5$COS)$_2$	S	S	1685		185.3		[4, 244]
(4-CH$_3$C$_6$H$_4$COS)$_2$	S	S	1706, 1692		185.8		[244]
(2-CH$_3$OC$_6$H$_4$COS)$_2$	S	S	1661, 1641		185.7		[244]
(C$_6$H$_5$COSe)$_2$	Se	Se	1740, 1694	380 (νSb-S)	187.3	615.3	[246]
(2-CH$_3$OC$_6$H$_4$COSe)$_2$	Se	Se	1676, 1651	375 (νSb-S)	185.0	681.13	[246]
(2-CHO$_3$C$_6$H$_4$COTe)$_2$	Te	Te	1676, 1651		185.5	806.2	[135, 247]
(C$_6$H$_5$COS)$_2$S	S	S	1692, 1675		185.5		[170]
(4-CH$_3$C$_6$H$_4$COS)$_2$S	S	S	1700, 1685		185.6		[156]
(C$_6$H$_5$COS)$_2$Se	Se	S	1691, 1690, 1685		186.9		[5, 156]
(4-CH$_3$C$_6$H$_4$COS)$_2$Se	Se	S	1690		186.6		[156]
(C$_6$H$_5$COS)$_2$Te	Te	S	1675		186.9		[156, 296]
(4-CH$_3$C$_6$H$_4$COS)$_2$Te	Te	S	1685		189.2		[156]
CF$_3$COSCl	Cl	S	1761				[266, 228]
C$_6$H$_5$COSCl	Cl	S	1688				[149, 154]
CF$_3$COSBr	Br	S	1761	1751			[266, 228]
C$_6$H$_5$COSBr	Br	S	1686				[147, 154]
CF$_3$COSI	I	S					[228]
C$_6$H$_5$COSI	I	S	1667	1727			[146, 154]

[a] KBr disc; [b] In CDCl$_3$; [c] In n-C$_6$H$_{12}$; [d] 3,5-Lut=3,5-Lutidine; [e] Ref.=[165]; [f] Ref.=[163]; [g] In CH$_3$CN.

4
Applications

Some of the acyl organyl and diacyl disulfides **119** have been found to be effective as antifungicides against *Histoplasma capslasma* [66, 272] and as anti-irradiation agents [65, 272]. Acetyl trichloromethyl disulfide is effective against German roach and milkweed bug and just as effective as Bordeaux against the fungus *Alteneria Sclertia fructicola* [93]. Liquid crystal properties of compounds bearing the RC(O)SOR- moiety have been reported [302]. Thiocarboxylic acid Pb, Cd and Ni complexes have been used as plasticizing agents for natural rubber [312]. A variety of metal thiocarboxylates, such as Ca [194, 293], Sr [194, 293], Ba [194, 312], Ga [289, 338], In [289, 338], Sn [257], Pb [257], Zn [41, 205, 249, 250, 258], Cd [249, 250], Hg [249], Sb [205], and Co [41] have been used as effective precursors to nano-crystalline metal sulfide materials. The facile preparation of high purity coin metal chalcogens by the thermolysis or alcoholysis of coin metal chalcogenocarboxylates has been reported [169]. Chalcogenocarboxylato complexes with transition metals such as Pd and Pt have been used effectively as additives for the high sensitization of silver halide photographic films [330–335].

Acknowledgements The authors are indebted to Professor Jugo Kohketsu of Chubu University, Japan, for his generous collection of numerous literature sources.

References

1. Adkins H, Thompson QE (1949) J Am Chem Soc 71:2242
2. Ainscough EW, Brodie AM, Coll RK, Coombridge BA, Waters JM (1998) J Organomet Chem 556:197
3. Al'fonsov AJ, Pudovik DA, Litvinov LA, Katsyuba SA, Filippova EA, Pudovik, MA, Kharullin VK, Batyeva SÉ, Pudovik AN (1987) Dokl Akad Nauk SSSR 296:103
4. Arabamudan G, Subrahmanyan T, Seshasayee M, Rao GVNA (1983) Polyhedron 2:1025
5. Arduini AL, Takats J (1981) Inorg Chem 20:2480
6. Armstrong DR, Banister AJ, Glegg W, Gill WD (1986) J Chem Soc Chem Commun 1672
7. Arora RS, Hari SC, Bhalla MS, Multani RK (1980) J Chin Chem Soc–Taip 27:65; Chem Abstr (1980) 93:204800
8. Arora RS, Multani RK (1983) J Inst Chem (India) 55:93
9. Baranovskii IB, Golubnichaya MA, Mazo GY, Shchelohov RN (1975) Zh Neorg Khim 20:844
10. Barrow MJ, Edsworth AV, HuntlyCM, Rankin DWH (1982) J Chem Soc Dalton 1131
11. Barton DHR, McCombie SW (1975) J Chem Soc Perk T 1574
12. Barton DHR, Chen C, Wall GM (1991) Tetrahedron 47:6127
13. Bauer W, Kühlein K (1985) In: Falbe J (ed) Methoden der organischen chemie. Georg Thieme, Stuttugart, Band 5, Teil 1, p 832
14. Bergmann M, Bloch I (1920) Ber Deut Chem Ges 53:977
15. Bergmann M (1920) Ber Deut Chem Ges 53:979
16. Bernard MA, Borel MM (1972) CR Acad Sci C 274:1743

17. Bernard MA, Borel MM, Ledesert MA (1973) B Soc Chim Fr 2194
18. Binder H, Brellochs B, Frei B, Simon A, Hettich B (1989) Chem Ber 122:1049
19. Bloch I, Bergmann M (1920) Ber Dtsch Chem Ges 53:961; Chem Abstr (1920) 14:3414
20. Böhme H, Clement M (1952) Liebigs Ann Chem 576:61
21. Böhme H, Zinner G (1954) Liebigs Ann Chem 585:142
22. Böhme H, Zinner G (1954) Liebigs Ann Chem 585:150
23. Böhme H, Freimuth F, Mundlos E (1954) Ber Deut Chem Ges 87:1661
24. Böhme H, Bessenberger H, Stachel HD (1957) Liebigs Ann Chem 602:1
25. Böhme H, Goubeaud HW (1959) Ber Deut Chem Ges 92:366
26. Bonamico M, Dessy G, Fares V (1969) J Chem Soc Chem Commun 697
27. Bonamico M, Dessy G, Fares V, Scaramuzza L (1976) J Chem Soc Dalton 67
28. Borel MM, Ledesert MA (1972) CR Acad Sci C 275:183
29. Borel MM, Ledesert MA (1973) CR Acad Sci C 276:181
30. Borel MM, Ledesert MA (1974) Acta Crystallogr B 30:2777
31. Borel MM, Ledesert MA (1975) Acta Crystallogr B 31:725
32. Borel MM, Dupriez G, Ledesert M (1975) J Inorg Nucl Chem 37:1533
33. Borel MM, Ledesert MA (1975) J Inorg Nucl Chem 37:2334
34. Borel MM, Geffrouais A, Ledesert MA (1976) Acta Crystallogr B 32:2385
35. Borel MM, Ledesert MA (1976) Acta Crystallogr B 32:2388
36. Brois SJ, Pilot JF, Barnum HW (1970) J Am Chem Soc 92:7629
37. Brükuner H, Härtel K (1958) Ger Patent 1 025 198; Chem Abstr (1960) 54:12468 h
38. Brunner H, Hollman A, Zabel A (2001) J Organomet Chem 630:169
39. Burnett TB, Dean PAW, Vittal J (1994) Can J Chem 72:1127
40. Carlsen L, Egsgaard H, Schumann E, Mrotzek H, Klein W-R (1980) J Chem Soc Perkin T 2 1557
41. Chandler CD, Roger C, Hampden-Smith MJ (1993) Chem Rev 93:1205
42. Coutts RSP, Wailes PC (1978) Aust J Chem 21:373
43. Coucouvanis DD (1970) Prog Inorg Chem 11:233
44. Cras JA, Willemes J (1987) In: Wilkinson G, Gillard RG, McCleverty JA (eds) Comprehensive coordination chemistry. Pergamon, Oxford, UK, vol 2, chap 16.4, p 571
45. Dabrowska U, Dabrowski J (1976) Chem Ber 109:1779
46. Darensbourg DJ, Rokicki A (1982) J Am Chem Soc 104:349
47. Davidson WE, Hill K, Henry MC (1965) J Organomet Chem 3:285
48. Dean PAW, Vittal JJ, Craig DC, Scudder ML (1998) Inorg Chem 37:1661
49. Deivaraj TC, Lai GX, Vittal JJ (2000) Inorg Chem 39:1028
50. Deivaraj TC, Dean PAW, Vittal JJ (2000) Inorg Chem 39:3071
51. Deivaraj TC, Vittal JJ (2000) Acta Crystallogr C 56:775
52. Denifl P, Bildstein B (1993) J Organomet Chem 453:53
53. Devy R, Vittal JJ, Dean PAE (1998) Inorg Chem 37:6939
54. Drake JE, Hemmings RT, Henderson HE (1976) Inorg Nucl Chem Lett 12:563
55. Dubs P, Stüssi R (1976) Helv Chim Acta 59:1307
56. Eikens W, Jäger S, Jones PG, Thöne C (1996) J Organomet Chem 511:67
57. El-Hinnawi MA, Al-Ajlouni AM (1987) J Organomet Chem 332:321
58. El-Hinnawi MA, Al-Ajlouni AM, Abu-Nasser JS, Powell AK, Vahrenkamp H (1989) J Organomet Chem 359:79
59. El-Hinnawi MA, Sumadi ML, Esmadi FT, Jibril I, Infof W, Huttner G (1989) J Organomet Chem 377:373
60. El-Hinnawi MA, El-Khateeb MY, Jibril I, Abu-Orabi ST (1989) Synth React Inorg Met 19:809
61. El-khateeb M, Younes A (2002) J Organomet Chem 664:228

62. Engelhardt A, Latschinoff P, Malyschew RZ (1868) Chem 358:455
63. Fanghänel E, Palmer T, Kersten J, Ludwigs R, Peter K, Schnering HG (1994) Synthesis 1067
64. Faust J, Mayer R (1976) J Prakt Chem 318:161
65. Field L, Buckman JD (1967) J Org Chem 32:3467
66. Field L, Hanley WS, McVeigh I, Evans Z (1971) J Med Chem 202
67. Field L, Hanley WS, McVeigh IZ (1971) J Org Chem 19:2735
68. Fromm E, Schmoldt P (1907) Ber Deut Chem Ges 40:2861
69. Fromm E, de Seixas Palma J (1906) Ber Deut Chem Ges 38:3317
70. Fromm E, Forster A (1912) Ann Chem Pharm 394:338
71. Fujisawa T, Kobayashi N (1971) J Org Chem 6:3546
72. Fujiwara S, Asai A, Shin-ike T, Kambe N, Sonoda N (1998) J Org Chem 63:1724
73. Furlani C, Luciani ML, Candori R (1968) J Inorg Nucl Chem 3121
74. Furuta K (1984) Bachelors thesis, Gifu University, Gifu, Japan
75. George TA, Lappert MF (1968) J Organomet Chem 14:327
76. Gilbert JD, Wilkinson G (1967) J Chem Soc A 1749
77. Gilman H, Lichtenwalter GD (1960) J Org Chem 25:1064
78. Gingerich RGW, Anglici RJ (1979) J Am Chem Soc 101:5604
79. Gladysz JA, Wong VK, Jick BS (1978) J Chem Soc Chem Commun 838
80. Gladysz JA, Wong VK, Jick BS (1979) Tetrahedron 35:2329
81. Glidewell C, Leslie EJ (1977) J Chem Soc Dalton 527
82. Goerdeler J, Yunis M (1985) Chem Ber 118:851
83. Goodfellow JA, Stephenson TA, Cornock MC (1978) J Chem Soc Dalton 1195
84. Goyal KC, Khosla BD (1981) Curr Sci India 50:128
85. Gould RO, Stephenson TA, Thomson MA (1978) J Chem Soc Dalton 769
86. Gronowicz GA, Ryan JW (1966) J Org Chem 31:3439
87. Haiduc I, King RB, Newton MG (1994) Chem Rev 94:301
88. Hall M, Sowerby DB (1980) J Chem Soc Dalton 1292
89. Hall M, Sowerby DB, Falshaw CP (1986) J Organomet Chem 315:321
90. Hausen HD (1972) Z Naturforsch B 57:82
91. Hausen HD, Guder HJ (1973) J Organomet Chem 57:243
92. Hawley RS, Kittleson AR (1951) US Patent 2 553 777; Chem Abstr (1951) 45:7742
93. Heap R, Saunder BC, Stacey GJ (1949) J Chem Soc 2983
94. Heimer NE, Field L, Neal RA (1981) J Org Chem 46:1374
95. Henry MC, Davidson WE (1962) J Org Chem 27:2252
96. Henry MC, Davidson WE (1963) Can J Chem 41:1276
97. Hieber W, Brück (1952) Z Anorg Allg Chem 269:13
98. Hieber W, Gscheidmeier M (1966) Chem Ber 99:2312
99. Hieber W, Rohm W (1969) Chem Ber 102:2787
100. Hildebrand U, Taraz K, Budzikiewicz H (1985) Tetrahedron Lett 26:4349
101. Hildebrand U, Taraz K, Budzikiewicz H (1985) Z Naturforsch B 40:1563
102. Hildebrand U, Hübner J, Budzikiewicz H (1986) Tetrahedron 42:5969
103. Hirabayashi Y, Mizuta M, Mazume T (1964) B Chem Soc Jpn 37:1002
104. Hirabayashi Y, Ishihara H, Echigo M (1987) Nippon Kagaku Kaishi 1430
105. Hirabayashi Y, Ishihara H, Ogi Y, Miyake T, Oida H, Kondo N, Goh H (1987) Nippon Kagaku Kaishi 1475
106. Hodson AGW, Thind RK, McPartlin M (2002) J Organomet Chem 664:277
107. Hölste G (1976) Anorg Allg Chem 425:57
108. Honnick WD, Zuckermann JJ (1979) J Organomet Chem 178:133
109. Hunter BK, Reevez LW (1968) Can J Chem 46:1399

110. Hunter JA, Lindsell WE, McCullough KJ, Parr RA, Scholes ML (1990) J Chem Soc Dalton 2145

111. Irgoric KJ (1990) In: Klamann D (ed) Methoden der organischen chemie, vol E12b. Georg Thieme, Stuttugart

112. Ishihara H, Kato S (1972) Tetrahedron Lett 13:3751

113. Ishihara H, Hirabayashi Y (1976) Chem Lett 203

114. Ishihara H, Kato S, Hirabayashi Y (1977) B Chem Soc Jpn 50:3007

115. Ishihara H, Muto S, Kato S (1986) Synthesis 128

116. Ishihara H, Matsunami N, Yamada Y (1987) Synthesis 371

117. Ishihara H, Muto S, Endo T, Komada M, Kato S (1989) Synthesis 929

118. Ishihara H, Yoshimura K, Kouketsu M (1998) Chem Lett 1287

119. Ishii A, Nakayama J (1995) In: Moody CJ (ed) Comprehensive org synthesis. Pergamon, vol 5, p 505

120. Jensen KA, Böje L, Hendriksen L (1972) Acta Chem Scand 26:1465

121. Jensen KA (1987) In: Klayman DL (ed) Organic selenium compounds: Their chemistry and biology. Wiley, New York, Ch 8, p 273

122. Jibril I, El-Hinnawi MA, El-kateeb M (1991) Polyhedron 10:2095

123. Jibril I, Esmadi F, Al-Masri H, Zsolnai L, Huttner G (1996) J Organomet Chem 510:109

124. Jibril I, Abu-Nimereh O (1996) Synth React Inorg Met 26:1409

125. Jibril I, Ali AK (1997) Indian J Chem A 36:987

126. Jibril I, Ali AK, Omar JT (1997) Polyhedron 16:3327

127. Jibril I, Abd-Alhadi EH (2000) Indian J Chem A 39:1055

128. Jibril I, Abd-Alhadi EH, Hamadh Z (2000) Transit Met Chem 25:407

129. Jones HO, Tasker HS (1909) J Chem Soc 95:1904

130. Kageyama H, Tsutsumi H, Murai T, Kato S (1989) Z Naturforsch B 44:1050

131. Kageyama H, Takagi H, Murai T, Kato S (1989) Z Naturforsch B 44:1519

132. Kageyama H, Kido K, Kato S, Murai T (1994) J Chem Soc Perkin T 1 1083

133. Kakigano T, Kanda T, Ishida M, Kato S (1987) Chem Lett 475

134. Kalinichenko NB, Kordoskii LE, Belyaeva VK, Marov IN, Hoyer EZ, Dietzsch W (1990) Zh Neorg Khim 35:1237; Chem Abstr (1990) 113:183584

135. Kanda T, Mizoguchi K, Koike T, Murai T, Kato S (1993) J Chem Soc Chem Commun 1631

136. Kanda T, Nakaiida S, Murai T, Kato S (1989) Tetrahedron Lett 30:1829

137. Kanda T, Mizoguchi K, Kagohashi S, Kato S (1994) Organometallics 17:1487

138. Kanda T, Koike T, Kagohashi S, Mizoguchi K, Murai T, Kato S (1995) Organometallics 14:4975

139. Kanda T, Mizoguchi K, Koike T, Murai T, Kato S (1994) Synthesis 282

140. Kanda T, Ibi M, Mochizuki K, Kato S (1998) Chem Lett 957

141. Katada T, Nishida M, Kato S, Mizuta M (1977) J Organomet Chem 129:189

142. Kato S, Akada W, Mizuta M (1972) Int J Sulfur Chem A 2:279

143. Kato S, Akada W, Mizuta M, Ishii Y (1973) B Chem Soc Jpn 46:244

144. Kato S, Watarai H, Katada T, Mizuta M, Miyagawa K, Ishida M (1981) Synthesis 370

145. Kato S, Hattori E, Sato H, Mizuta M, Ishida M (1981) Z Naturforsch B 36:783

146. Kato S, Hattori E, Mizuta M, Ishida M (1982) Angew Chem Int Edit 21:150, Angew Chem 91:148

147. Kato S, Miyagawa K, Kawabata S, Ishida M (1982) Synthesis 1013

148. Kato S, Komatsu K, Miyagawa K, Ishida M (1983) Synthesis 552

149. Kato S, Itoh K, Miyagawa K, Ishida M (1983) Synthesis 814

150. Kato S, Oguri M, Ishida M (1983) Z Naturforsch B 38:1585

151. Kato S, Kabuto H, Kimura M, Ishihara H, Murai T (1985) Synthesis 519

152. Kato S, Kabuto H, Ishihara H, Murai T (1985) Synthesis 520

153. Kato S (1986) Org Prep Proced Int 18:369
154. Kato S, Komatsu K, Murai T, Hamaguchi H, Okazaki R (1986) Chem Lett 935
155. Kato S, Ono Y, Miyagawa K, Murai R, Ishida M (1986) Tetrahedron Lett 27:4595
156. Kato S, Ono Y, Miyazaki T, Itoh Y, Aoyama T, Ishida M, Murai T (1987) Nippon Kagaku Kaishi 1457
157. Kato S, Shi Min, Beniya Y, Ishida M (1987) Tetrahedron Lett 28:6473
158. Kato S, Ishida M (1988) Sulfur Report 8:155
159. Kato S, Yasui E, Terashima K, Ishihara H, Murai T (1988) B Chem Soc Jpn 61:3931
160. Kato S, Ishihara H, Ibi K, Kageyama H, Murai T (1990) J Organomet Chem 386:313
161. Kato S, Kageyama H, Takagi K, Mizoguchi K, Murai T (1990) J Prakt Chem 332:893
162. Kato S, Kageyama H, Kanda T, Murai T, Kawamura T (1990) Tetrahedron Lett 31:3587
163. Kato S, Hori A, Mizuta M, Katada T, Ishihara H, Fujieda K, Ikebe Y (1991) J Organomet Chem 420:13
164. Kato S, Sasaki H, Yagihara M (1992) Phosphorus Sulfur 67:27
165. Kato S, Ibi K, Kageyama H, Ishihara H, Murai T (1992) Z Naturforsch B 47:558
166. Kato S, Kageyama H, Kawahara Y, Murai T, Ishihara H (1992) Chem Ber 125:417
167. Kato S, Niyomura O, Nakaiida S, Kawahara Y, Kanda T, Yamada R, Hori S (1999) Inorg Chem 38:519
168. Kato S, Niyomura O, Kawahara Y, Kanda T (1999) J Chem Soc Dalton 1677
169. Kato S (2003) Jpn Kokai Tokkyo Koho 2003-54948
170. Kato S (unpublished results)
171. Kawahara Y, Kato S, Kanda T, Murai T, Ishihara H (1993) J Chem Soc Chem Commun 277
172. Kawahara Y, Kato S, Kanda T, Murai T (1994) B Chem Soc Jpn 67:1881
173. Kawahara Y, Kato S, Kanda T, Murai T (1995) Chem Lett 87
174. Kawahara Y, Kato S, Kanda T, Murai T, Ebihara M (1995) B Chem Soc Jpn 68:3507
175. Kawahara Y, Kato S, Kanda T, Murai T, Miki K (1996) J Chem Soc Dalton 79
176. Kawamura S, Horii T, Nakabayashi T, Hamada M (1975) B Chem Soc Jpn 48:2993
177. Kawamura S, Sato A, Nakabayashi T, Hamada M (1975) Chem Lett 1231
178. Kekulé A (1854) Ann Chem Pharm 90:309
179. Kekulé A (1862) Ann Chem Pharm 123:273
180. Kerr KB, Walde AW (1955) US Patent 2,702,776; Chem Abstr (1955) 49:7816
181. Kimura M, Kimura T, Iwata A, Sugiura N, Ishida M, Kato S (2004) (in preparation)
182. Kimura M (1986) Masters thesis, Gifu University, Gifu, Japan
183. King RB (1963) J Am Chem Soc 85:1918
184. Kirmse R, Dietzsch W, Stach J, Olk RM, Hoyer E (1987) Z Anorg Allg Chem 548:133; Chem Abstr (1987) 107:227830
185. Köster R, Kucznierz R, Seidel G, Betz P (1992) Chem Ber 125:1023
186. Koch KN, Mloston G, Senning A (1999) Eur J Org Chem 83
187. Kojima Y, Ibi K, Ishihara H, Murai, Kato S (1993) B Chem Soc Jpn 66:990
188. Komatsu Y (1985) Masters thesis, Gifu University, Gifu, Japan
189. Krebs AW, Henry MC (1963) J Org Chem 28:1911
190. Krebs H, Weber EF, Fassbender H (1954) Z Anorg Allg Chem 267:128
191. Kricheldorf HR, Leppert E (1971) Synthesis 435
192. Kricheldorf HR, Leppert E (1972) Makromol Chem 158:223
193. Krüger HJ, Holm RH (1990) J Am Chem Soc 112:2955
194. Kunze K, Bihry B, Atanasova P, Hampden-Smith MJ, Duesler E (1996) Chem Vapor Depos 2:105
195. Kym O (1899) Ber Deut Chem Ges 32:3533
196. Latif KA, Chakraborty PK (1967) Tetrahedron Lett 8:971

197. Lee LK, Teo SG, Fong WF, Teo SB, Fun HK (1994) J Coord Chem 31:103
198. Legrand L, Lozach N (1969) B Soc Chim Fr 37:1170
199. Legrand L, Lozach N (1969) B Soc Chim Fr 37:1173
200. Leriverend C, Metzner T (1994) Synthesis 761
201. Lesser R, Weiss R (1924) Ber Deut Chem Ges 57:1077
202. Levedev EP, Baburian VA, Mizhiritskii MDG (1979) J Gen Chem 49:940; (1979) Z Obsh Khim 49:1081
203. Litvinov IA, Zuev MB, Kataeva ON, Naumov VA (1991) J Gen Chem 61:1017; (1991) Zh Obsch Khim 61:1120
204. Luciani ML, Furlani C (1971) Inorg Chem 10:2614
205. Ma L, Payne DA (1994) Chem Mater 6:875
206. Malhotra KC, Puri JK (1975) Indian J Chem 13:184
207. Malhora HS, Multani RK (1984) J I Chem (India) 56:135
208. Mamedov KS, Alekperov RE, Magerramov AI, Amiraslanov IR, Musaev FN, Movsumov EM (1977) Dokl Akad Nauk Azelbaijan SSSR 33:31; Chem Abstr (1977) 87:209740 g
209. Marcos CF, Polo C, Rakitin OA, Rees CW, Trroda T (1997) Angew Chem Int Edit 36:281
210. Martel B, Duffaut N (1966) CR Acad Sci C 263:74
211. Masumoto H, Tsutsumi H, Kanda T, Komada M, Murai T, Kato S (1989) Sulfur Lett 10:103
212. Matsuhashi Y, Tokitoh N, Okazaki R (1993) Organometallics 12:2573
213. Mattes R, Meschede W, Stork W (1975) Chem Ber 108:1
214. Mattes R, Stork W, Kahlenberg J (1975) Spectrochim Acta 33A:643
215. Mattes R, Weber H (1979) J Organomet Chem 178:191
216. Matz C, Mattes R (1978) Z Naturforsch B 33:461
217. Mayer R, Scheithauer S (1985) In: Falbe J (ed) Methoden der organischen chemie. Georg Thieme, Stuttugart, Band 5, Teil 1, p 785
218. Mayer R, Scheithauer S (1985) In: Falbe J (ed) Methoden der organischen chemie. Georg Thieme, Stuttugart, Band 5, Teil 2, p 891
219. Mazurkiewicz W, Senning A (1983) Sulfur Lett 1:127
220. Mehrotra RC, Srivastava G, Vasanta EN (1980) Inorg Chim Acta 47:125
221. Melson GA, Greene GA, Bryan RF (1970) Inorg Chem 9:1116
222. Melson GA, Crawford NP, Geddes BJ (1970) Inorg Chem 9:1123
223. Mereiter K, Mikenda W, Steinwender E (1993) J Cryst Spectrosc 23:397
224. Meschede W, Mattes R (1976) Chem Ber 109:2510
225. Metzner P, Pham TN (1988) J Chem Soc Chem Commun 390
226. Mikolajczyk M, Kielbasinski P (1976) J Chem Soc Perkin T 1 564
227. Milligan B, Saville B, Swan JM (1961) J Chem Soc 4850
228. Minkwitz R, Sawatzki J (1988) Z Anorg Allg Chem 566:151
229. Mizuta M, Katada T, Itoh E, Kato S, Miyagawa K (1980) Synthesis 721
230. Mueller WH, Butler PE (1967) J Org Chem 32:2925
231. Mueller WH, Oswald AA (1997) US Patent 4,003,939; Chem Abstr (1977) 86:170853y
232. Murai T, Oida S, Min S, Kato S (1986) Tetrahedron Lett 27:4593
233. Murai T, Kimura T, Miwa A, Kurachi D, Kato S (2002) Chem Lett 914
234. Musaev FN, Movsumov EM, Mamedov KS, Amiraslanov IR, Magerramov AI, Shnulin AN (1977) J Struct Chem (USSR) 18:865; (1977) Zh Struct Khim 18:1089
235. Musaev FN, Mamedov KS, Movsumov EM, Shnulin AN, Amiraslanov IR (1978) J Struct Chem (USSR) 19:440; (1978) Zh Struct Khim 19:508; Chem Abstr (1978) 89:189300z
236. Musaev FN, Movsumov EM, Mamedov KS, Amiraslanov I (1978) Sov J Coord Chem 4:956
237. Musaev FN, Mamedov KS, Movsumov EM, Amiraslanov IR (1979) Koord Khim 5:119; Chem Abstr (1979) 90:145153v

238. Nair KP, Nair CGR (1971) Indian J Chem A 9:706
239. Nakagawa K, Shiba S, Horikawa M, Sato K, Nakamura H, Harada N, Harada F (1980) Synth Commun 10:305
240. Nakata N (2000) Masters thesis, Gifu University, Gifu, Japan
241. Neo YC, Vittal JJ, Hor TSA (2001) J Organomet Chem 637–639:757
242. Nishiyama Y, Katsuura A, Okamoto Y, Hamanaka S (1989) Chem Lett 1825
243. Nishiyama Y, Katsuura A, Negoro A, Hamanaka S (1991) J Org Chem 56:3776
244. Niyomura O, Kitoh Y, Nagayama K, Kato S (1999) Sulfur Lett 22:195
245. Niyomura O, Kato S, Kanda T (1999) Inorg Chem 38:507
246. Niyomura O, Tani K, Kato S (1999) Heteroatom Chem 10:373
247. Niyomura O, Kato S, Inagaki S (2000) J Am Chem Soc 122:2132
248. Noble P Jr, Tarbell DS (1963) Org Synth IV:924
249. Nyman MD, Hamden-Smith MD, Duesler EN (1997) Inorg Chem 36:2218
250. Nyman M, Hampden-Smith MJ, Duesler EN (1996) Adv Mater CVD 2:171
251. Ogawa A, Sonoda N (1995) In: Moody CJ (ed) Comprehensive organic synthesis. Pergamon, Oxford, vol 5, p 231
252. Oguri M (1985) Masters thesis, Gifu University, Gifu, Japan
253. Oro LA, Beltran FG, Tejel FJ (1972) An Quim 68:1461
254. Osuka A, Ohmasa N, Uno Y, Suzuki H (1983) Synthesis 68
255. Osuka A, Uno Y, Horiuchi H, Suzuki H (1984) Synthesis 145
256. Otera J, Okawara R (1969) J Organomet Chem 17:353
257. Ouhadi T, Hubert AJ, Teyssie P, Deroane EG (1973) J Am Chem Soc 95:449
258. Ouhadi T, Bioul JP, Stevens C, Warin R, Hocks L, Teyssie P (1976) Inorg Chim Acta 19:203
259. Pertlik F, Mikenda W, Steinwender E (1993) J Cryst Spectrosc 23:389
260. Poller RC, Ruddick JNR (1973) J Organomet Chem 60:87
261. Preisenberger M, Schier A, Schmidbaur H (1999) J Chem Soc Dalton 1645
262. Raasch MS (1972) J Org Chem 37:3820
263. Reid DH (1970–1981) Organic sulphur, selenium and tellurium, vols 1–6. Royal Society of Chemistry, London
264. Reid EE (1962) Organic chemistry of bivalent sulfur, vol IV. Chemical Publishing Co, Inc., New York, p 11
265. Reifenberg GH, Considine WJ (1969) Brit Patent 1 173 466
266. Rochat WV, Gard GL (1969) J Org Chem 34:4173
267. Rout GC, Seshasayee T, Subrahmanyan T, Aravamudan G (1983) Acta Crystallogr C 39:1387
268. Sampanthar JT, Vittal JJ, Dean PAW (1999) J Chem Soc Dalton 3153
269. Sampanthar JT, Deivaraj TC, Vittal JJ, Dean PAW (1999) J Chem Soc Dalton 4419
270. Sato T, Otera J, Nozaki H (1990) Tetrahedron Lett 31:3595
271. Savant VV, Gopalakrishnam J, Patel CC (1970) Inorg Chem 9:748
272. Schard LT, Werner RH, Dillon L, Field L, Tate CE (1975) J Med Chem 18:344
273. Schauble JH, Williams JD (1972) J Org Chem 37:2514
274. Scheithauer S, Mayer R (1979) In: Senning A (ed) Topics in sulfur chemistry. Georg Thieme, Stuttgart, vol 4
275. Schumann H, Thom KF, Schmidt M (1963) Angew Chem 75:138
276. Schumann H, Thom KF, Schmidt M (1963) J Organomet Chem 1:167
277. Schumann H, Thom KF, Schmidt M (1964) J Organomet Chem 2:97
278. Schumann H, Schmidt M (1965) Angew Chem 77:1049
279. Schumann H, Thom KF, Schmidt M (1965) J Organomet Chem 4:22
280. Schumann H, Thom KF, Schmidt M (1965) J Organomet Chem 4:28
281. Segi M, Takahashi T, Ichinose H, Li GM, Nakajima T (1992) Tetrahedron Lett 51:7865

282. Senning A (1979) Angew Chem Int Edit 18:941
283. Senning A, Jensen B (1983) Sulfur Lett 1:147
284. Senning A, Jensen B (1984) Sulfur Lett 2:11
285. Senning A, Abdel-Megeed MF, Mazurkiewicz W, Chevallier M-A (1985) Sulfur Lett 3:123
286. Severengiz T, du Mont W-W, Lenoir D, Voss H (1985) Angew Chem Int Edit 24:1041
287. Severengiz T, du Mont W-W (1987) J Chem Soc Chem Commun 820
288. Seyferth D, Kiwan AM (1985) J Organomet Chem 286:219
289. Shang G, Hampden-Smith MJ, Duesler EN (1996) J Chem Soc Chem Commun 1733
290. Shi Min (1988) Masters thesis, Gifu University, Gifu, Japan
291. Singh P, Bhattacharya S, Gupta VD, Nöth H (1996) Chem Ber 129:1093
292. Singh P, Singh S, Gupta VD, Nöth H (1998) Z Naturforsch B 53:1475
293. So JH, Boudjouk P (1990) Inorg Chem 29:1592
294. Speier G, Fülöp V, Argay G (1991) Transit Met Chem 16:576
295. Steele DF, Stephenson TA (1973) Inorg Nucl Chem Lett 9:777
296. Subrahmanyan T, Aravamudan G, Rout GC, Seshasayee M (1984) J Cryst Spectrosc 14:239
297. Sudha BP, Dixit NS, Patel CC (1978) Indian J Chem A 16:856
298. Suzuki H, Inamasu T, Ogawa T, Tani H (1990) J Chem Res–S 56
299. Szperl L (1930) Rocz Chem 10:652; Chem Abstr (1931) 25:938
300. Szperl L, Pendolank J, Wiorogorski W, Tomaszewska M, Podwysocka L (1931) Rocz Chem 11:753; Chem Abstr (1961) 26:5552
301. Szperl L, Wiorogorski W (1932) Rocz Chem 12:71
302. Takatoh K, Sakamoto M (1991) B Chem Soc Jpn 64:2770
303. Tani K, Kato S, Kanda T, Inagaki S (2000) Org Lett 3:655
304. Tani K, Matsuyama K, Kato S, Yamada K, Mifune H (2000) B Chem Soc Jpn 73:1243
305. Tani K, Hanabusa S, Kato S, Mutoh S, Suzuki S, Ishida M (2001) J Chem Soc Dalton 518
306. Tani K, Yamada R, Kanda T, Kato S, Murai T (2002) Organometallics 21:1487
307. Tarugi N (1895) Gazz Chim Ital 25:341
308. Teo SG, Teo SB, Lee LK, Arifin Z, Ng SW (1995) J Coord Chem 34:253
309. Tohru Y, Nishigaki M, Seko T, Kanefusa E, Maekawa E (1985) Synthesis 878
310. Tohru Y, Yamada Y, Oshima T, Imao S, Maekawa E (1987) Nippon Kagaku Kaishi 1442
311. Tsipis CA, Hadjikostas CC, Sigalas MP (1980) Spectrochim Acta 32:41
312. Tsvetkov AI, Grinberg AE, Makarova IM, Prashchikina AS, Makeeva AR (1960) USSR Patent 132801; Chem Abstr (1961) 55:8913a
313. Uemura S, Tanaka S, Okano M (1977) B Chem Soc Jpn 50:220
314. Ulrich UC (1859) Ann Chim Phys 56:236
315. Usacheva GM, Kamai GK (1968) Izv Akad Nauk SSSR Ser Khim 413; Chem Abstr (1968) 69:52239d
316. Usacheva GM, Kamai GK (1968) Izv Akad Nauk SSSR Ser Khim 1878; Chem Abstr (1969) 70:4239c
317. Usacheva GM, Kamai GK (1970) J Gen Chem 40:1298, (1970) Zh Obsch Khim 40:1306
318. Wang J-X, Cui W-F, Hu Y-L, Zang S-S (1990) J Chem Res–S 230
319. Vittal JJ, Dean P AW (1993) Inorg Chem 32:791
320. Vittal JJ, Dean P AW (1996) Acta Crystallogr C 52:1180
321. Vittal JJ, Dean P AW (1996) Inorg Chem 35:3089
322. Vittal JJ, Dean P AW (1997) Acta Crystallogr C 53:409
323. Vittal JJ, Dean P AW (1998) Acta Crystallogr C 54:319
324. Vittal JJ, Dean P AW (1998) Polyhedron 17:1937
325. Voss J (1979) In: Patai S (ed) Supplement B: The chemistry of carboxyric acid derivatives. Wiley, New York, p 1021
326. Weber H, Mattes R (1979) Chem Ber 112:95

327. Weidlein J (1971) J Organomet Chem 32:181
328. Weigert F (1903) Ber Deut Chem Ges 36:1007
329. Winter G (1980) Rev Inorg Chem 2:253
330. Yagihara M, Sasaki H, Mimatsu H, Kato S (1992) Jpn Patent 2 729 728
331. Yagihara M, Sasaki H, Mimatsu H, Kato S (1992) Jpn Patent 3 079 405
332. Yagihara M, Sasaki H, Mimatsu H, Kato S (1995) Jpn Patent 3 038 458
333. Yagihara M, Sasaki H, Mimatsu H, Kato S (1993) Eur Patent 585 787 B1
334. Yagihara M, Sasaki H, Mimatsu H, Kato S (1993) Eur Patent 661 589 B1
335. Yagihara M, Sasaki H, Mimatsu H, Kato S (1995) Jpn Kokai Tokkyo Koho Hei 7–98483
336. Yousif NM (1989) Phosphorus Sulfur 46:79
337. Zemlyanski NI, Chernaya NM, Turkevich VV (1967) J Gen Chem 37:464, (1967) Zh
     Obsch Khim 37:495
338. Zhou HS, Sasahara H, Honma I, Komiyama H (1994) Chem Mater 6:1534

Top Curr Chem (2005) 251: 87–140
DOI 10.1007/b101007
© Springer-Verlag Berlin Heidelberg 2005

Thio-, Seleno-, and Telluro-Carboxylic Acid Esters

Shin-ichi Fujiwara[1] (✉) · Nobuaki Kambe[2]

[1] Department of Chemistry, Osaka Dental University, Hirakata, Osaka 573–1121, Japan
fujiwara@cc.osaka-dent.ac.jp
[2] Department of Molecular Chemistry, Graduate School of Engineering, Osaka University,
Suita, Osaka 565–0871, Japan
kambe@chem.eng.osaka-u.ac.jp

Abstract This chapter reviews studies into the preparation of thio-, seleno-, and telluro-carboxylic acid esters (referred to thiol, selenol, and tellurol esters, respectively) and into their use in organic synthesis. The studies reviewed here span the last ten years, although some important and original findings reported further back than this are also included.

Keywords Chalcogeno carboxylic acid ester · Thiol ester · Selenol ester · Tellurol ester

1
Introduction

Chalcogeno esters (thiol, selenol and tellurol esters) are useful synthetic intermediates that have been employed, for example, as acylating reagents, building blocks for heterocyclic compounds, precursors of acyl radicals and anions, and in asymmetric aldol reactions. This chapter deals with the transformations and recently-developed synthetic reactions of chalcogeno esters. Chalcogeno esters that are naturally occurring or intermediates in biological metabolism process – for example, derivatives of Coenzyme A, which play critical roles in biochemical transformations – are out of the scope of this chapter.

2
Thio-Carboxylic Acid Esters (Thiol Esters)

In spite of the growing interest in organic transformations of thio-carboxylic acid esters (thiol esters, see Fig. 1) 1, until the mid-1980s, methods for preparing them were limited to those based on conventional methodologies (such as the condensation of the thiols with the parent carboxylic acids in the presence of an activating agent, and substitution of acid chlorides or acid anhydrides with metal thiolates; see for example [1]). Nowadays, in addition to these methods, thiol esters 1 can now be synthesized by (i) transition metal-catalyzed carbonylation, (ii) reaction of acyllithiums with disulfides, (iii) hydration of thioacetylenes, (iv) Tishchenko-type reactions, and (v) couplings of thiol chloroformate with organotin compounds. The thiol esters 1 have been used in asymmetric Michael additions and Mannich reactions as well as in asymmetric aldol reactions. They can be converted to aldehydes and ketones and also utilized as precursors of acyl radicals and as protecting groups for thiols. Methods of synthesizing lactones and performing decarbonylative addition to alkynes have also been developed.

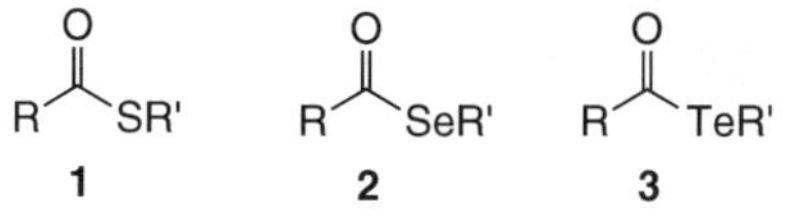

Fig. 1 Thiol (1), selenol (2), and tellurol (3) esters

2.1
Syntheses

2.1.1
Carbonylation of Thiols and Disulfides with Carbon Monoxide

In 1960, it was reported that reacting thiols, disulfides, and sulfides with carbon monoxide gave thiol esters in the presence of $Co_2(CO)_8$ or metal oxide catalysts at 250–300 °C and under 100–1000 atm, as shown in Eqs. 1 to 3 [2]. Both aliphatic and aromatic thiols undergo carbonylation to give thiol esters in yields up to 46% with conversions up to 73%. No reaction took place between carbon monoxide and benzenethiol at 70 °C in the presence of $Co_2(CO)_8$, nor at 275 °C without the catalyst. Reaction conditions for carbonylation of disulfides and sulfides were similar to those used for thiols, but yields were low (up to 30%).

$$2\ RSH\ +\ CO\ \xrightarrow[\text{benzene}]{\text{cat. } Co_2(CO)_8}\ RC\overset{\displaystyle O}{\overset{\|}{S}}R\ +\ H_2S \tag{1}$$

$$RSSR\ +\ 2\ CO\ \xrightarrow[\text{benzene}]{\text{cat. } Co_2(CO)_8}\ RC\overset{\displaystyle O}{\overset{\|}{S}}R\ +\ COS \tag{2}$$

$$RSR\ +\ CO\ \xrightarrow[\text{benzene}]{\text{cat. } Co_2(CO)_8}\ RC\overset{\displaystyle O}{\overset{\|}{S}}R \tag{3}$$

In 1985, Alper and co-workers reported the $Co_2(CO)_8$-catalyzed desulfurization and carbonylation of benzylic and aromatic thiols to carboxylic esters with carbon monoxide (58 atm) in aqueous alcohols at 190 °C (Eq. 4) [3]. However, desulfurization products are formed when the reaction is performed in ben-

$$RSH\ +\ CO\ +\ R'OH\ \xrightarrow[\substack{H_2O,\ 190\ °C \\ 58\text{-}61\ atm}]{\substack{Co_2(CO)_8\ (5 \\ mol\ \%)}}\ RC\overset{\displaystyle O}{\overset{\|}{O}}R'\ +\ H_2S \tag{4}$$

R	R'	yield, %
4-CH$_3$C$_6$H$_4$CH$_2$	Et	75
	Me	83
	i-Pr	74
	t-Bu	28
Ph	Et	32
CH$_3$(CH$_2$)$_7$	Et	trace

zene instead of aqueous alcohols [4]. It is also interesting that thiol esters are also obtained in good to excellent yields when 2,3-dimethyl-1,3-butadiene (DMBD) or 2,3-dimethoxy-1,3-butadiene (DMEBD) is added (Eq. 5) [5].

$$\text{RSH} + \text{CO} \xrightarrow[\substack{\text{diene (1.0-1.2 equiv)} \\ \text{185-190 °C, 55-61 atm} \\ \text{PhH (30 mL), H}_2\text{O (2 mL)}}]{\text{Co}_2(\text{CO})_8 \text{ (5 mol \%)}} \overset{\displaystyle O}{\overset{\|}{\text{RCSR}}} + \text{COS} \qquad (5)$$

10 mmol

R	diene	yield, %
Ph	DMEBD	87
4-CH$_3$C$_6$H$_4$	DMBD	61
	DMEBD	82
2-naphtyl	DMBD	63
	DMEBD	85

The yield of the thiol ester depends upon the diene employed, as shown in Eq. 5. Under similar conditions, isoprene (13%), indene (37%), and 1,3-cyclohexadiene (8%) are much less efficient (R=4-CH$_3$C$_6$H$_4$). The carbonylation reaction is applicable to a variety of aromatic and benzylic thiols, but aliphatic thiols are inert under similar reaction conditions.

In the reaction of disulfides with carbon monoxide using a catalytic amount of Co$_2$(CO)$_8$, diaryl disulfides give the corresponding thiol esters in better yields than dibenzyl disulfide [6].

2.1.2
Carbonylative Addition of Diphenyl Disulfide and Thiols to Acetylenes with Carbon Monoxide

Ogawa, Sonoda, and co-workers revealed that the Pd(PPh$_3$)$_4$-catalyzed addition of diaryl disulfides to terminal acetylenes takes place stereoselectively, giving high yields of (Z)-1,2-bis(arylthio)-1-alkenes **4**.

When the reaction is performed under pressurized carbon monoxide, carbonylative addition occurs, affording the (Z)-1,3-bis(arylthio)-2-alken-1-ones **5** (Scheme 1), where carbon monoxide is regioselectively incorporated at the terminal carbon of the acetylenes [7]. Higher pressure is essential to suppress the generation of the direct adduct of the disulfide **4**.

They also reported an interesting finding; that acetylenes react with benzenethiol and carbon monoxide to give different carbonylation products depending on the catalysts used (Scheme 2). For example, when RhH(CO)(PPh$_3$)$_4$ is employed, the carbon monoxide and phenylthio group are introduced selectively into the terminal and internal positions of acetylenes, giving **6** [8]. However, when Pt(PPh$_3$)$_4$ is used instead of RhH(CO)(PPh$_3$)$_4$, a dramatic changeover of the regioselectivity of CO introduction is observed, and hydrothiocarboxylation of the acetylenes (a hydride and a thiocarboxyl group are introduced into

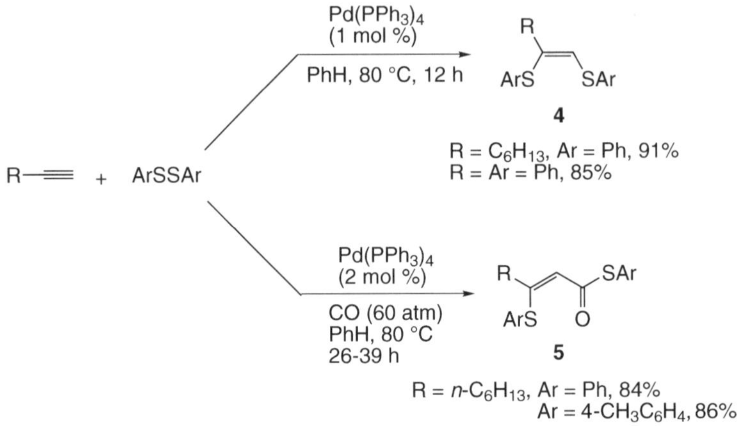

Scheme 1 Addition of disulfide to alkyne with and without carbon monoxide

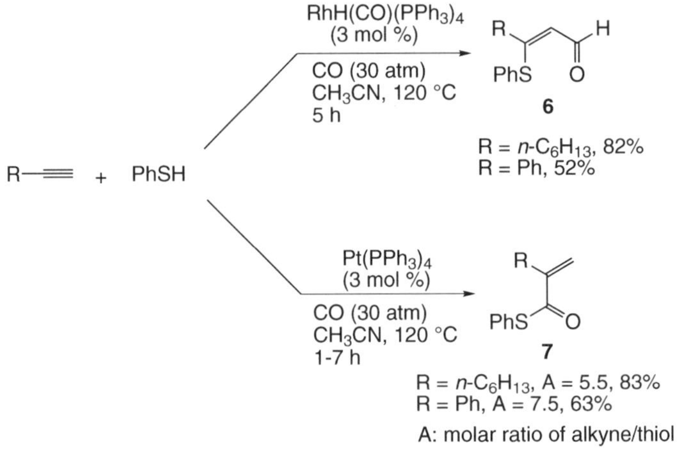

Scheme 2 Dramatic changeover of the regioselectivity of carbon monoxide insertion

the terminal and internal positions of acetylenes, respectively) takes place to afford conjugated thiol esters [9]. Product **7** may undergo conjugate addition of PhSH during the reaction, but the use of excess acetylene affords **7** predominantly. The procedure for this hydrothiocarboxylation can be used with a variety of aromatic and aliphatic thiols, giving α,β-unsaturated thiol esters in good to excellent yields. In particular, the hydrothiocarboxylation with aliphatic thiols proceeds smoothly without using an excess amount of acetylene.

2.1.3
Carbonylation of *N,S*-Acetals

Carbonylations of a series of *N,S*-acetals **8** were carried out in dry benzene under pressurized CO at 140 °C for 24 h in the presence of 1 mol% of [Rh(COD)Cl]$_2$. Thiol esters **9** were obtained as the only products, in 68–92% yields (Eq. 6) [10].

$$\text{(6)}$$

R	R'	yield, %
Et	Et	74
i-Bu	*i*-Bu	82
cyclohexyl	cyclohexyl	92
Me	cyclohexyl	81
allyl	cyclohexyl	68
Me	Bn	81

The substrates applicable to this reaction are limited to benzylsulfides. Neither alkyl nor aryl sulfides underwent carbonylation, and the starting material was recovered along with some decomposition products. This reaction is regioselective, and no CO insertion into the C–N bond was observed, even when benzylamine units were involved. Such regioselectivity seems unusual, since CO incorporation between carbon-heteroatom bonds may become less favorable in the order of C–N>C–S>C–O [11]. It was also reported that carbon monoxide was inserted into the C–N bond of the thiazolidine rather than its C–S bond [12]. In the present reaction, the presence of an N atom at the β-position from the S atom is essential for the insertion of CO into the C–S bond. It is likely that chelation of the substrate to Rh in a bidentate manner is important. Compounds with nitrogen atoms in their heterocyclic rings are not effective. Introduction of a phenyl group at the carbon between the sulfur and nitrogen atoms alternates the regioselectivity, leading to the insertion of carbon monoxide into the C–S bond of the same side of the nitrogen exclusively (Eq. 7).

$$\text{(7)}$$

R	R'	yield, %
Et	Et	83
Me	Bn	46
-CH$_2$CH$_2$OCH$_2$CH$_2$-		80

2.1.4
Trapping of Acyllithums Derived from Alkyllithiums and Carbon Monoxide with Disulfides or Elemental Sulfur

Seyferth and co-workers reported the nucleophilic acylation of carbon electrophiles such as aldehydes [13a], ketones [13a,b], esters[13a,b], lactones [13c], and trialkylchlorosilanes [13d] with acyl- and aroyllithiums generated in situ from the RLi/CO system at very low temperatures (–110 °C to –135 °C). They have also applied this system to the acylation of disulfides, as shown in Eq. 8 [14].

$$
\text{RLi/CO + R'SSR'} \xrightarrow[\substack{\text{THF/Et}_2\text{O/}n\text{-pentane} \\ = 13/13/4 \\ -110\,°\text{C, 2 h, then rt}}]{} \underset{\text{O}}{\overset{\text{O}}{\underset{\|}{\text{RCSR'}}}} + \text{R'SLi} \qquad (8)
$$

R	R'	yield, %
t-Bu	Me	66
Bu	Me	78
t-Bu	i-Pr	71
s-Bu	i-Pr	74

A typical procedure is as follows. The disulfide and a mixed solvent of dry THF, Et$_2$O, and n-pentane are charged to a flask. The solution is cooled to –110 °C, and carbon monoxide is bubbled in for 30 min. Then a solution of alkyllithium is added using a syringe pump. After the addition is complete, the yellow reaction mixture is stirred at –110 °C under CO for 2 h. A reaction temperature of –110 °C is suitable and reacting at lower temperatures or –78 °C usually gives lower yields of products. When the acyl halide corresponding to the acyllithium formed is available, the resulting BuSLi can be converted to the same thiol ester, giving rise to an improvement in the thio ester synthesis (Eq. 9).

$$
t\text{-BuLi/CO + Bu}_2\text{S}_2 \xrightarrow{-110\,°\text{C}} t\text{-BuCSBu} + \text{BuSLi} \qquad (9)
$$

Besides these nucleophilic acylations of disulfides, they also developed acylation of elemental sulfur. For example, the generation of t-Bu(CO)Li from t-BuLi and CO in the presence of elemental sulfur at –78 °C, followed by the addition of MeI, gave t-BuC(O)SMe with a 57% yield. s-BuLi also affords the product in a similar yield, but this reaction is not successful when BuLi is used.

2.1.5
Hydration of Thioacetylenes

Braga and co-workers developed a convenient method of synthesizing thiol esters using the acid catalyzed hydration of thioacetylenes **10** in CH_2Cl_2 in the presence of silica gel (Eq. 10) [15].

$$R\!\!-\!\!\!\equiv\!\!\!-\!SR' \quad \xrightarrow[\substack{\text{silica (1 g), CH}_2\text{Cl}_2\\ 40\,°C}]{\text{acid (1.1 equiv)}} \quad R\overset{O}{\underset{}{\diagup\!\!\diagdown}}SR' \tag{10}$$

10, 1 mmol

R	R'	acid	time, h	yield, %
Ph	Me	TFA	2	86
		p-TsOH	10	86
		$ClSO_3H$	6	85
		$HClO_4$	140	80
		HCl	15	80
		AcOH	140	–
t-Bu	Me	TFA	16	85

Strong acids such as p-TsOH, trifluoroacetic acid (TFA), $ClSO_3H$, $HClO_4$, and HCl proved to be effective, but no reaction took place in acetic acid. Other solvents such as DMF, $CHCl_3$, benzene, Et_2O, and THF were not suitable for efficient hydrolysis. The reaction does not proceed satisfactorily in the absence of silica containing water in natural contents. The use of additional amounts of water neither improves the yields nor increases the rates of the reactions.

This protocol is applicable to the synthesis of selenol esters starting from selenoacetylenes **11** (Eq. 11) [15b, 16].

$$Ph\!\!-\!\!\!\equiv\!\!\!-\!SeMe \quad \xrightarrow[\substack{\text{silica (1 g), CH}_2\text{Cl}_2\\ 40\,°C}]{\text{acid (1.1 equiv)}} \quad Ph\overset{O}{\underset{}{\diagup\!\!\diagdown}}SeMe \tag{11}$$

11, 1 mmol

acid	time, h	yield, %
TFA	48	92
p-TsOH	70	88
$ClSO_3H$	56	48
$HClO_4$	96	70
HCl	96	40
AcOH	140	5

The difference between the sulfur and selenium analogs was their hydration reaction time. The rate of hydration of alkynyl selenides is very slow compared to their sulfur counterparts. For example, hydration of a selenoacetylene (R=Ph, R′=Me) with TFA needed 48 h to complete, whereas hydration of the corresponding thioacetylenes completed within 2 h. The first step (addition of the acid to the chalcogeno acetylenes) was found to have comparable rates in both cases, but the rate of the second step leading to the chalcogeno esters was slower for selenium (Scheme 3).

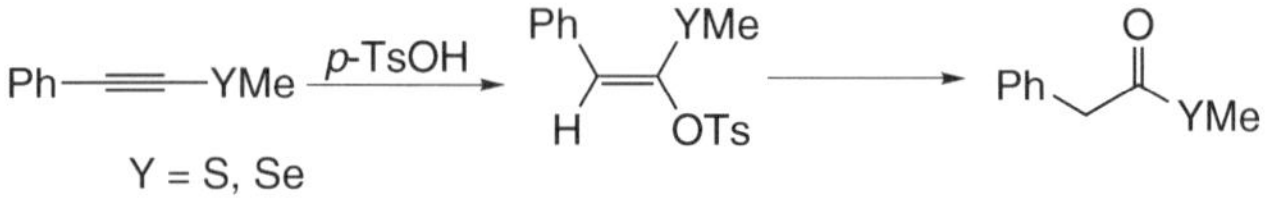

Scheme 3 Reaction pathway for hydrolysis of chalcogenoacetylenes

2.1.6
Tishchenko-Type Reactions of Aldehydes with *i*-Bu$_2$AlSR

A new method of synthesizing thiol, selenol, tellurol esters from aldehydes via a Tishchenko-type reaction using diisobutylaluminum chalcogenoate (*i*-Bu$_2$AlYR, Y=S, Se, Te) has been reported; it is shown in Eq. 12 [17].

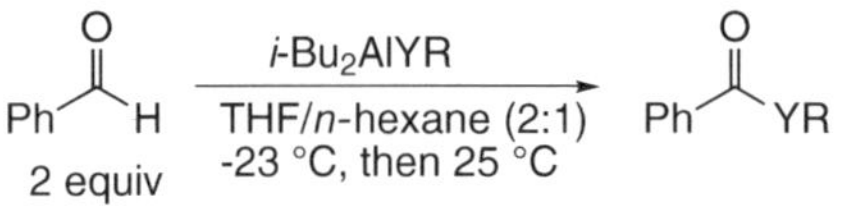

$$(12)$$

YR	time, h	yield, %
SBu	2	83
SeBu	2	79
TeBu	1	52 (71*)
SPh	20	61
SePh	20 (10)	29 (44*)
SBn	2	83

* In the presence of Et$_2$AlCl

This reaction proceeded efficiently to give thiol and selenol esters when *i*-Bu$_2$AlSR and *i*-Bu$_2$AlSeR were used, respectively. Thiol and selenol esters were obtained in good yields from both aromatic and aliphatic aldehydes except in the case of pivalaldehyde. *S*-and *Se*-Ph esters were formed in moderate to good yields. *Te*-Bu esters were obtained in satisfactory yields from aromatic aldehydes. Addition of Et$_2$AlCl improved the yields. When arenetelluroate (*i*-Bu$_2$Al-TeAr) was employed, reduction of aldehydes to alcohols predominated, and the *Te*-Ar esters could not be obtained.

2.1.7
Dehydration of Carboxylic Acids with Thiols by Solid Supported Carbodiimide

The dehydration of carboxylic acid with thiols by the use of polymer supported ethyl (dimethylaminopropyl)carbodiimide (EDAC) **12** [18] was reported by Adamczyl and co-worker (Eq. 13) [19].

$$(13)$$

R	time, h	yield, %
Ph	16	86
C_6F_5	16	74
Bn	16	77
i-Pr	72	63[*]
t-Bu	72	34[*]

[*] Ten equiv of thiol was used

The condensation of benzenethiol with a slight excess amount of carboxylic acid proceeds faster than that of benzylic thiol. This rate difference is in accord with previous observations using a soluble carbodiimide in Et_2O [20]. The more sterically hindered secondary and tertiary thiols (isopropyl and t-butyl thiols) formed only small amounts of the desired products. This method can be extended to the preparation of various α-amino thiol esters.

2.1.8
Cross Coupling of (α-(Acyloxy)alkyl)tributylstannanes with Thiol Chloroformate

Falck and co-workers report cross-coupling of (α-(acyloxy)alkyl)tributylstannanes with thiol chloroformate using a catalytic amount of copper salts. For example, (α-(acetyloxy)benzyl)tributylstannane **13** was treated with thiol chloroformate **14** in the presence of 8 mol% of CuCN in toluene at 75 °C for 21 h to give the thiol ester **15** with 98% yield (Eq. 14) [21].

$$(14)$$

2.2
Reactions

2.2.1
Asymmetric Aldol Reaction between Silyl Enol Ethers of Thiol Esters and Aldehydes

The titanium tetrachloride ($TiCl_4$)-mediated aldol reaction of silyl enol ethers with aldehydes was first reported by Mukaiyama and co-workers [22]. Following this report, several other Lewis acids such as $BF_3 \cdot Et_2O$, and $SnCl_4$, or fluoride anions such as Bu_4F were found to be effective promoters or catalysts in this reaction.

In 1989, a highly enantioselective aldol reaction of achiral silyl enol ethers of thiol esters with achiral aldehydes was developed by using a novel chiral promoter system consisting of chiral diamine-coordinated tin(II) triflate and tributyltin fluoride (or dibutyltin diacetate) [23]. When the silyl enol ether **16** of S-ethyl ethanethioate was treated with PhCHO in the presence of stoichiometric amounts of tin(II) triflate, (S)-1-methyl-2-[(piperidin-1-yl)-methyl]-pyrrolidine (**18**), and tributyltin fluoride, the aldol reaction proceeded at –78 °C to afford the corresponding adduct **17** in 78% yield with 82% ee (Scheme 4).

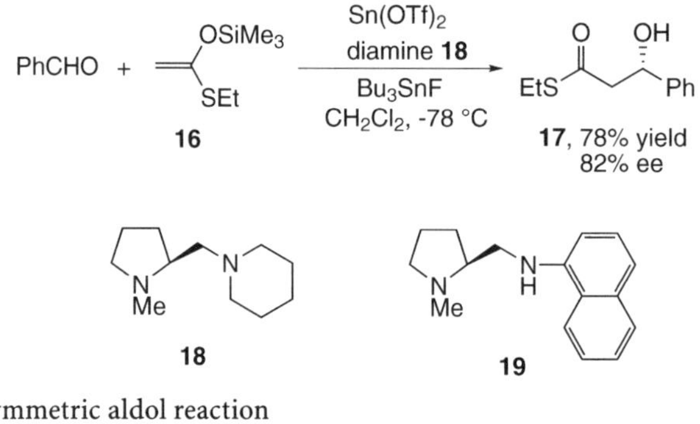

Scheme 4 Asymmetric aldol reaction

The importance of the combination of chiral diamine-coordinated tin(II) triflate and tributyltin fluoride is obvious from the result that no enantiomeric selection was observed when using only chiral diamine-coordinated tin(II) triflate or a combination with other metal salts. Tributyltin fluoride was found to be the most effective among the several tin(IV) fluorides examined. The maximum ee was obtained (92%) when (S)-1-methyl-2[(N-1-naphthylamino)methyl]pyrrolidine (**19**) was employed as the chiral diamine.

The authors studied the further possible extension of this enantioselective reaction to the synthesis of syn-α-methyl-β-hydroxy thiol esters and found that the reaction of PhCHO with enol silyl ether **20** in the presence of stoichiometric amounts of tin(II) triflate, chiral diamine **19**, and tributyltin fluoride afforded only syn aldol **21** with excellent enantioselectivity (Eq. 15).

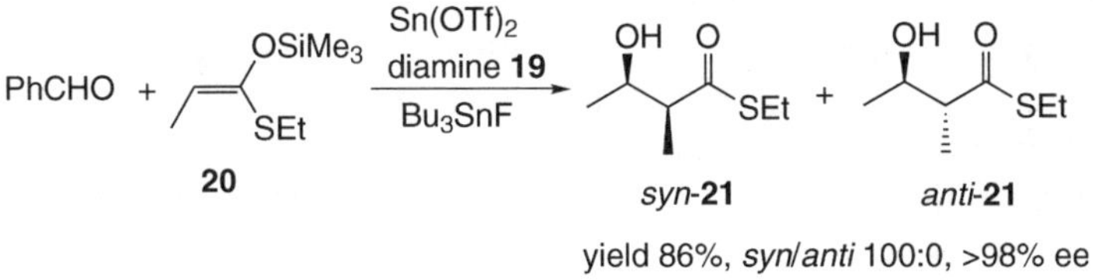

$$\text{yield 86\%, }syn/anti\text{ 100:0, >98\% ee}$$

These findings brought about extensive studies of asymmetric aldol reactions using silyl enol ethers of thiol esters that employed other chiral diamines [24] or organometals such as Cu [25], Zr [26], Ti [27], B [28], Sc [29], and Pr [30]. An imino aldol reaction catalyzed by a cation-exchange resin [31], and uncatalyzed aldol addition of enoxysilacyclobutanes to aldehydes [32] have also been reported.

2.2.2
Asymmetric Michael Addition of Silyl Enol Ethers of Thiol Esters to Enones

In 1974, Mukaiyama and co-workers reported the first examples of Lewis acid-catalyzed Michael reactions between silyl enolates and α,β-unsaturated carbonyl compounds [33]. Evans and co-workers developed a catalytic asymmetric Michael reaction of silyl enol ethers of thiol esters to alkylidene malonates. For example, the reaction of alkylidene malonate **23** with 2.2 equiv of silyl enol ether **22** was carried out in the presence of 10 mol % of catalyst **25** and 2 equiv of hexafluoro-2-propanol (HFIP) in PhMe/CH$_2$Cl$_2$ (3:1) at –78 °C to give the expected adduct **24** in 93% ee (Scheme 5) [34]. Borane complex-catalyzed asymmetric Michael addition has also been reported [35].

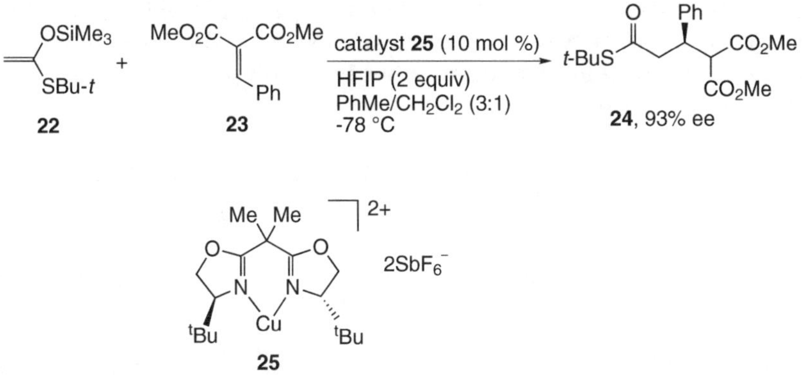

Scheme 5 Asymmetric Michael addition

2.2.3
Asymmetric Mannich-Type Reactions of Silyl Enol Ethers of Thiol Esters

Kobayashi and co-workers discovered catalytic enantioselective Mannich-type reactions of silyl enol ethers of thiol esters with aldimines using a novel zirconium catalyst. For example, in the presence of 10 mol% of catalyst **28**, the silyl enol ether **20** was treated with aldimine **26** in CH_2Cl_2 at –45 °C in the co-existence of *N*-methylimidazole (NMI) as an additive to afford the adduct **27** in 78% yield with 88% ee (Scheme 6) [36].

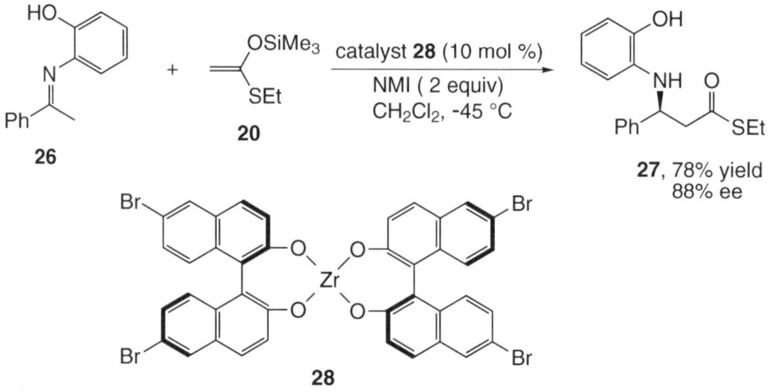

Scheme 6 Asymmetric Mannich-type reaction

2.2.4
Ir-Catalyzed Substitution of Propargylic-Type Esters with Silyl Enol Ethers of Thiol Esters

Matsuda and co-workers found that silyl enol ethers of thiol esters are good nucleophiles in the reaction with propargylic-type esters to form β-alkynyl thiol esters in the presence of a catalytic amount of [Ir(cod)(POPh$_3$)$_2$]OTf **32** [37]. The acetoxy group of **29** was readily substituted at 25 °C to form thiol ester **31** with 93% yield by simply stirring together **29**, 4 equiv of silyl enol ether **30**, and **32** (5 mol%) (Eq. 16).

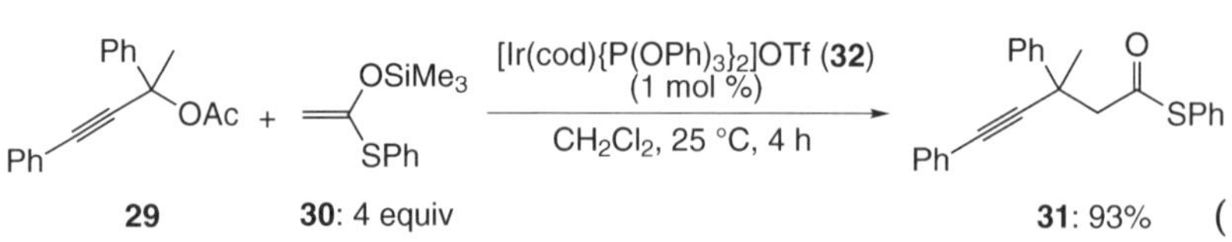

$$(16)$$

2.2.5
Reduction to Aldehydes

The reduction of thiol esters to aldehydes using Raney nickel was reported to proceed in 70–80% ethanol under reflux more than five decades ago. [38] Fujita and co-workers developed a low temperature (–20 to –50 °C) reduction of thio esters by treating with diisobutylaluminum hydride (DIBALH) in hexane-dichlorometane (Eq. 17) [39]. The authors also reported that treatment of thiol esters with $NaBH_4$ in aqueous THF at room temperature gave alcohols in high yields (Eq. 18).

$$ \text{(17)} $$

R	yield, %
Ph	93
$n\text{-}C_{15}H_{31}$	79
$n\text{-}C_9H_{19}$	72
$n\text{-}C_5H_{11}$	54
PhCH=CH *(trans)*	64

$$ \text{(18)} $$

R	yield, %
Ph	92
$n\text{-}C_{15}H_{31}$	99
$n\text{-}C_9H_{19}$	98
$n\text{-}C_5H_{11}$	96
PhCH=CH *(trans)*	92

In 1990, Fukuyama and co-workers established a palladium-mediated reduction of thiol esters to aldehydes with Et_3SiH (Eq. 19) [40].

This reduction proceeds at room temperature and functional groups such as azide and nitro groups are incompatible. The practical utility of this reaction is demonstrated by its application to the total synthesis of complex natural products such as (+)-neothramycin A methyl ether [41]. Recently, an improved protocol using odorless dodecanethiol has been reported [42, 43].

Penn and co-worker report the reduction of thiol esters with lithium. Treatment of aryl and alkyl thiol esters with lithium wire at –78 °C for 4–5 h, followed by quenching with MeOH, resulted in the exclusive formation of aldehydes (Eq. 20). The reaction mechanism seems to involve electron-transfer from lithium

$$
\text{R-C(=O)-SEt} \xrightarrow[\substack{10\% \text{ Pd/C (2-3mol \%)} \\ \text{acetone, rt}}]{\text{Et}_3\text{SiH (2-3 equiv)}} \text{R-C(=O)-H} + \text{Et}_3\text{SiSEt}
$$

(19)

MeO–C$_6$H$_4$–COSEt   91%

MeO–C$_6$H$_4$–O–(CH$_2$)$_3$–COSEt   94%

Ph–CH(OAc)–COSEt   84%

PhS–CH$_2$–COSEt   75%

$$
\text{R-C(=O)-S-Ph} \xrightarrow[\text{2) MeOH}]{\text{1) Li/THF, -78 °C}} \text{R-C(=O)-H}
$$

(20)

R	yield, %
Ph	98
$n\text{-C}_{10}\text{H}_{21}$	99
cyclohexyl	100
t-Bu	98

to a thiol ester, which undergoes bond cleavage to generate a radical and an anion. The radical formed is further reduced to yield thiolate or acyl anion which is trapped with electrophiles. The formation of an acyl anion is supported by evidence that acetophenone is produced by successfully quenching of the benzoyl anion with methyl iodide [44].

2.2.6
Substitution with Organometallic Reagents Leading to Ketones

The reactions of thiol esters with organocopper and Grignard reagents to ketones were revealed in 1974 by Anderson and co-workers (Eq. 21) [45–46], and Mukaiyama and co-workers (Eq. 22) [47], respectively.

$$
n\text{-C}_9\text{H}_{19}\text{-C(=O)-SEt} \xrightarrow[\text{Et}_2\text{O, 1.5-2 h}]{\text{R}_2\text{CuLi (1.0-1.2 equiv)}} n\text{-C}_9\text{H}_{18}\text{-C(=O)-R}
$$

(21)

R	temp, °C	yield, %
Me	-78	75
Bu	-40	87
Ph	-40	74

$$(22)$$

R	yield, %
Ph	quant
Me	91
$n\text{-}C_5H_{11}$	86

Fukuyama and co-workers disclosed the palladium-catalyzed synthesis of ketones by the reaction of thio esters with organozinc reagents (Eq. 23) [48]. Treatment of thio esters with a catalytic amount of $PdCl_2(PPh_3)_2$ (5 mol%) and EtZnI (1.5 eq) in toluene at room temperature for 5 min affords ketones in high yields. Ethyl zinc iodide as well as i-butyl-, benzyl-, phenyl-, β-phenethyl-, and vinylzinc iodides react smoothly to afford the corresponding ketones. The geometry of the olefins on the alkyl chains is retained in the reaction. Alkyl, aryl, and α,β-unsaturated thio esters can be converted into ketones in good to excellent yields. It should be noted that a variety of sensitive functional groups including ketones, α-acetates, sulfides, aromatic bromides, chlorides and even aldehydes are compatible with this protocol. This remarkable chemoselectivity indicates that ketone formation is much faster than the oxidative addition of palladium to aromatic bromide or nucleophilic additions of zinc reagents to aldehydes.

$$(23)$$

99%   75%   91%   69%

Srogel, Liesbeskind and co-worker reported the palladium-catalyzed cross-coupling of the thiol esters **32** with the boronic acids **33** (Eq. 24) [49].

$$(24)$$

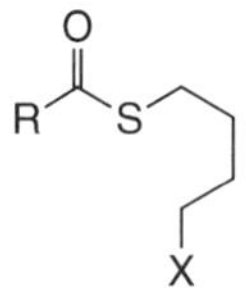

32a: R = phenyl, X = Br
32b: R = methyl, X = Br
32c: R = undecyl, X = Br

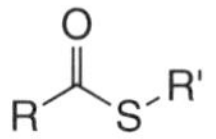

35a: R = 4-MeC$_6$H$_4$, R' = Me
35b: R = undecyl, R' = Et

Fig. 2 4-Halo-*n*-butyl and normal thiol esters

The coupling of stable organosulfur compounds with boronic acids is of great synthetic importance, since both reaction partners are readily available, low toxicity, and are stable toward many reagents. 4-Halo-*n*-butyl and normal thio esters (Fig. 2) were prepared and investigated as substrates for palladium-catalyzed cross-coupling with boronic acids.

In the presence of 5–10% palladium catalyst **34**, 20–45% NaI, and 4–6 equiv K$_2$CO$_3$, various boronic acids underwent cross-coupling with thio esters **32** in dimethylacetamide at 90–95 °C for 12–24 h to give the desired ketones in moderate to high yields. Tetrahydrothiophene, detected by GC/MS, was formed by intramolecular nucleophilic attack at the terminal carbon by a sulfur atom. Interestingly, in the absence of NaI, the bromobutyl thiol ester **32a** gave only a trace amount of benzophenone when treated with phenylboronic acid in the presence of **34** in DMA at 95 °C for 22 h. In the presence of NaI, the bromide atom at the terminal carbon is replaced by an iodide atom, leading to sulfonium formation via intramolecular *S*-alkylative activation.

Only a trace of *p*-tolyl phenyl ketone was formed when the simple thiol ester **35a** was treated with phenylboronic acid in the presence of NaI under similar reaction conditions. Although less efficient than the intramolecular system, intermolecular alkylative activation was also feasible. For example, **35b** was transformed into undecyl phenyl ketone (48%) when treated with PhB(OH)$_2$ and 1,4-dibromobutane in the presence of 6 mol% of **34**, 30% NaI, and 4.1 equiv of K$_2$CO$_3$ in DMF at 90 °C for 18 h. Use of 1,2-dibromoethane and 1-bromohexane in place of 1,4-dibromobutane did not promote this reaction.

Liebeskind and Srogl developed Pd-catalyzed cross-coupling of simple thiol esters with boronic acids to give ketones in the presence of Cu(I) thiophene-2-carboxylate (CuTC) under nonbasic conditions (Eq. 25) [50].

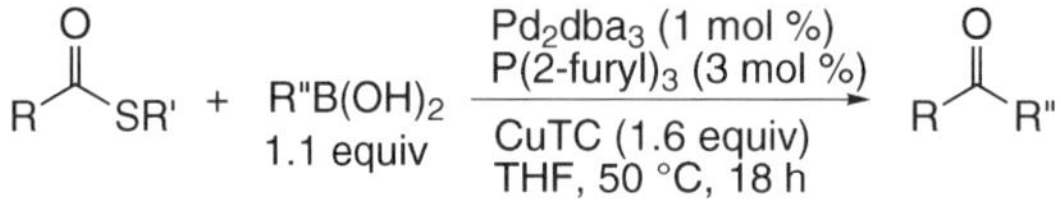

$$\text{R} \overset{\text{O}}{\underset{}{\text{C}}}\text{SR}' + \underset{1.1\ \text{equiv}}{\text{R''B(OH)}_2} \xrightarrow[\substack{\text{CuTC (1.6 equiv)} \\ \text{THF, 50 °C, 18 h}}]{\substack{\text{Pd}_2\text{dba}_3\ (1\ \text{mol \%}) \\ \text{P(2-furyl)}_3\ (3\ \text{mol \%})}} \text{R} \overset{\text{O}}{\underset{}{\text{C}}}\text{R''} \tag{25}$$

R	R'	R"	yield, %
Ph	CH$_2$CONMe$_2$	2-MeOC$_6$H$_4$	88
		3-MeOC$_6$H$_4$	83
		2-naphthyl	79
n-C$_{11}$H$_{23}$	4-MeC$_6$H$_4$	3-NO$_2$C$_6$H$_4$	79
CF$_3$		(*E*)-β-stryl	93

Unlike the usual Suzuki-Miyaura cross-coupling of boronic acids with halides [51], bases were detrimental to the present reaction. Addition of 1 equiv of K_2CO_3 to the reaction significantly suppressed the yield of ketone. On the other hand, ketones were formed in moderate yields in acetic acid. It was proposed that CuTC coordinates to the sulfur atom to activate the acylpalladium-thiolate complex toward transmetalation with a boron reagent (Scheme 7).

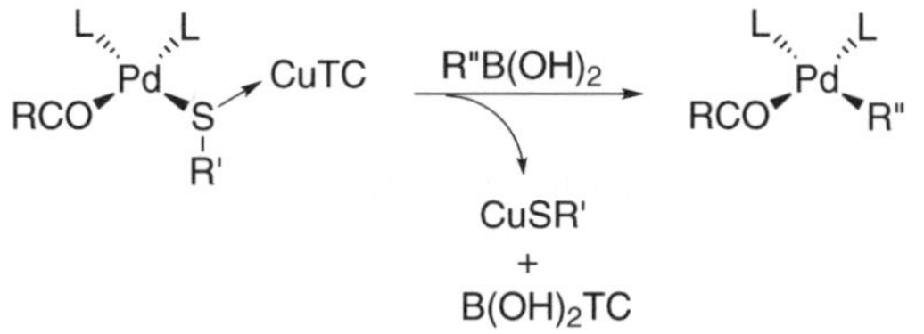

Scheme 7 Activation of acylpalladium-thiolate by CuTC

α,β-Acetylenic ketones are useful synthetic precursors of compounds such as chiral propargyl alcohols [52], α,β-unsaturated ketones [53], as well as a variety of Michael addition products, and have therefore served as crucial intermediates for the synthesis of natural products [54]. Yamaguchi and co-workers reported a mild and efficient synthesis of α,β-acetylenic ketones from the reaction of thiol esters with trimethylsilylacetylenes in the presence of $AgBF_4$. The reaction takes place in CH_2Cl_2 at room temperature and completes within 30 min (Eq. 26) [55]. Among the catalysts examined, $AgBF_4$ proved to be the most effective. Other catalysts such as AgF and $CuCl_2$ gave no product and $SnCl_4$ gave a yield of less than 10%.

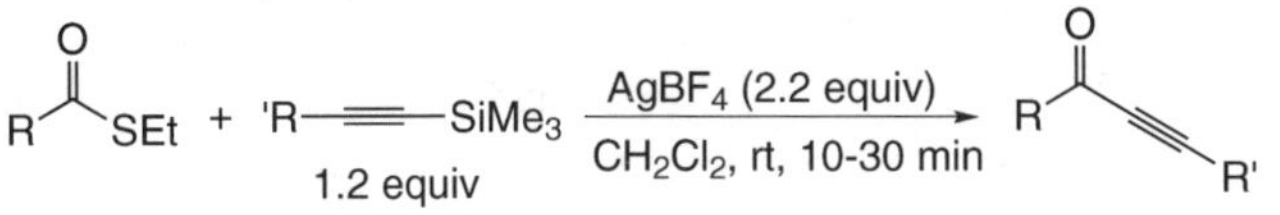

$$(26)$$

R	R'	yield, %
Pr	$n\text{-}C_6H_{13}$	96
Pr	Ph	88
i-Pr	$n\text{-}C_6H_{13}$	97
i-Pr	Ph	97
Ph	Ph	92

In 2003, Sonogashira-type palladium-catalyzed coupling of thiol esters with 1-alkynes was found to give α,β-acetylenic ketones (Eq. 27) [56].

When thiol esters were subjected to the conventional Sonogashira conditions, the expected coupling products were obtained, although the yields were poor (~17%). However, the yield was dramatically improved by adding phos-

$$\text{(27)}$$

Ar = 4-MeOC$_6$H$_4$, 94%
  = 4-BrC$_6$H$_4$, 80%   80%   64%

phine ligands such as tri-2-furylphosphine. Among the catalysts examined, PdCl$_2$(dppf) proved to be the most effective.

2.2.7
Synthesis of β-Lactones by the Reaction of Lithium Enolates of Thiol Esters with Carbonyl Compounds

Danheiser and co-worker describe a convenient one-step preparation of β-lactones based on the addition of thiol ester enolates to carbonyl compounds. Treatment of thiol esters with 1 equiv of LDA in THF at –78 °C for 30 min furnishes lithium enolates, which smoothly combine with ketones and aldehydes at –78 °C. Subsequent gradual warming results in the formation of β-lactones in good to high yields (Eq. 28) [57].

$$\text{(28)}$$

R	yield, %
H	86
Me	92
i-Pr	48
MeO	78

2.2.8
Generation of Acyl Radicals from Thiol Esters by Intramolecular Homolytic Substitution at Sulfur

For transformations based on stannane-mediated radical chain sequences, selenol esters are the precursors of choice owing to their ease of preparation, stabilities, and ability to accomplish S$_H$2 reactions (see Sect. 3.2.1). However, replacing the selenol esters by thiol esters in these tributyltin hydride- or allyl-tributylstannane-mediated chain reactions would be attractive from a synthetic viewpoint.

Unfortunately, the radical chain reaction of simple *S*-phenyl thiol esters was not efficiently triggered by trialkyltin radicals [58]. The homolytic bond dissociation of *S*-2-naphthyl and other thiol esters by photolysis also seems to be a low quantum yield process (Φ=0.08–0.10) [59].

Crich and co-worker report an attractive solution to this problem by a combination of efficient iodine abstraction by tributyltin or tris(trimethylsilyl)silyl radicals from an aryl iodide to form an aryl radical and successive intramolecular homolytic attack at sulfur by the aryl radicals to generate acyl radicals (Scheme 8) [60].

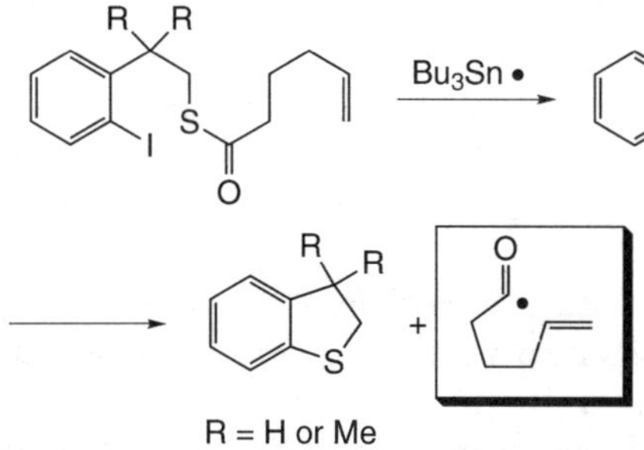

Scheme 8 Generation of acyl radical by intramolecular hemolytic substitution at sulfur

Reactions of thiol esters with the three most commonly used reagents – tributylstannane, tris(trimethylsilyl)silane (TTMSS), and allyltributylstannane – were conducted using AIBN as an initiator. The reaction of **36** with 1.3 equiv of Bu₃SnH in refluxing benzene with AIBN for 1 h led to complete consumption of the substrate and clean formation of the cyclization and reduction products (**39** and **40**, respectively, see Fig. 3) in a ratio of 96:4. As anticipated, dihydrobenzothiophene **41** was also formed in this reaction. It is worth noting that **40**

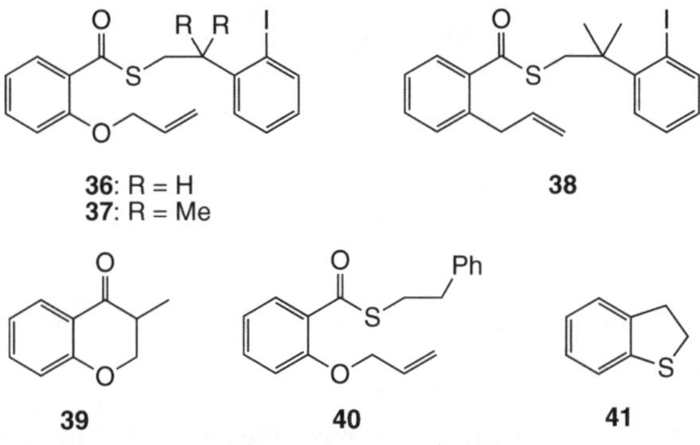

36: R = H
37: R = Me

38

39

40

41

Fig. 3 Compounds 36–41

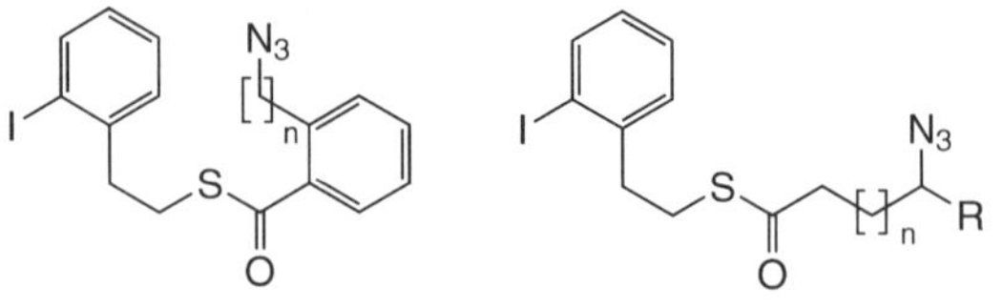

Fig. 4 Thiol esters **42** and **43**

was only formed in very low yields, suggesting that the intramolecular attack of an aryl radical at a sulfur atom is faster than the intermolecular hydrogen atom abstraction from the stannane. As may be expected, use of the poorer hydrogen atom donor TTMSS in place of Bu₃SnH as the reducing agent completely suppressed the formation of **40**. Hydrogen transfer to the aryl radical could also be completely suppressed by use of **37** and **38**, where two methyl groups enhance cyclization.

Benati, Spagnolo, Strazzari and co-workers applied this Crich's method to the generation of acyl radicals bearing an azido group in the side chain from thiol esters **42** and **43** (see Fig. 4) [61].

Interestingly, the acyl radicals formed show different chemical behaviors depending on their structures. Aroyl radicals from **42** were highly prone to performing an unprecedented intramolecular five- and six-membered cyclization via attack onto the azido moiety (Scheme 9), whereas acyl redicals from **43** undergo rapid decarbonylation.

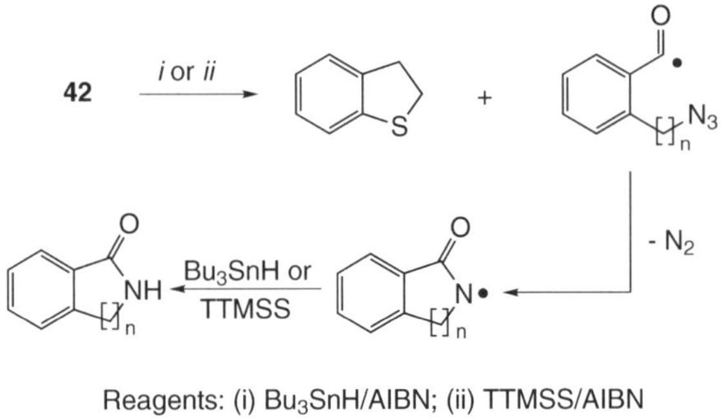

Reagents: (i) Bu₃SnH/AIBN; (ii) TTMSS/AIBN

Scheme 9 Azidoacyl radical cyclization

2.2.9
Reduction of Thiol Esters with SmI₂ and Synthetic Application as Synthons of Unstable Acyl Radicals

The kinetic stability of acyl radicals is strongly controlled by the decarbonylation reaction, which is influenced by the nature of substituent X (Eq. 29). For example, acyl radicals possessing a heteroatom substituent at the α-position undergo rapid decarbonylation due to the radical stabilizing effect of substituent X [62].

$$X \overset{O}{\diagdown}\!\!\!\diagup\! \bullet \;\; \rightleftharpoons \;\; X\!-\!\bullet \;+\; CO \qquad\qquad (29)$$

Skrydstrup and co-workers reported a SmI_2-mediated generation of formal acyl radical equivalents from amino acid thiol esters, and their synthetic application. Trapping of the intermediates formed in this way with a variety of acrylamides provides a simple way to access γ-ketoamides and -esters, which are incorporated in small peptide strands of a series of potential protease inhibitors (Eq. 30) [63].

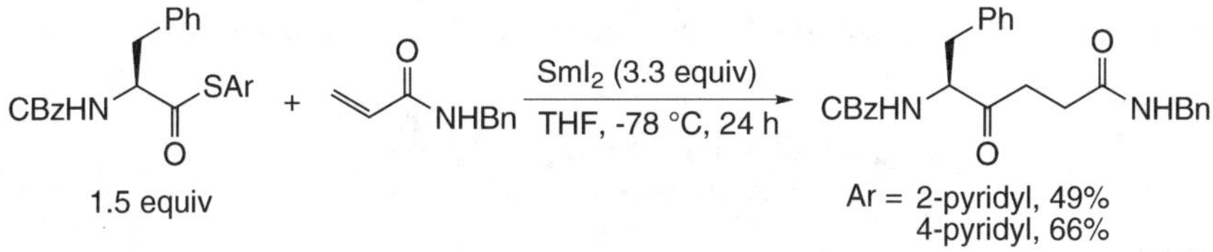

$$\qquad\qquad (30)$$

For example, a solution of acrylamide and thiol ester (1.5 equiv) in THF was added dropwise to a 0.1 M solution of SmI_2 cooled to -78 °C. After the mixture was stirred for 1 h, followed by oxidation of the excess SmI_2 with oxygen, γ-keto-amide was formed in 49% yield. That decarbonylation (as shown in Eq. 29) is not observed implies that the reacting species is not a free acyl radical. It is proposed that electron transfer into the carbonyl group could generate a ketyl radical anion equivalent to the structure **44** coordinating to samarium(III) (Scheme 10).

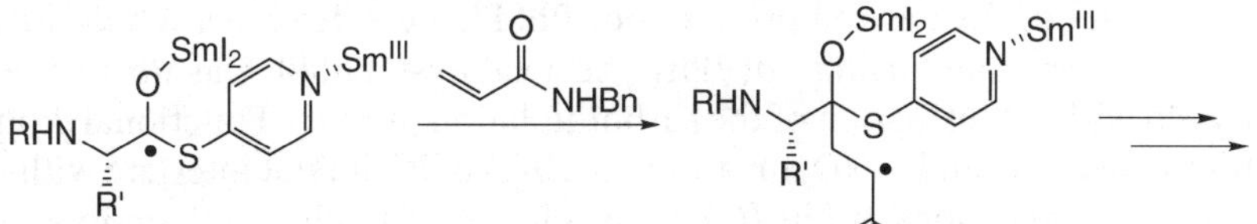

Scheme 10 Proposed reaction pathway for SmI_2 mediated addition of thiol ester

2.2.10
Decarbonylation of Thiol Esters to Sulfides and Carbothiolation of Alkynes Catalyzed by Pd

T. Yamamoto, A. Yamamoto and co-worker report palladium-catalyzed decarbonylation of thiol esters to sulfides (Eq. 31) [64].

$$\text{Ar}\overset{\text{O}}{\underset{}{\|}}\text{SAr} \quad \xrightarrow[\text{toluene, 100 °C, 3 h}]{\text{Pd(PCy}_3)_2 \text{ (5 mol \%)}} \quad \text{ArSAr} \tag{31}$$

$$100\% \text{ (Ar = Ph)}$$
$$98\% \text{ (Ar = 4-MeC}_6\text{H}_4)$$

Symmetrical diaryl sulfides are obtained from the corresponding thiol esters by heating them in the presence of $Pd(PCy_3)_2$. Other zero-valent Ni and Pd complexes such as $Pd(PPh_3)_4$, $Pd(dpe)_2$, $Ni(PPh_3)_4$, and $Ni(PEt_3)_4$ are less effective than $Pd(PCy_3)_2$. A key intermediate of the reaction seems to be an aryl Pd(II) compound carrying a thiolate ligand $ArPd(SR)L_2$ generated via decarbonylation of the aroyl Pd(II) formed by the oxidative addition of thiol ester to Pd(0).

Kuniyasu and co-workers report regio- and stereoselective additions of carbon and sulfur functionalities to alkynes by utilizing a similar intermediate generated in situ from thiol ester and Pt(0), which undergoes insertion of alkyne into the S–Pt bond and subsequent reductive elimination (Eq. 32) [65].

$$R\overset{\text{O}}{\underset{}{\|}}\text{SPh} + \text{=}\text{—R}' \quad \xrightarrow[\substack{\text{toluene, reflux}\\10\text{-}15 \text{ h}}]{\text{Pt(PPh}_3)_4 \text{ (5 mol \%)}} \tag{32}$$

$$1.2 \text{ equiv}$$

R	R'	yield, %
Ph	n-C$_6$H$_{13}$	77
4-MeC$_6$H$_4$		70
4-BrC$_6$H$_4$		81
	Ph	71
	Me$_3$Si	75

For example, the reaction of PhC(O)SPh with 1-octyne carried out in the presence of $Pt(PPh_3)_4$ under toluene reflux for 10 h produced the anticipated (Z)-2-(phenylthio)-1-phenyl-1-octene with 77% yield. Neither regio- and stereoisomers of the desired product nor PhSPh were detected. When the reaction was performed using $Pd(PPh_3)_4$ as a catalyst, PhSPh was formed with 40% yield without formation of the carbothiolation product. Functional groups such as 4-Me, 4-Br, and 4-NO$_2$ on aromatic rings of R' did not interfere with the reactions. The reactions of PhC(O)SBu or PhC(O)SCH$_2$Ph with 1-octyne were very sluggish. This platinum-catalyzed arylthiolation of alkynes was not so sensitive to electronic effects of the aromatic rings in R. The group R can be replaced by a vinyl group to afford the vinylthiolation product with the stereochemistry of the vinyl moiety retained.

They also extend their reaction to the carboselenation of alkynes with selenol esters under similar conditions employed for the carbothiolation (Eq. 33) [66].

$$R\overset{\text{O}}{\underset{}{\|}}\text{SeR}' + \text{=}\text{—R}'' \quad \xrightarrow{\text{Pt(PPh}_3)_4} \tag{33}$$

Selenol esters bearing a 4-MeOC$_6$H$_4$ group on Se underwent aroylselenation under CO (10 atms) to give the corresponding product with 42% yield along with a trace amount of the arylselenation product (Eq. 34).

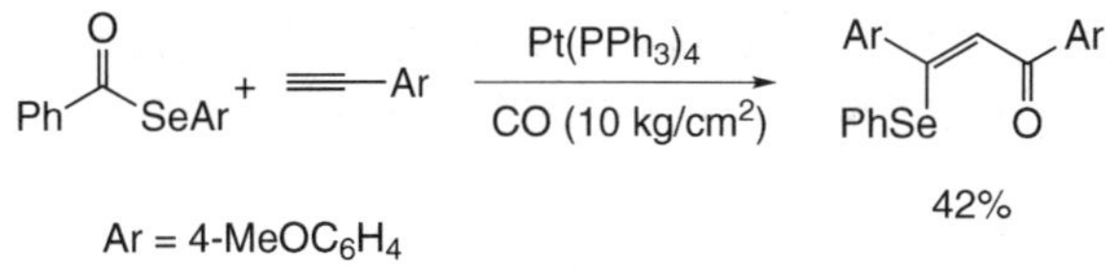

$$\text{(34)}$$

Ar = 4-MeOC$_6$H$_4$

2.2.11
As Protecting Groups for Thiols for Suzuki-Miyaura Cross-Coupling

In Suzuki-Miyaura cross-coupling of boranes or boronic acids, the thiol group poses a special problem arising from its strong affinity toward late transition metals. Due to the poisoning of the catalyst, more than stoichiometric amounts of palladium would be necessary to obtain a coupling reaction. Sulfides and thio-heterocycles are applicable to Suzuki-Miyaura coupling. Although sulfides seem to be suitable protecting groups for thiols, deprotection due to the cleavage of S–C bonds is problematic. The commonly employed reaction conditions for their cleavage are too harsh for a number of functional groups.

Terfort and co-workers focused their efforts on the use of acyl groups, since these are easily attached to the thiol moiety, forming thiol esters, which in turn are easily cleaved by aqueous base. Reaction of thiol esters with p-tolylboronic acid anhydride in the presence of Pd(PPh$_3$)$_4$ followed by saponification afforded the coupling product in high yield (Scheme 11) [67].

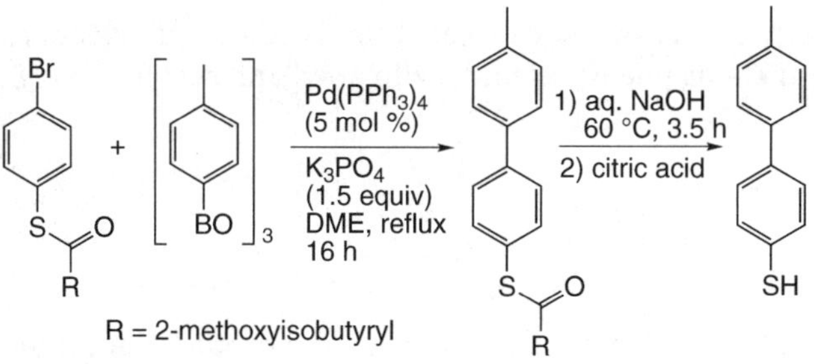

Scheme 11 Thiol ester as a protecting group for thiol

3
Seleno-Carboxylic Acid Esters (Selenol Esters)

Seleno-carboxylic acid esters (selenol esters) **2** are also prepared like thiol esters by the reaction of acyl halides with selenols, or diselenides as well as their alkali metal salts. Most of the preparative methods for thiol esters shown in Sect. 2.1 are applicable to the synthesis of selenol esters **2**. Unique synthetic methods for selenol esters are the following carbonylation reactions: (1) group transfer radical carbonylation reaction; (2) carbonylation of organolithium compounds with Se and CO; (3) Pd-catalyzed coupling of stannyl selenide, aryl iodides, and CO. Selenol esters **2** have been widely used as precursors of acyl radicals. They can also be utilized as acylating reagents and dienophiles in Diels-Alder reactions, and can be converted to aldehydes, α-seleno ketones, and vinyl selenides.

3.1
Syntheses

3.1.1
Reactions of Selenolate Anions with Acid Chlorides

A common method of synthesizing selenol esters is to react selenolate anions with acyl chlorides. Selenolate anions can be easily generated through the reductive cleavage of the Se–Se bond using various reducing agents, such as sodium borohydride, sodium, and lithium aluminum hydride. Reaction of Grignard and organolithium reagents with elemental selenium is also a convenient and practical method [68].

Zhang and co-workers report novel syntheses of selenol esters by using SmI_2 [69a], $Sm/TMSCl/H_2O$ [69b], $TiCl_4$-Sm [69c], $Sm/CrCl_3$ [69d], and $Sm/CoCl_2$ [69e] as reducing agents.

3.1.2
Reaction of Thallous Selenide and Bis(organoseleno)mercury with Acid Chlorides

In 1980, Detty and co-worker revealed the synthesis of selenol esters by the reaction of thallous phenyl selenide with aroyl and acyl halides (Eq. 35) [70].

$$\underset{R=Ph,\,Me}{\underset{R}{\overset{O}{\parallel}}\!\!C\!-\!Cl} \;+\; PhSeTl \;\xrightarrow[\substack{1\text{-}2.5\,h}]{Et_2O,\,rt}\; \underset{>\,97\%}{\underset{R}{\overset{O}{\parallel}}\!\!C\!-\!SePh} \qquad (35)$$

Thallous phenyl selenide is prepared by the dropwise addition of thallous ethoxide to a solution of benzeneselenol in hexanes-Et_2O, and is obtained as an orange solid in quantitative yields. Acyl and aroyl halides react readily with thallous phenyl selenide to give the corresponding selenol esters in >97% isolated yields.

Recently, Silveira and co-workers reported the synthesis of thiol, selenol and tellurol esters based on the reaction of bis(organochalcogenyl)mercury compounds with acyl chlorides in the presence of a catalytic amount of Bu_4NX (Eq. 36) [71].

The starting mercurials, $Hg(SPh)_2$, $Hg(SePh)_2$, and $Hg(TeBu)_2$, are stable solids that are easily prepared by the reaction of metallic Hg with PhSH,

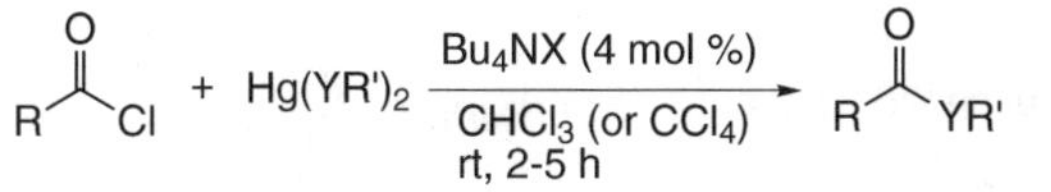

$$\text{(36)}$$

R	SePh (yield, %)	SPh (yield, %)	TeBu (yield, %)
Ph	99	98	39
4-ClC$_6$H$_4$	74	99	44
C$_6$H$_5$CH$_2$	94	99	35
t-Bu	81	94	7

PhSeSePh, or BuTeTeBu. The yield decreases by changing Y from S to Se or Te. The addition of the ammonium halide catalyst greatly improves the efficiency of the reaction. The reaction between benzoyl chloride and Hg(SePh)$_2$ was monitored by IR spectroscopy to obtain an insight into the reaction pathway. At the start of the reaction, the C=O stretching frequency of benzoyl chloride barely changed ($\approx$5 cm^{-1}). Unexpectedly, however, no sign of the product (1680 cm^{-1}) was detected. The characteristic band of PhCOSePh only began to appear after 1.5 h. Instead, strong absorptions at 1530, 1475, and 1425 cm^{-1} were observed, which could not be attributed to Hg(SePh)$_2$ or ClHgSePh. Therefore, this transformation might follow the pathway depicted in Scheme 12.

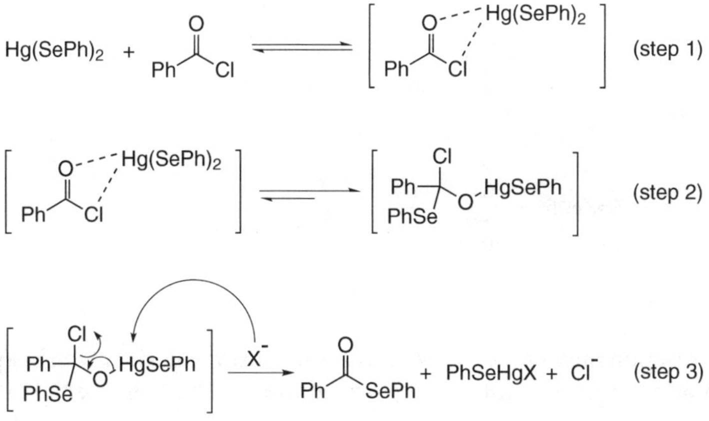

Scheme 12 Proposed mechanism for the reaction of bis(organochalcogenyl)mercury with acyl halide

3.1.3
Metal-Catalyzed Arylselenation of Acyl Chlorides

Nishiyama, Sonoda and co-workers revealed that the Pd(PPh$_3$)$_4$-catalyzed reaction of PhSeSnBu$_3$ with acyl or aroyl chlorides afforded selenol esters in moderate to excellent yields (Eq. 37) [72].

$$\underset{R}{\overset{O}{\big|\big|}}C-Cl \ + \ PhSeSnBu_3 \quad \xrightarrow[\text{toluene, 25 °C, 6 h}]{Pd(PPh_3)_4 \ (1 \ mol \ \%)} \quad \underset{R}{\overset{O}{\big|\big|}}C-SePh \tag{37}$$

R	yield, %
Ph	97
4-MeC$_6$H$_4$	83
4-ClC$_6$H$_4$	91
Pr	85
t-Bu	51

For example, when PhSeSnBu$_3$ was allowed to react with benzoyl chloride in the presence of a catalytic amount of Pd(PPh$_3$)$_4$ at 25 °C for 6 h, the corresponding selenol ester was formed with 97% yield. 4-Methyl, 4-methoxy, 4-chloro, 4-nitro, and 4-cyanobenzoly chlorides also afforded selenol esters with 71–91% yields. This coupling reaction proceeds efficiently when acyl chlorides with linear alkyl chains or benzylic groups are employed, although α-methyl- and α,α-dimethyl-substituted acyl chloride gives moderate yields of products.

Beletskaya and co-workers reported the reactions of acyl and aroyl chlorides with PhSeSnBu$_3$, which was generated in situ by the reaction of PhSeSePh with Bu$_3$SnSnBu$_3$ in daylight [73]. However, Nishiyama and co-workers reported that the reaction of isolated PhSeSnBu$_3$ with benzoyl chloride under similar reaction conditions to those of Beletskaya's report afforded the product with only 24% yield.

Ranu and co-workers discovered a simple reaction between acyl chlorides and PhSeSePh mediated by indium(I) iodide at room temperature to produce selenol esters (Eq. 38) [74].

$$\underset{R}{\overset{O}{\big|\big|}}C-Cl \ + \ PhSeSePh \quad \xrightarrow[\substack{CH_2Cl_2, \ rt \\ 40\text{-}45 \ min}]{InI \ (50 \ mol \ \%)} \quad \underset{R}{\overset{O}{\big|\big|}}C-SePh \tag{38}$$

$$R = Ph, \ 72\%$$
$$Me, \ 52\%$$

This reaction might proceed by the reaction of acyl chlorides with bis(phenylseleno)iodoindium(III), InI(SePh)$_2$, formed from InI and PhSeSePh [75].

3.1.4
Methylselenation of Esters with Dimethylaluminum Methylselenoate

Kozikowski and co-worker reported the conversion of O-alkyl esters to selenol esters using dimethylaluminum methylselenoate (Me$_2$AlSeMe) (Eq. 39) [76].

Me$_2$AlSeMe is conveniently prepared by heating a toluene solution containing Me$_3$Al with selenium powder. The transformations of a variety of methyl and ethyl esters to their corresponding selenol esters were found to complete within 1 h (30 min at 0 °C, with an additional 30 min at room temperature). The

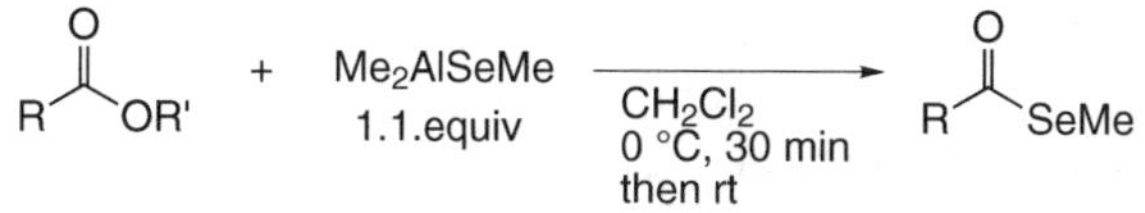

$$\text{(39)}$$

R	R'	yield, %
$n\text{-}C_6H_{13}$	Me	95
Ph	Me	99
cyclohexyl	Me	93
cyclopropyl	Et	96

ethyl ester of cyclopropanecarboxylate gave the corresponding selenol esters without ring opening. The *t*-butyl ester was converted to its selenol ester only in low yield, even after prolonged heating. While δ-valerolactone (**45**) is transformed to the δ-hydroxyselenol ester **46**, γ-butyrolactone (**47**) is recovered unchanged (see Fig. 5).

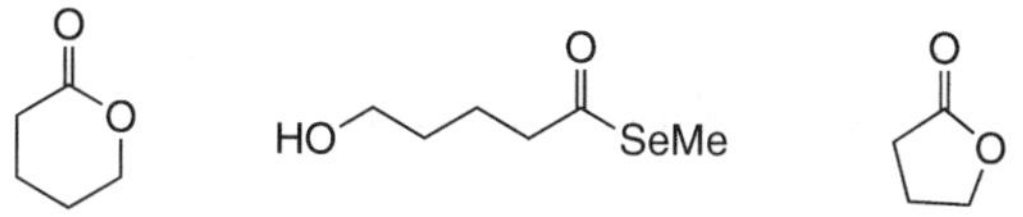

45 **46** **47**

Fig. 5 Cyclic lactones **45** and **47**, and δ-hydroxyselenol ester **46**

3.1.5
Oxidative Selenation of Hydrazines with Benzeneseleninic Acid

Back and co-workers revealed that the oxidations of *N*-acyl and *N*-aroyl hydrazines **48** with benzeneseleninic acid **49** produced selenol esters (Eq. 40) [77].

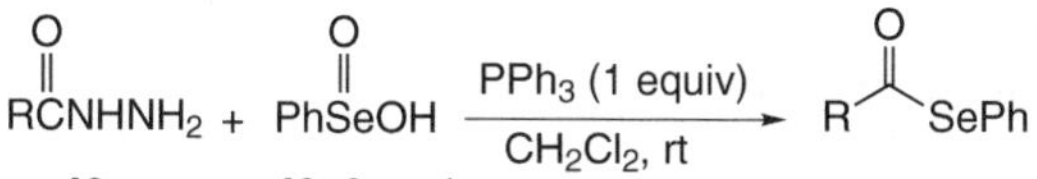

$$\text{(40)}$$

R	yield, %
cyclohexyl	86
Ph	77
t-Bu	83
$PhCH_2$	73

Benzeneseleninic acid is a stable, readily available, odorless solid which can be used to oxidize various organic substrates. The reaction shown in Eq. 40 was conducted by slowly adding a mixture of the hydrazide and PPh_3 to benzeneseleninic acid in CH_2Cl_2 solution.

The initial oxidation of the hydrazide **48** is expected to generate the corresponding acyl- or aroyldiazene **50** and benzeneselenenic acid. Further reaction of these species produces the hypothetical intermediate **51**, followed by nitrogen extrusion leading to selenol ester formation (Scheme 13).

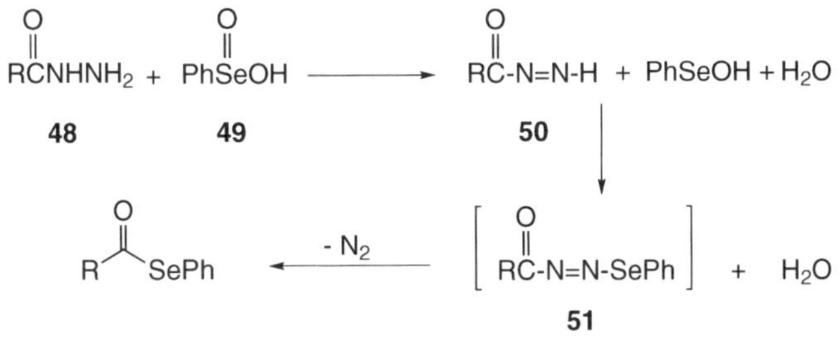

Scheme 13 Reaction pathway of the oxidation of hydrazine with $PhSeO_2H$

3.1.6
Tandem Alkylation-Acylation of a Se^{2-} Equivalent

Segi, Sonoda and co-workers reported a one-pot synthesis of selenol esters by tandem alkylation-acylation of bis(trimethylsillyl) selenide **52** as a Se^{2-} equivalent (Scheme 14) [78].

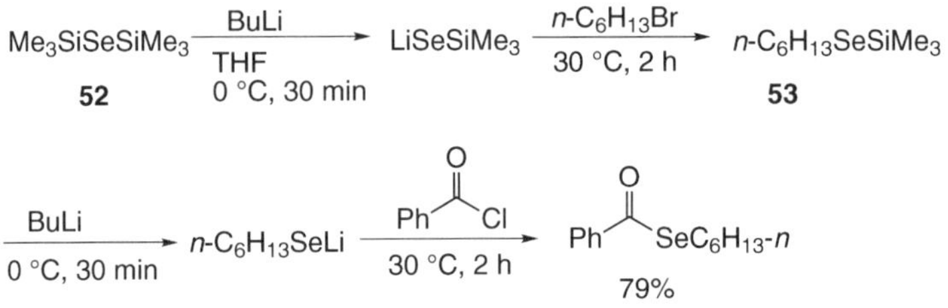

Scheme 14 Tandem alkylation and acylation of Se^{2-} equivalent

The reaction of **52** with BuLi followed by alkylation with alkyl halides gave the silyl selenides **53**. The repetition of a similar operation with BuLi and acyl chlorides afforded selenol esters in good yields. Selenol ester was also obtained in a good yield when the order of the alkylation/acylation sequence was changed.

Elemental selenium can be reduced by carbon monoxide and water in the presence of tertiary amine to produce an amine salt of hydrogen selenide ($[HSe^-]\,[R_3NH^+]$) [79], which can be trapped by acid halides and alkyl halides to give selenol esters (Scheme 15) [80]. A variety of acyl or aroyl chlorides can be used as the substrates. However, the bulkiness of the alkyl halides influences the yields of the products.

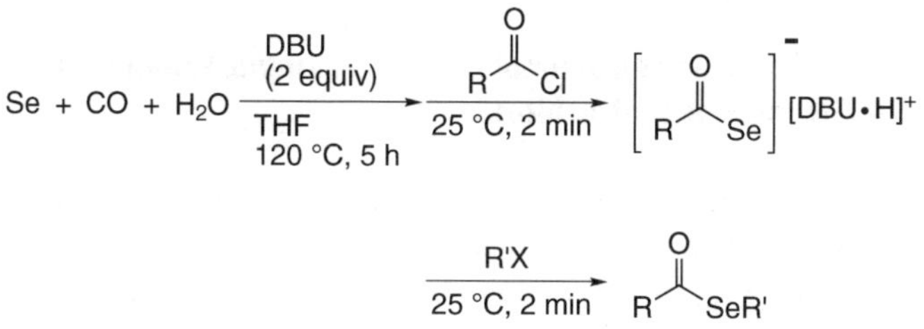

Scheme 15 Synthesis of selenol ester by the use of Se/CO/H$_2$O reaction system

R	R'	yield, %
Ph	Me	96
	PhCH$_2$	93
2-furyl		80
Bu		91
t-Bu		70

3.1.7
Desulfurization of *Se*-Aryl Acylmethanesulfenoselenoates

Kato and co-workers described a method of preparing selenol esters by desulfurization of *Se*-aryl acylmethanesulfenoseleoates **54** with PPh$_3$ (Eq. 41) [81].

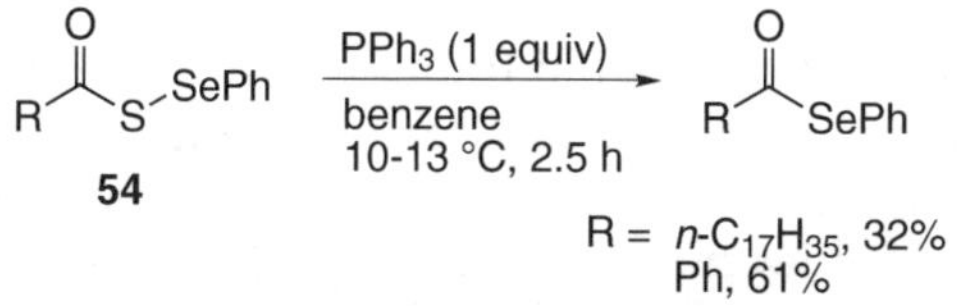

(41)

$$R = n\text{-}C_{17}H_{35}, 32\%$$
$$Ph, 61\%$$

54 can be synthesized conveniently either by the reaction of the acylsulfenyl bromides **55** (see Fig. 6) with ArSeSeAr or by that of the metal thiocarboxylates **56** with ArSeBr. The desulfurization reactions are clean and complete at room temperature within 3 h. Attempts to prepare Te analogs by a similar desulfurization technique from **57** failed. However, the deselenation of the aryl aroyl diselenides **58** proceeds to give selenol esters [82].

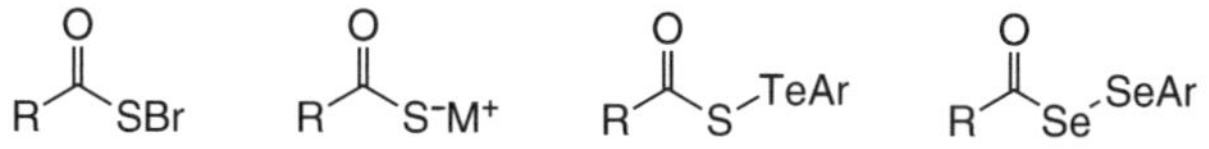

Fig. 6 Compounds **55–58**

3.1.8
Synthesis of α,β-Unsaturated Selenol Esters and Vinylic Selenol Esters by the Use of Zirconocene Hydrochloride

α,β-Unsaturated selenol esters serve as dienophiles in Diels-Alder reactions with a variety of 1,3-dienes. They can also be prepared by other methods shown in this chapter. In 1999, Huang and co-worker reported a new method for the synthesis of α,β-unsaturated selenol esters by the reaction of acylzirconene chlorides **60** with electrophilic selenium bromides (Scheme 16) [83].

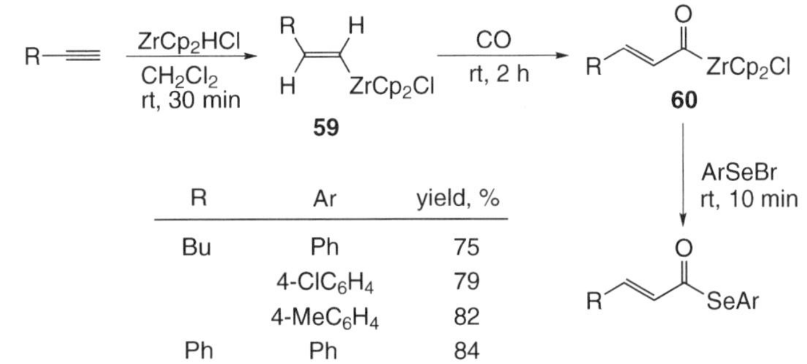

R	Ar	yield, %
Bu	Ph	75
	4-ClC$_6$H$_4$	79
	4-MeC$_6$H$_4$	82
Ph	Ph	84

Scheme 16 Synthesis of α,β-unsaturated selenol esters by the use of zirconene hydrochloride

Acylzirconocene chlorides **60** can be prepared through sequential treatment of alkynes with zirconene hydrochloride and carbon monoxide [84]. This method affords a variety of α,β-unsaturated selenol esters in good yields at room temperature. Diaryl diselenides can also be used in place of ArSeBr [85a]. α,β-Unsaturated tellurol esters can be prepared by the same protocol shown in Scheme 16 [85b].

Vinylic selenol esters were formed by using the same zirconocene intermediate **59**. The insertion of elemental selenium into the (sp^2)C–Zr bond of alkenylchlorozirconocenes **59** affords the (E)-vinylseleno zirconocenes **61**,

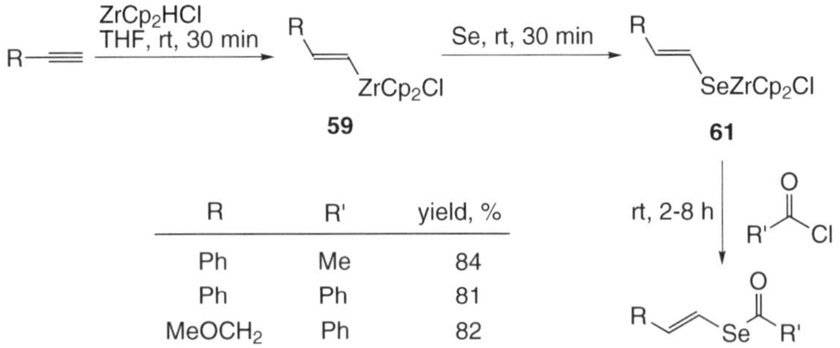

R	R'	yield, %
Ph	Me	84
Ph	Ph	81
MeOCH$_2$	Ph	82

Scheme 17 Synthesis of vinylic selenol ester

which were trapped by acyl chlorides giving (*E*)-vinylic selenol esters in good yields (Scheme 17) [86].

3.1.9
Carbophilic Addition of Organocopper Reagents to Carbonyl Selenide (SeCO)

Carbonyl sulfide (SCO) reacts with Grignard reagents at the carbonyl carbon [87], and this has been used to introduce a thiocarboxyl unit into organic molecules. For example, thiocarboxylations of enolates [88], phosphorous ylides [89], and an acyllithium [90] have been reported. Selenocarboxylation to organocopper reagents with carbonyl selenide (SeCO), followed by alkylation, gave selenol esters (Eq. 42) [91].

$$SeCO \ + \ RCu(CN)Li \xrightarrow[\substack{THF,\ -78\ °C \\ 10\ min}]{} \xrightarrow[\substack{-78\ °C,\ 30\ min, \\ then\ 20\ °C}]{MeI} R\overset{O}{\underset{}{\diagdown}}SeMe \qquad (42)$$

R	yield, %
Ph	86
4-MeOC$_6$H$_4$	70
4-MeC$_6$H$_4$	65
4-ClC$_6$H$_4$	84

The reaction of SeCO with PhLi resulted in selenophilic attack with concomitant elimination of CO, giving rise to PhSeBn after trapping with BnBr. A similar result was also obtained with PhMgBr. In marked contrast to these results, when a cyanocuprate, PhCu(CN)Li, was employed, carbophilic attack took place successfully to give the corresponding selenol ester in 86% yield after trapping with MeI. In a similar manner, selenol esters were obtained in high yields from lithium cyanoarylcuprates with an electron-releasing or -withdrawing substituent at the *para* position.

3.1.10
Co$_2$(CO)$_8$-Catalyzed Carbonylation of Diaryl Diselenides

Uemura and co-workers reported that diaryl diselenides reacted with carbon monoxide (5–100 atm) at 100–200 °C for 1–4 h in the presence of Co$_2$(CO)$_8$ to give the corresponding selenol esters (Eq. 43) [92].

$$ArSeSeAr \ + \ \underset{\substack{20\text{-}100 \\ atm}}{CO} \xrightarrow[\substack{CH_3CN \\ 150\text{-}200\ °C,\ 1\ h}]{\substack{Co_2(CO)_8 \\ (40\ mol\ \%)}} Ar\overset{O}{\underset{}{\diagdown}}SeAr \qquad (43)$$

$$Ar = Ph,\ 90\%$$
$$4\text{-}MeOC_6H_4,\ 90\%$$

Similar treatment of diphenyl ditelluride affords telluro analogs in lower yields. Didodecyl diselenide also gives selenol ester, while selenol esters were not produced from dibenzyl diselenide and didodecyl ditelluride.

The reactivity of PhMMPh (M=S, Se, Te) was found to decrease in the order Te>Se>S. The carbonylation of diaryl diselenides proceeds catalytically with $Co_2(CO)_8$ in the presence of PPh_3.

3.1.11
Carbonylative Addition of Diphenyl Diselenide to Acetylenes with Carbon Monoxide

The procedure for the carbonylative addition of diaryl disulfide to acetylene shown in Sect. 2.1.2 (Scheme 1) is also applicable to the diselenide/acetylene reaction system (Eq. 44) [7].

$$R-\!\!\!\equiv\!\!\!-\; +\; ArSeSeAr \quad \xrightarrow[\substack{CO\ (15\ kg/cm^2)\\ benzene,\ 80\ °C}]{\substack{Pd(PPh_3)_4\\(2\ mol\ \%)}} \quad \begin{array}{c} R \diagdown\;\diagup SeAr \\ ArSe \quad O \end{array} \tag{44}$$

R	Ar	yield, %
$n\text{-}C_6H_{13}$	$4\text{-}MeC_6H_4$	80
	$4\text{-}CF_3C_6H_4$	89
$(CH_2)_3OH$	Ph	76

This carbonylation is completely regioselective and highly stereoselective; the carbonyl group is introduced only at the terminal carbon of the acetylene, and Z-isomers are obtained. Acetylenes bearing a C=C double bond react chemoselectively at the triple bond. The hydroxyl group does not interfere with the carbonylative addition. In contrast, propargylamine affords a complicated mixture.

When the reaction of **62** with carbon monoxide is conducted in the presence of $Pd(PPh_3)_4$, no CO-incorporated products are formed (Eq. 45). This result indicates that **62** is not a precursor for the carbonylation.

$$\begin{array}{c} R \\ PhSe \diagup\;\diagdown SePh \end{array} +\; CO \xrightarrow{Pd(PPh_3)_4/PhSeSePh} \!\!\!/\!\!\!/ \quad \begin{array}{c} R \diagdown\;\diagup SePh \\ PhSe \quad O \end{array} \tag{45}$$

62

3.1.12
Group Transfer Carbonylation

Ryu, Sonoda and co-workers reported a three component coupling reaction, based on group transfer carbonylation, that involves methyl α-(phenylseleno)-acetate **63** (and related derivatives), terminal alkenes, and CO (Scheme 18) [93].

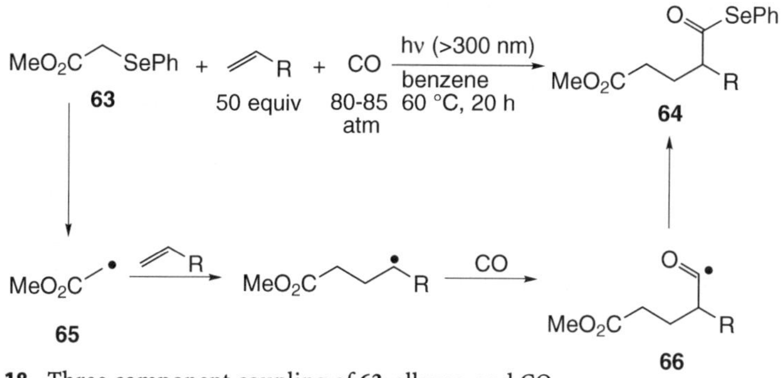

Scheme 18 Three component coupling of **63**, alkyne, and CO

For example, when a benzene solution of methyl α-(phenylseleno)acetate **63** and 1-octene (50 equiv) was irradiated using a 500 W xenon lamp (>300 nm) for 20 h at 60 °C under 80 atm of CO pressure, selenol ester **64**, the three-component coupling product, was formed with 58% yield. In order to irradiate under CO pressure, an autoclave equipped with quartz windows was employed in this work. The reaction pathways shown in Scheme 18 may account for the formation of the selenol esters: (1) the photoinduced homolysis of methyl α-(phenylseleno)acetate **63** takes place; (2) the (methoxycarbonyl)methyl radical **65** generated adds to 1-octene; (3) the addition of the resulting alkyl radical to CO leads to an acyl radical **66**, and; (4) an S_H2 reaction on the selenium of **63** prompted by the attack of the acyl radical yields the selenol ester **64** and regenerates **65**.

3.1.13
Carbonylation of Organolithium Compounds with Selenium and Carbon Monoxide

In 1996, it was revealed that benzylic and allylic organolithium compounds underwent carbonylation with CO aided by selenium, yielding selenol esters after trapping with alkyl halides [94].

Organolithium compounds are known to react with selenium to give lithium selenolates. Indeed, when the (9-methylfluorenyl)lithium **68** generated from **67** and BuLi was allowed to react with selenium at 20 °C, the corresponding selenide **69** was obtained in 93% yield after quenching with MeI. However, when CO was introduced at 20 °C into a THF solution of selenolate **68**, a stoichiometric amount of CO was absorbed within 90 min. Addition of MeI gave the selenol ester **70** in 93% yield (Scheme 19).

Addition of HMPA accelerated the CO absorption, and carbonylation proceeded even at −23 °C. Fluorenes with a butyl or cyclohexyl group, triphenylmethane, and pentamethylcyclopentadiene also underwent carbonylation to give the corresponding selenol esters.

Lithium enolates of ketones and aldehydes also undergo carbonylation with carbon monoxide aided by selenium to yield the β-keto and β-formyl selenol

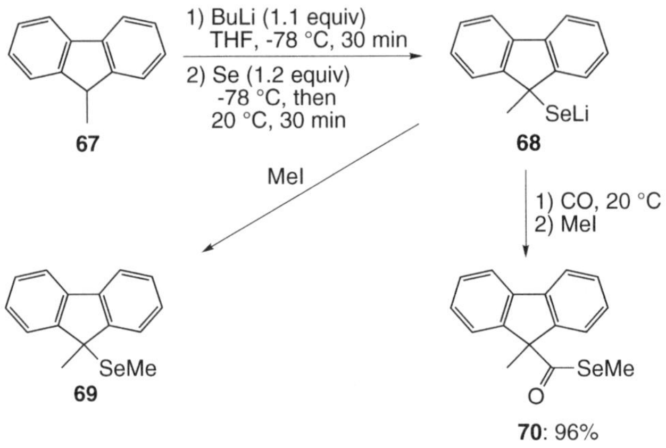

Scheme 19 Carbonylation of organolithium compound with Se and CO

esters **72**. A unique carbonylation mechanism involving the selenocarbonate intermediates **71** has been suggested on the basis of control experiments (Scheme 20) [95].

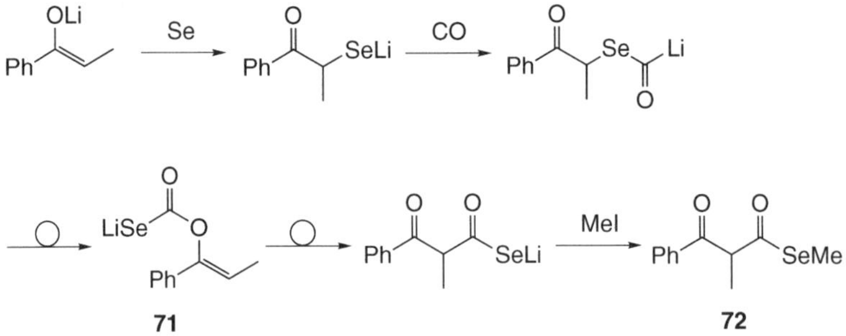

Scheme 20 Carbonylation of lithium enolate of ketone and aldehyde

3.1.14
Pd(PPh$_3$)$_4$-Catalyzed Coupling of Phenyl Tributylstannyl Selenide with Aryl Iodides and Carbon Monoxide

Pd(PPh$_3$)$_4$ catalyzes the three-component coupling of phenyl tributylstannyl selenide **73** with aryl iodides and carbon monoxide to afford the corresponding selenol esters (Eq. 46) [96].

When phenyl tributylstannyl selenide **73** (see Fig. 7) was allowed to react with iodobenzene in the presence of a catalytic amount of Pd(PPh$_3$)$_4$ in toluene solution under carbon monoxide (5 atm) at 80 °C for 5 h, selenol ester was formed with 89% yield. Chloro- and bromobenzenes did not give the selenol ester under the same reaction conditions, but aryl iodides with various substituents on

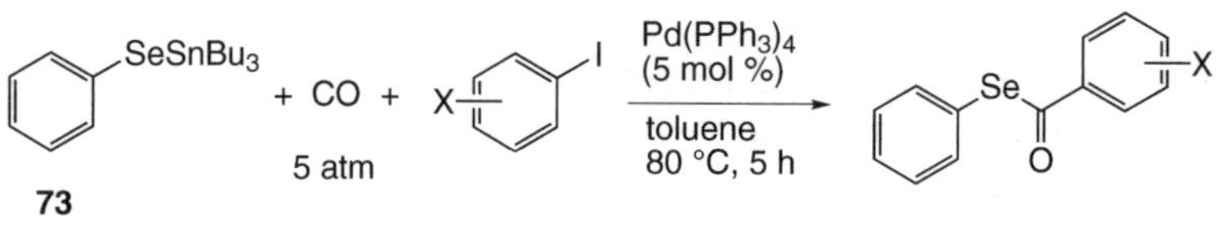

$$\text{(46)}$$

X	yield, %
H	89
4-Me	73
4-MeO	70
4-Cl	85

the ring can be used in this reaction. The first step of this reaction would be the oxidative addition of the aryl iodide to the low-valent palladium species **74**, followed by the insertion of carbon monoxide into the palladium-carbon bond to form the aroyl palladium complex **75**. The following transmetalation of **75** with PhSeSnBu$_3$ generates the intermediate **76**. The subsequent reductive elimination from **76** affords the selenol ester and the low-valent palladium species.

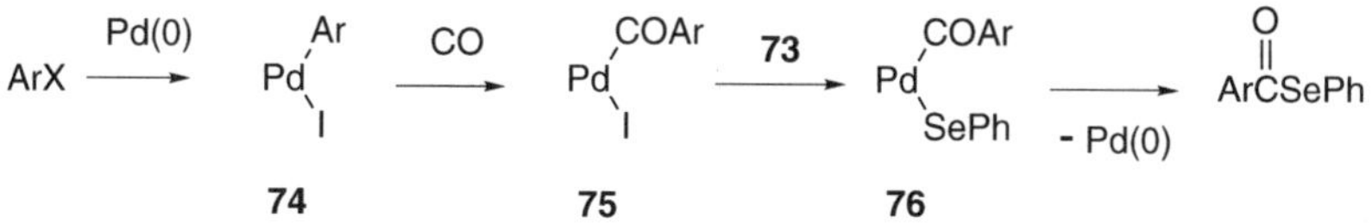

Fig. 7 Palladium species **74–76**

3.2
Reactions

3.2.1
Precursors of Acyl Radicals

In 1980, Graf and co-workers reported that phenyl selenol esters underwent reduction to the corresponding aldehydes and alkanes (reduction and decarbonylation) in the presence of trialkyltin hydrides and a free-radical initiator through generation of acyl radicals (Scheme 21) [97].

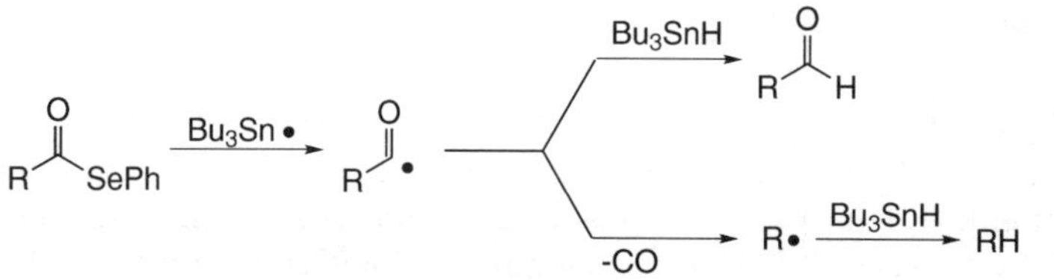

Scheme 21 Generation of acyl radical from selenol ester

In 1988, Boger and co-worker found that the intramolecular free-radical cyclization reactions of acyl radicals generated from selenol esters proceeded efficiently, suppressing competing reduction and decarbonylation (Eq. 47) [98].

$$(47)$$

X	yield, %
CO$_2$Me	84
H	76
MeO	64

They also revealed that primary alkyl-, vinyl-, and aryl-substituted acy radicals generated by treating selenol esters with Bu$_3$SnH can be utilized in intermolecular alkene addition reactions (Eq. 48) [99, 98b]. Acyl radicals exhibit nucleophilic character and react efficiently with alkenes bearing electron-withdrawing or radical-stabilizing groups.

$$(48)$$

R	X	yield, %
4-MeOC$_6$H$_4$	CO$_2$Bn	60
	CN	64
	Ph	71
Ph	CO$_2$Me	58
Me	Ph	51

Since these findings, selenol esters have been frequently used as precursors of acyl radicals in organic synthesis, especially for the total synthesis of natural products [62, 100, 101].

3.2.2
Utilization as Acylating Reagents

The ability of the selenol ester to serve as a potent acyl-transfer agent was demonstrated by the conversion of selenol ester to its corresponding acid, ester, or amide. This was accomplished by simply stirring the selenol ester with H$_2$O/CH$_3$CN, MeOH/CH$_3$CN, or cyclohexylamine/CH$_3$CN, respectively, at room temperature for 1 h in the presence of HgCl$_2$ and CaCO$_3$ (Scheme 22) [76a].

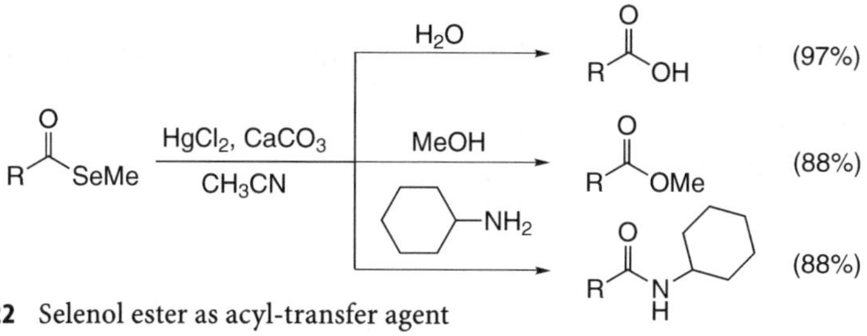

Scheme 22 Selenol ester as acyl-transfer agent

Kozikowski and co-worker also reported that the copper(I) triflate-benzene complex, $(CuOTf)_2PhH$, promoted Friedel-Crafts acylations of aromatic compounds (Eq. 49) [102]. Other metal complexes such as $HgCl_2$, $AgNO_3$, CuCl, and Cu_2O were not effective. The reaction proceeds efficiently when arenes with electron-donating groups are employed. Toluene, used as the solvent, was only acylated in low yields in this Cu(I)-selenol ester system, implying that more reactive arenes are needed for this reaction. Methylthiol esters were sluggish under these conditions. Heterocyclic compounds such as furan, thiophene, pyrrole, and N-methylindole readily underwent acylation.

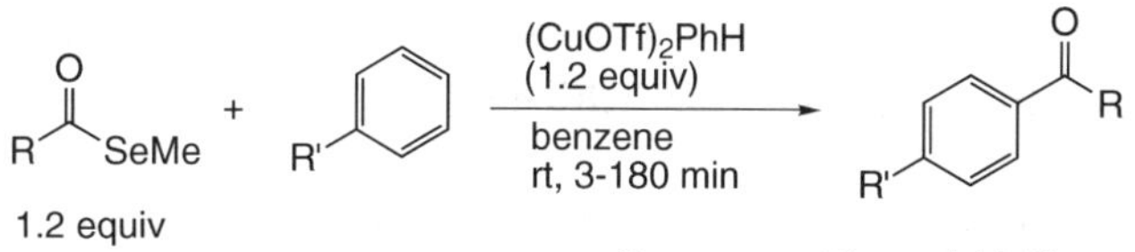

$$(49)$$

R	R'	yield, %
$n\text{-}C_6H_{13}$	MeO	81
$CH_2{=}CH(CH_2)_2$		55
$CH_3C(O)(CH_2)_2$		60
$n\text{-}C_6H_{13}$	Me	28

Oxazoles can also be synthesized by simply stirring selenol ester and isonitrile together at room temperature for 6–20 h in the presence of 1.5 equiv of Et_3N or DBU and 1.5 equiv of anhydrous cuprous oxide (Eq. 50).

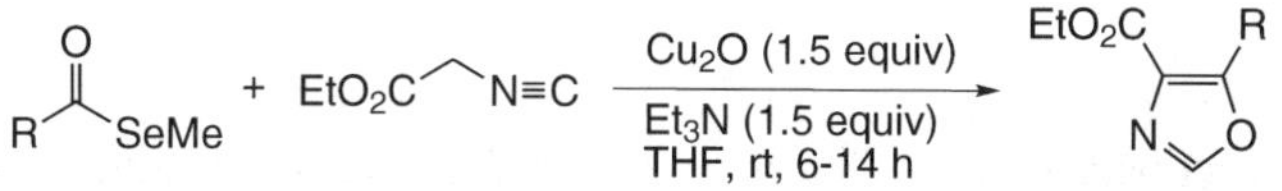

$$(50)$$

R	yield, %
Ph	60
cyclopropyl	61
$n\text{-}C_6H_{13}$	85

3.2.3
Homologation of Selenol Esters to α-Seleno Ketones with Diazomethane

Back and co-worker reported diazomethane reacts with a series of selenol esters in the presence of CuI, CuSePh or Cu powder, producing α-seleno ketones in moderate yields (Eq. 51) [103].

$$
\underset{R}{\overset{O}{\parallel}}\text{SePh} + \underset{\text{excess}}{\text{CH}_2\text{N}_2} \xrightarrow[\text{Et}_2\text{O, rt, 4-12 h}]{\text{Cu (or CuI)}} \underset{R}{\overset{O}{\parallel}}\text{SePh} \tag{51}
$$

R	yield, %
Ph	56
Me	65
PhCH$_2$	56
2-furyl	42

The thiol ester failed to provide the corresponding α-thio ketone when treated with diazomethane. No insertion was observed when ethyl diazoacetate (EtOC(O)CHN$_2$) or trimethylsilyldiazomethane (Me$_3$SiCHN$_2$) were used with the selenol esters.

3.2.4
Pd(PPh$_3$)$_4$-Catalyzed Reduction to Aldehydes with Bu$_3$SnH

Ogawa, Sonoda and co-workers reported the Pd(PPh$_3$)$_4$-catalyzed chemoselective and siteselective reduction of (Z)-1,3-bis(arylseleno)-2-alken-1-ones 77 with Bu$_3$SnH to provide the corresponding aldehydes in high yields (Eq. 52) [104].

$$
\textbf{77} + \text{Bu}_3\text{SnH} \xrightarrow[\substack{\text{benzene} \\ 25\,°\text{C, 3-17 min}}]{\substack{\text{Pd(PPh}_3)_4 \\ (0.4 \text{ mol \%})}} \tag{52}
$$

R	yield, %
n-C$_6$H$_{13}$	91
Ph	90
HO(CH$_2$)$_2$	87

In this reaction the carbon-carbon double bond was not reduced and no decarbonylative product was formed. The reduction was retarded by the additon of PPh$_3$ (8 mol%). When the AIBN-initiated reaction system was applied to 77, the same product was obtained, but isomerization of the C=C double bond also took place.

3.2.5
Addition to Alkynes

Reaction of selenol esters with phenylpropiolonitrile in the presence of K_2CO_3 in refluxing THF-containing water affords (Z)-3-arylselenocinnamonitriles **78** stereoselectively in high yields (Eq. 53) [105].

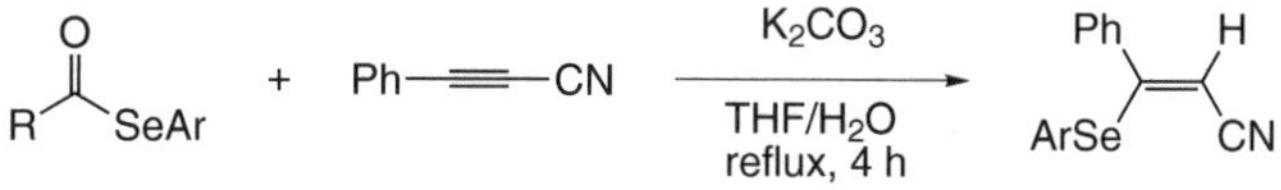

$$
\begin{array}{ccc}
 & & K_2CO_3 \\
R\text{-CO-SeAr} + Ph\text{---}CN & \xrightarrow{\quad\quad\quad} & \text{ArSe-C(Ph)=CH-CN} \\
 & & \text{THF/H}_2\text{O} \\
 & & \text{reflux, 4 h}
\end{array}
\qquad (53)
$$

78

R	Ar	yield, %
Me	Ph	89
Me	4-MeOC$_6$H$_4$	93
Ph	4-ClC$_6$H$_4$	96

Selenol esters add to terminal alkynes in the presence of CuX (X=Cl, I) and trimethylamine hydrochloride at 80–90 °C for 6–8 h to give (Z)-β-seleno-α,β-unsaturated ketones **79** in high yields (Eq. 54) [106].

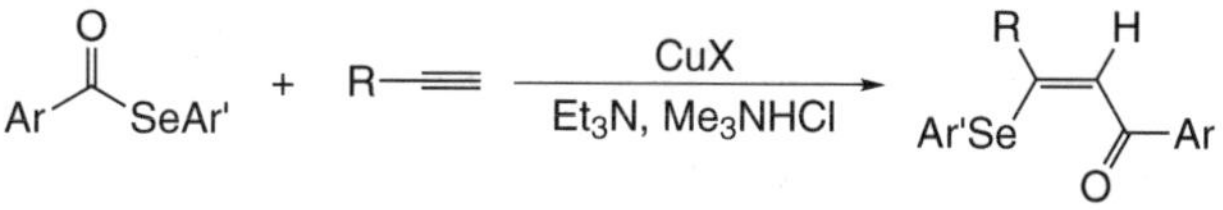

$$
\text{Ar-CO-SeAr'} + R\text{---} \xrightarrow[\text{Et}_3\text{N, Me}_3\text{NHCl}]{\text{CuX}} \text{Ar'Se-C(R)=CH-CO-Ar}
\qquad (54)
$$

79

The proposed reaction pathways are shown in Scheme 23. Alkynylcopper **80** formed from alkyne and cuprous halide in the presence of a base attacks selenol ester at the carbonyl carbon to form α,β-alkynone **81** and **82**. Then arylselenol generated from **82** adds to α,β-alkynone to produce the (Z)-β-arylseleno-α,β-unsaturated ketone **79**.

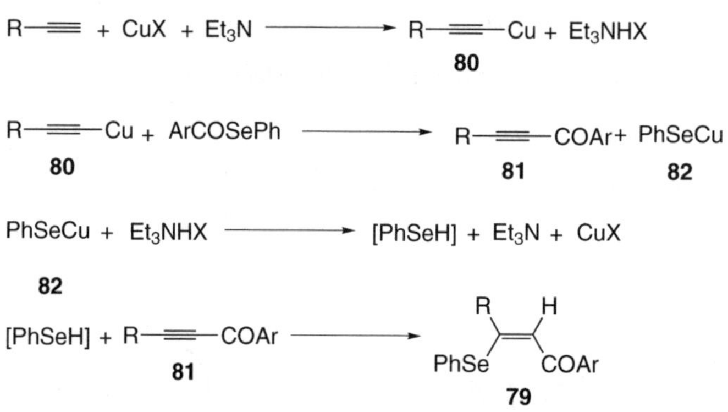

$$
R\text{---} + \text{CuX} + \text{Et}_3\text{N} \longrightarrow R\text{---Cu} + \text{Et}_3\text{NHX}
$$

80

$$
R\text{---Cu} + \text{ArCOSePh} \longrightarrow R\text{---COAr} + \text{PhSeCu}
$$

80   **81**   **82**

$$
\text{PhSeCu} + \text{Et}_3\text{NHX} \longrightarrow [\text{PhSeH}] + \text{Et}_3\text{N} + \text{CuX}
$$

82

$$
[\text{PhSeH}] + R\text{---COAr} \longrightarrow \text{PhSe-C(R)=CH-COAr}
$$

81   **79**

Scheme 23 Reaction pathway of addition of selenol ester to alkyne

Tellurol esters also add to alkynes under similar reaction conditions (Eq. 55) [107].

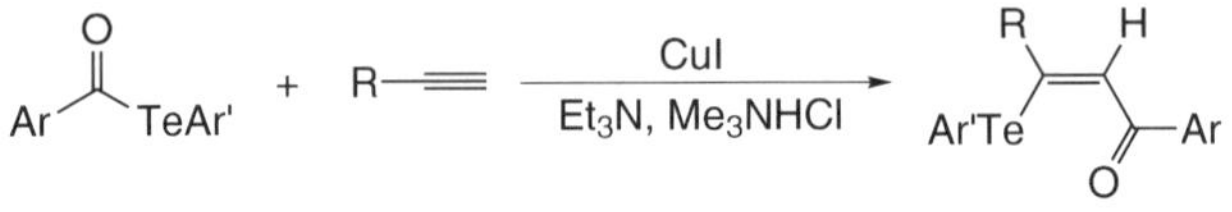

$$\text{(55)}$$

64-91%

3.2.6
Diels-Alder Reactions of α,β-Unsaturated Selenol Esters

α,β-Unsaturated selenol esters react with a variety of 1,3-dienes to give Diels-Alder products. Good levels of regioselectivity are obtained with unsymmetrical dienes when Lewis acid promoters are used [108].

The results in Scheme 24 show that the thermal reaction of selenol ester with isoprene is non-regioselective. Regioselectivity was improved by using a strong Lewis acid at low temperatures. The solid-supported Lewis acid facilitates the reaction but regioselectivity was not improved. The use of thiol esters affords similar results. Selenol and thiol esters are more reactive than the corresponding parent esters.

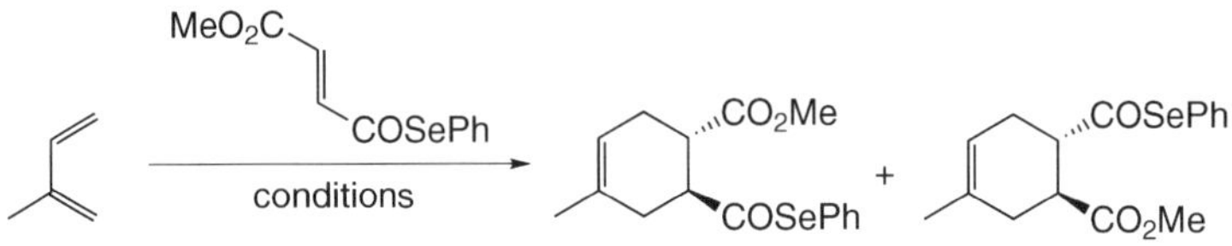

83   **84**

conditions	yield	ratio (**83**:**84**)
190–195 °C	81%	55:45
TiCl$_4$, -60 °C – rt	77%	100:0
SiO$_2$-Et$_2$AlCl, rt	88%	46:54

Scheme 24   Diels-Alder reaction of α,β-unsaturated selenol ester

Reactions of α,β-unsaturated selenol esters with cyclohexanone enamines provide bicycle[3.3.1]nonane-2,9-diones **85** (Eq. 56) [109].

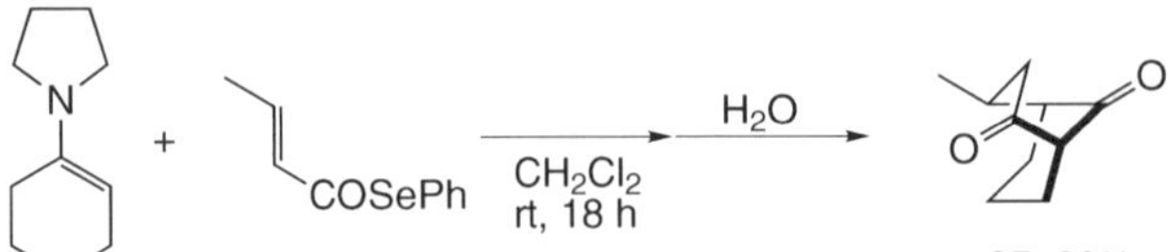

$$\text{(56)}$$

85, 66%

3.2.7
[2+2] Cycloaddition of Silyl *O,Se*-Ketene Acetals of Selenol Esters to Aldehydes

Abe and co-workers revealed that the photochemical [2+2] cycloaddition (Paternò-Büchi reaction) of silyl enol ethers **86** of selenol esters to aromatic aldehydes affords the 3-selanyl-3-siloxyoxetanes **87** (*trans:cis*=65:35–85:15) regioselectively in good to high yields (Eq. 57) [110].

$$\text{(57)}$$

3.2.8
Deoxygenative Methylenation (Witting-Type Reaction) of Selenol Ester with Dimethyl Titanocene

Petasis and co-worker reported that methylenation of selenol and thio esters with dimethyl titanocene led to the formation of the corresponding vinyl selenides and sulfides (Eq. 58) [111]. Dimethyl titanocene can be easily prepared from titanocene dichloride and methyllithium [112]. The methylenation reactions involve simply heating a mixture of dimethyl titanocene and chalcogeno esters at 60–75 °C.

$$\text{(58)}$$

R = H, Y = Se 77%
R = Me, Y = S 70%

4
Telluro-Carboxylic Acid Esters (Tellurol Esters)

Syntheses of tellurol esters **3** are limited in comparison to thiol and selenol esters. A conventional preparative method is the reaction of tellurolate anions with acid chlorides or acid anhydride [113]. $Co_2(CO)_8$-mediated carbonylation of dichalcogenides (described in Sect. 3.1.10) is applicable to diphenyl ditelluride, but the yields of tellurol esters are poor. They can also be obtained by transesterification of esters with i-$Bu_2AlTeBu$. Reactions of tellurol esters **3** have not been extensively explored yet. However, generation of acyllithiums by Li-Te exchange is characteristic. They can be employed as acyl radical precursors in tin free radical reactions.

4.1
Syntheses

4.1.1
Alkylation of Tellurocarboxylates

Tellurocarboxylates (RCOTe⁻) are considered to be one of the most fundamental starting compounds for the synthesis of tellurol esters. However, due to their instability, tellurocarboxylates were not described in the literature until the mid-80s. Kato and co-workers synthesized ammonium or potassium tellurocarboxylates by reacting bis(acyl) telluride **88** with primary or secondary amines or potassium ethoxide, and trapped them with methyl iodide, giving rise to tellurol esters (Eq. 59) [114].

$$\text{(59)}$$

The authors also synthesized tellurol esters by successive acylation and alkylation of sodium telluride (Na_2Te) via tellurocarboxylates [115]. Suzuki, Tani and co-workers succeeded in the synthesis of tellurol esters by a similar method using sodium hydrogen telluride (NaTeH) (Eq. 60) [116].

$$\text{(60)}$$

4.1.2
Reaction of Trimethylsilyl Benzenetellurolate with Acyl Chloride

Ogura and co-workers reported that $PhTeSiMe_3$ reacted with aroyl chlorides in THF at room temperature for 3 h to give tellurol esters (Eq. 61) [117].

$$\text{(61)}$$

X	yield, %
H	87
Cl	87
CN	83
Me	91

Schiesser and co-worker conducted the same reaction as shown in Eq. 61 in the presence of a catalytic amount of $Pd(PPh_3)_4$ [118]. Zhang and co-worker reported the samarium diiodide-induced reductive cleavage of the Te-Si bond of $PhTeSiMe_3$ and the subsequent trapping with acyl halides to give tellurol esters in good to high yields [119].

4.1.3
Transesterification of Esters with *i*-Bu$_2$AlTeBu

A variety of esters can be converted to the corresponding tellurol esters in moderate to good yields by reacting with i-Bu$_2$AlTeBu (Eq. 62) [120].

$$\text{(62)}$$

R	R'	yield, %
Ph	Et	80
	i-Pr	3
	Ph	62
4-MeC$_6$H$_4$	Et	71
4-ClC$_6$H$_4$	Et	92

For example, the reaction of ethyl benzoate with 1.3 equiv of i-Bu$_2$AlTeBu, prepared in situ by Ogura's method from BuTeTeBu and 2 mol equiv of DIBALH at 25 °C in toluene in the presence of a catalytic amount of PPh$_3$ for 20 h [121], affords the tellurol ester in 82% yield. This transesterification proceeds smoothly in nonpolar solvents such as toluene and n-hexane. Aryl esters react efficiently in comparison with alkyl esters. Addition of a catalytic amount of PPh$_3$ or ZnCl$_2$ accelerates the reaction and improves the yields. Substituents on the oxygens of the esters exert significant steric effects on this reaction, but the reaction is insensitive to the steric bulkiness of substituents on the carbonyl carbon.

4.2
Reactions

4.2.1
Synthesis of Ketones, Carboxylic Acids, and Esters

Like selenol esters, tellurol esters can also serve as acyl transfer reagents, giving rise to the corresponding ketones, carboxylic acids, and esters. For example, tellurol esters react with lithium dialkylcuprates at −78 °C to give the corresponding ketones in high yields (Eq. 63) [117].

Carboxylic acids were obtained in good yields from tellurol esters in the presence of copper(II) chloride dihydrate in acetone at room temperature for 5 min (Eq. 64) [122].

$$\text{Ph}\overset{\text{O}}{\underset{}{\|}}\text{TePh} + \text{R}'_2\text{CuLi} \xrightarrow[\text{-78 °C, 15-30 min}]{\text{THF/Et}_2\text{O}} \text{Ph}\overset{\text{O}}{\underset{}{\|}}\text{R}' \qquad (63)$$

$$\begin{array}{cc} \text{R}' = \text{Me, 87\%} \\ \text{Bu, 97\%} \end{array}$$

$$\text{R}\overset{\text{O}}{\underset{}{\|}}\text{TeBu-}n \xrightarrow[\substack{\text{acetone} \\ \text{rt, 5 min}}]{\text{CuCl}_2\text{·2H}_2\text{O}} \text{R}\overset{\text{O}}{\underset{}{\|}}\text{OH} \qquad (64)$$

R	yield, %
Ph	96
PhCH$_2$	65
t-Bu	73

Selenol esters were also hydrolyzed to carboxylic acids, but thiol esters are inert under similar conditions. The nature of the copper salt has a dramatic influence on the reaction course. No hydrolysis products were obtained when tellurol esters were treated with copper sulfate or copper(I) chloride.

When the reaction was performed in dry ethanol or methanol at room temperature, both tellurol and selenol esters were transformed into the corresponding esters (Eq. 65).

$$\text{R}\overset{\text{O}}{\underset{}{\|}}\text{YBu} \xrightarrow[\text{CuCl}_2\text{, rt}]{\text{R'OH}} \text{R}\overset{\text{O}}{\underset{}{\|}}\text{OR}' \qquad (65)$$

$$\text{Y = Te, Se} \qquad 69\text{-}88\%$$

4.2.2
Generation of Acyl- and Aroyllithiums via Lithium-Tellurium Exchange

The most straightforward method for nucleophilic introduction of acyl and aroyl groups into organic molecules is the use of carbonyl anions, represented by acyl-and aroyllithiums, as nucleophiles. Their synthetic utility, however, has been severely limited for a long time because of difficulties both in their generation and in controlling their reaction courses. A practically useful route to acyl- and aroyllithiums has been developed, based on the efficient lithium-tellurium exchange of tellurol esters [123].

When BuLi was added to aryltellurol esters in THF/Et$_2$O at low temperatures in the presence of pinacolone as an electrophile, α-hydroxy ketones **90** were obtained in high yields via aroyllithiun **89** together with dibutyl telluride (Eq. 66) [124].

Although the corresponding selenol esters are candidates to be the precursors of carbonyl anions, reacting PhC(O)SeBu under similar conditions did not give the desired adduct.

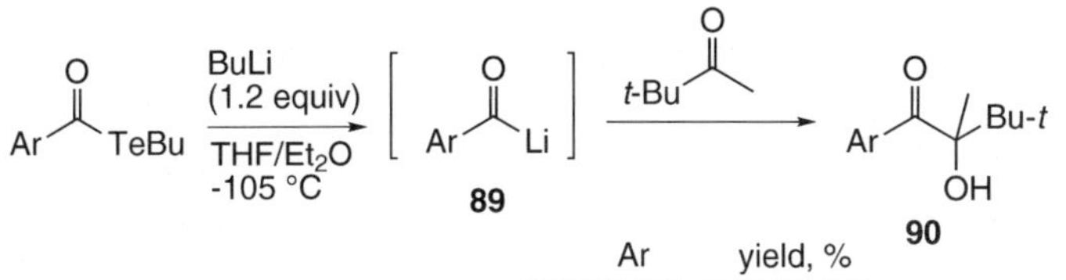

$$\text{(66)}$$

Ar	yield, %
Ph	85
4-CF$_3$C$_6$H$_4$	60
2,6-F$_2$C$_6$H$_3$	74
t-Bu (-78 °C)	66

Under similar conditions, tellurol esters bearing acidic α-hydrogenes gave complex results, but when **91** was treated with 2 equiv of BuLi in the presence of 2 equiv of Me$_3$SiCl at –105 °C, the enol silyl ether **92** of acyl silane was obtained quantitatively with high Z-stereoselectivity (Eq. 67) [125].

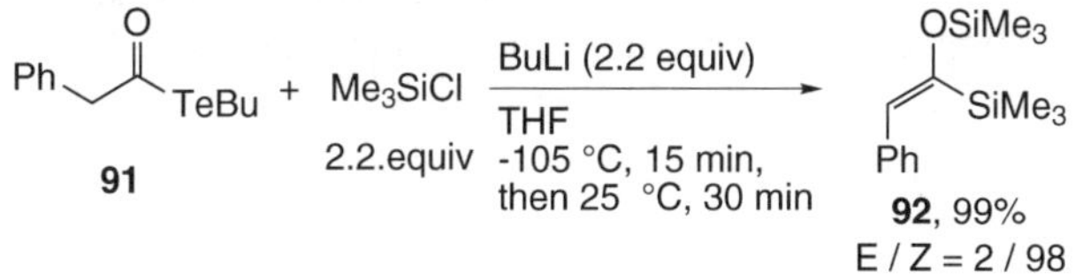

$$\text{(67)}$$

Ethanetelluroates with other anion stabilizing groups at the α carbon, such as aryl, phenylthio, and benzyloxy groups, gave enol silyl ethers of the corresponding acylsilanes in good to high yields. This reaction was stereoselective, and Z-isomers were obtained as sole or major products from a variety of chlorosilanes. Control experiments revealed that the reaction comprises the following consecutive processes, as shown in Scheme 25: (i) α-proton abstraction to give the enolate **93**; (ii) chlorosilane-trapping of **93** to give **94**; (iii) lithium-tellurium exchange of **94** to form (α-siloxyvinyl)lithium **95**; (iv) 1,2-silicon shift, and; (v) chlorosilane-trapping of the resulting acylsilane enolate **92**.

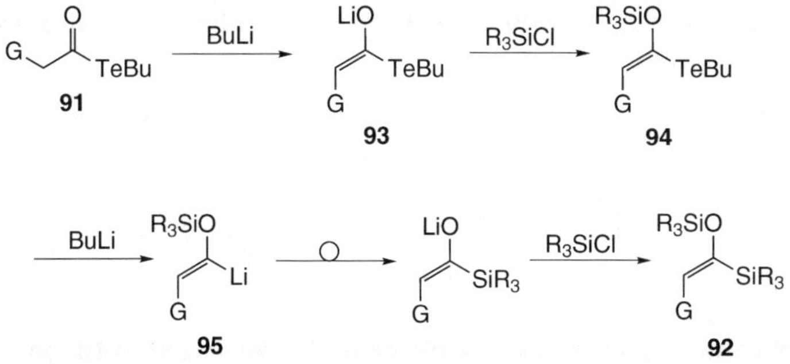

Scheme 25 Reaction pathway for the formation of silyl enol ether

4.2.3
Generation of Acyl Radicals and ArTe Group Transfer

Selenol esters are excellent precursors for the generation of acyl radicals used to construct radical chain reactions in the presence of Bu_3SnH. This reaction involves hydrogen atom abstraction from the Bu_3SnH by a carbon radical to form a stannyl radical which then abstracts the seleno moiety from the selenol ester to generate acyl radicals (the essential chain propagation steps). Crich and co-workers developed a new method of generating acyl radicals by photolysis of tellurol esters. Combination of this process with their inter- or intramolecular trapping and group transfer of the ArTe moiety provides an alternative chain sequence without a hydrogen transfer process [126]. For example, the photolysis of **96** (see Fig. 8) under an inert atmosphere at 8 °C in benzene in the presence of 1 equiv of diphenyl diselenide gave the selenol ester **97** quantitatively. A similar reaction using diphenyl disulfide in place of diphenyl diselenide afforded the corresponding thiol ester **98** with 85% yield, although a longer irradiation time was required. Thermal reactions can also take place; for example, refluxing a benzene solution of **96** and diphenyl diselenide under nitrogen in the dark gave 80% of **97** after 2 h, while **98** was obtained with only 16% yield after 16 h when diphenyl disulfide was employed.

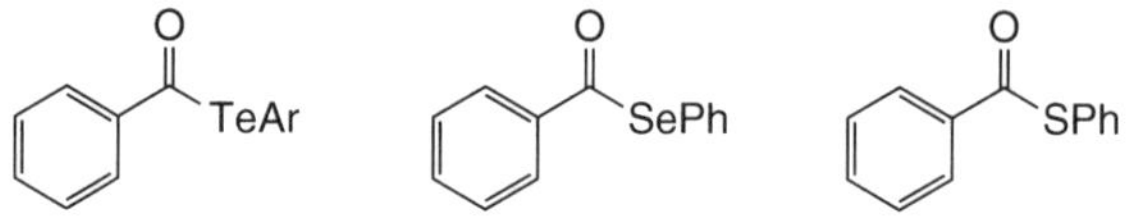

96: Ar = 4-F-C_6H_4 **97**   **98**

Fig. 8 Chalcogeno esters **96–98**

These observations are best interpreted in terms of either photochemically- and/or thermally-initiated homolytic cleavage of the acyl-tellurium bonds, followed by the attack of the acyl radicals formed in this way on the Se or S atom of the diselenide or disulfide.

Irradiation of **99** in benzene at reflux gave **100** in high yields as the products of acyl radical cyclization with ArTe group transfer, along with an elimination product **101** (Eq. 68).

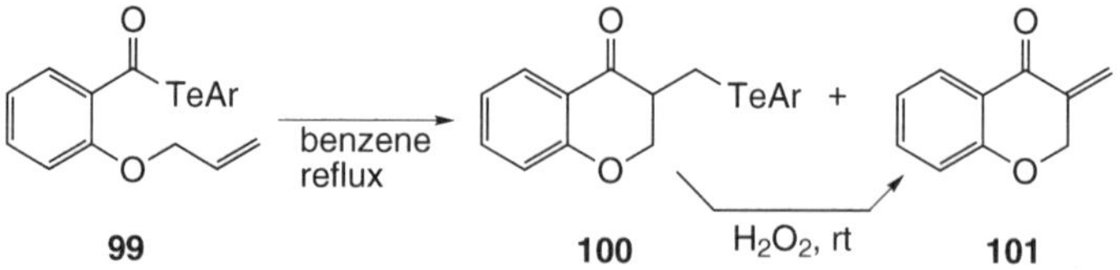

$$ \text{(68)} $$

99   **100** H_2O_2, rt   **101**

The tellurium transfer product **100** could be converted to **101** in excellent yield by simply stirring it at room temperature with hydrogen peroxide in THF.

Yamago, Yoshida and co-workers reported that tellurol esters bearing alkyl, aryl, alkenyl carbon residues react with 2,6-xylyl isonitrile at 100 °C for 6 h to give the α-acyl-substituted imidoyl tellurides **102** in good to high yields (Eq. 69) [127].

$$(69)$$

87%   78%   73%   48%

Decarbonylation from acyl radicals is a serious side reaction in the reaction of tellurol esters bearing secondary and tertiary alkyl carbon units as R. However, decarbonylation can be avoided by carrying out the reaction under CO pressure. In addition, electrochemical oxidation of the product **102** in the presence of water afforded the corresponding α-acyl amide in quantitative yields.

4.2.4
Reaction with Phosphoranes and Arylpropiolates

Tellurol esters react with phosphoranes at room temperature to give α-acylphosphiranes **103** (Eq. 70) [128]. The reaction with arylpropiolates gives (Z)-β-aryltelluro-α,β-unsaturated esters **104** with high stereoselectivity (Eq. 71) [129].

$$(70)$$

Ar	R	yield, %
Ph	Et	51
Ph	Me	44
4-BrC$_6$H$_4$	Et	53

$$(71)$$

Ar	Ar'	yield, %
Ph	Ph	87
4-MeOC$_6$H$_4$	Ph	88
Ph	4-ClC$_6$H$_4$	91

References

1. Mukaiayama T, Takeda T, Atsumi K (1974) Chem Lett 187; Yamada S, Yokoyama Y, Shioiri T (1974) J Org Chem 39:3302; Masamune S, Kamata S, Diakur J, Sugihara Y, Bates GS (1975) Can J Chem 53:3693
2. Holmquist HE, Carnahan JE (1960) J Org Chem 25:2240
3. Shim SC, Antebi S, Alper H (1985) J Org Chem 47:149
4. Shim SC, Antebi S, Alper H (1985) Tetrahedron Lett 26:1935
5. Antebi S, Alper H (1986) Organometallics 5:596
6. Antebi S, Alper A (1985) Tetrahederon Lett 26:2609
7. Kuniyasu H, Ogawa A, Miyazaki S, Ryu I, Kambe N, Sonoda N (1991) J Am Chem Soc 113:9796
8. Ogawa A, Takeba M, Kawakami J, Ryu I, Kambe N, Sonoda N (1995) J Am Chem Soc 115:7564
9. Ogawa A, Kawakami J, Mihara M, Ikeda T, Sonoda N, Hirao T (1997) J Am Chem Soc 119:12380
10. Khumtaveeporn K, Alper H (1994) J Org Chem 59:1414
11. Roberto D, Alper H (1989) J Am Chem Soc 111:7539; Wang MD, Calet S, Alper H (1989) J Org Chem 54:20
12. Khumtaveeporn K, Alper H (1994) J Am Chem Soc 116:5662
13. Seyferth D, Weinstein RM, Wang W-L, Hui RC (1983) Tetrahederon Lett 24:4907; Seyferth D, Weinstein RM, Wang W-L (1983) J Org Chem 48:1144; Weinstein RM, Wang W-L, Seyferth D (1983) J Org Chem 48:3367; Seyferth D, Weinstein RM (1982) J Am Chem Soc 104:5534
14. Seyferth D, Hui RC (1984) Organometallics 3:327
15. Braga AL, Rodrigues OED, de Avira E, Silveira CC (1998) Tetrahedron Lett 39:3395; Braga AL, Martins TLC, Silveira CC, Rodrigues OED (2001) Tetrahedron 57:3297
16. Sheng S, Liu X (2002) Org Prep Proc Int 34:489
17. Inoue T, Takeda T, Kambe N, Ogawa A, Ryu I, Sonoda N (1994) J Org Chem 59:5824
18. Desai MC, Stramiello LMS (1993) Tetrahedron Lett 34:7685
19. Adamczyk M, Fishpaugh JR (1996) Tetrahedron Lett 37:4305
20. Grunwell R, Forest L (1976) Synth Commun 6:453
21. Falck JR, Bhatt RK, Ye J (1995) J Am Chem Soc 117:5973
22. Mukaiyama T, Narasaka K, Banno K (1973) Chem Lett 1011; Mukaiyama T, Banno K, Narasaka K (1974) J Am Chem Soc 96:7503
23. Kobayshi S, Uchiro H, Fujishita Y, Shiina I, Mukaiyama T (1991) J Am Chem Soc 113:4247; Kobayshi S, Mukaiyama T (1989) Chem Lett 297; Mukaiyama T, Uchiro H, Kobayashi S (1989) Chem Lett 1001; Mukaiyama T, Uchiro H, Kobayashi S (1989) Chem Lett 1757; Mukaiyam T, Kobayashi S, Uchiro H, Shiina I (1990) Chem Lett 129; Mukaiyama T, Uchiro H, Shiina I, Kobayashi S (1989) Chem Lett 1019; Mukaiyama T, Shiina I, Kobayashi S (1991) Chem Lett 1901
24. Kobayashi S, Horibe M (1994) J Am Chem Soc 116:9805; Kobayashi S, Horibe M, Saito Y (1994) Tetrahedron 50:9629; Evans DA, Murry JA, Kozlowski MC (1996) J Am Chem Soc 118:5814; Evans DA, MacMillan DWC, Campos KR (1997) J Am Chem Soc 119:10859
25. Evans DA, Kozlowski MC, Burgey CS, MacMillan DWC (1997) J Am Chem Soc 119:7893; Evans DA, Kozlowski MC, Murry JA, Burgey CS, Campos KR, Connell BT, Staples RJ (1999) J Am Chem Soc 121:669; Evans DA, Burgey CS, Kozlowski MC, Tregay SW (1999) J Am Chem Soc 121:686; Orlandi S, Mandoli A, Pini D, Salvadori P (2001) Angew Chem Int Edit 40:2519

26. Ishitani H, Yamashita Y, Shimizu H, Kobayashi S (2000) J Am Chem Soc 122:5403; Yamashita Y, Ishitani H, Shimizu H, Kobayashi S (2002) J Am Chem Soc 124:3292
27. Mikami K, Matsukawa S (1994) J Am Chem Soc 116:4077; Keck GE, Krishnamurthy D (1995) J Am Chem Soc 117:2363
28. Parmee ER, Hong Y, Tempkin O, Masamune S (1992) Tetrahedron Lett 33:1729
29. Evans DA, Masse CE, Wu J (2002) Org Lett 4:3375; Evans DA, Wu J, Masse CE, MacMillan DWC (2002) Org Lett 4:3379
30. Hamada T, Manabe K, Ishikawa S, Nagayama S, Shiro M, Kobayashi S (2003) J Am Chem Soc 125:2989
31. Shimizu M, Itohara S, Hase E (2001) Chem Commun 2318
32. Denmark SE, Griedel BD, Coe DM, Schnute ME (1994) J Am Chem Soc 116:7026
33. Narasaka K, Soai K, Mukaiyama T (1974) Chem Lett 1223
34. Evans DA, Rovis T, Kozlowski MC, Tedrow JS (1999) J Am Chem Soc 121:1994; Evans DA, Rovis T, Kozlowski MC, Downey CW, Tedrow JS (2000) J Am Chem Soc 122:9134; Evans DA, Scheidt KA, Johnston JN, Willis MC (2001) J Am Chem Soc 123:4480; Sibi MP, Chen J (2002) Org Lett 4:2933
35. Harada T, Iwai H, Takatsuki H, Fujita K, Kubo M, Oku A (2001) Org Lett 3:2101
36. Ishitani H, Ueno M, Kobayashi S (1997) J Am Chem Soc 119:7153; Ishitani H, Ueno M, Kobayashi S (2000) J Am Chem Soc 122:8180; Ueno M, Ishitani H, Kobayashi S (2002) Org Lett 4:3395; Kobayashi S, Matsubara R, Nakayama Y, Kitagawa H, Sugiura M (2003) J Am Chem Soc 125:2507
37. Matsuda I, Komori K-I, Itoh K (2002) J Am Chem Soc 124:9072
38. Wolfrom ML, Karabinos JV (1946) J Am Chem Soc 68:1455
39. Nagao Y, Kawabata K, Fujita E (1978) J Chem Soc Chem Comm 330; McGarvey GJ, Williams JM, Hiner RN, Matsubara Y, Oh T (1986) J Am Chem Soc 108:4943; McGarvey GJ, Hiner RN, Williams JM, Matsubara Y, Poarch JW (1986) J Org Chem 51:3742
40. Fukuyama T, Lin S-C, Li L (1990) J Am Chem Soc 112:7050; Tokuyama H, Yokoshima S, Lin S-C, Li L, Fukuyama T (2002) Synthesis 1121
41. Kanda Y, Fukuyama T (1993) J Am Chem Soc 115:8451; Fujiwara A, Kan T, Fukuyama T (2000) Synlett 1667; Evans DA, Black WC (1993) J Am Chem Soc 115:4497; Evans DA, Ng HP, Rieger DL (1993) J Am Chem Soc 115:11446; Smith AB III, Chen SS-Y, Nelson FC, Reichert JM, Salvatore BA (1997) J Am Chem Soc 119:10935; Makino K, Henmi Y, Hamada Y (2002) Synlett 613
42. Miyazaki T, Han-ya Y, Tokoyama H, Fukuyama T (2004) Synlett 477
43. Node M, Kumar K, Nishide K, Ohsugi S, Miyamoto T (2001) Tetrahedron Lett 42:9207; Nishide K, Ohsugi S, Fudesaka M, Kodama S, Node M (2002) Tetrahedron Lett 43:5177
44. Penn JH, Owens WH (1992) Tetrahedron Lett 33:3737
45. Anderson RJ, Henrick CA, Rosenblum LD (1974) J Am Chem Soc 96:3654
46. Masamune S (1988) Pure Appl Chem 60:1587; Cama L, Christensen BG (1980) Tetrahedron Lett 21:2013
47. Araki M, Sakata S, Takei H, Mukaiyama T (1974) B Chem Soc Jpn 47:1777
48. Tokuyama H, Yokoshima S, Yamashita T, Fukuyama T (1998) Tetrahedron Lett 39:3189
49. Savarin C, Srogl J, Liebeskind LS (2000) Org Lett 2:3229
50. Liebeskind LS, Srogl J (2000) J Am Chem Soc 122:11260
51. Miyaura N, Suzuki A (1995) Chem Rev 95:2457
52. Helal CJ, Magriotis PA, Corey EJ (1996) J Am Chem Soc 118:10938
53. Trost BM, Schmidt T (1988) J Am Chem Soc 110:2301
54. Huguet J, Karpf M, Dreiding AS (1982) Helv Chim Acta 61:2413; Sayo N, Azuma K, Mikami K, Nakai T (1984) Tetrahedron Lett 25:565

55. Kawanami Y, Katsuki T, Yamaguchi M (1983) Tetrahedon Lett 24:5131
56. Tokuyama H, Miyazaki T, Yokoshima S, Fukuyama T (2003) Synlett 1512
57. Danheiser RL, Nowick JS (1991) J Org Chem 56:1176
58. Crich D, Fortt SM (1989) Tetrahedron 45:6581; Boger DL, Mathvink RJ (1988) J Org Chem 53:3377
59. Penn JH, Liu F (1994) J Org Chem 59:2608
60. Crich D, Yao Q (1996) J Org Chem 61:3566
61. Benati L, Leardini R, Minozzi M, Nanni D, Spagnolo P, Strazzari S, Zanardi G (2002) Org Lett 4:3079
62. Chatgilialoglu C, Crich D, Komatsu M, Ryu I (1999) Chem Rev 99:1991
63. Blakskjær P, Høj B, Riber D, Skrydstrup T (2003) J Am Chem Soc 125:4030
64. Osakada K, Yamamoto T, Yamamoto A (1987) Tetrahedron Lett 28:6321
65. Sugoh K, Kuniyasu H, Sugae T, Ohtaka A, Takai Y, Tanaka A, Machino C, Kambe N, Kurosawa H (2001) J Am Chem Soc 123:5108
66. Hirai T, Kuniyasu H, Kato T, Kurata Y, Kambe N (2003) Org Lett 5:3871
67. Zeysing B, Gosch C, Terfort A (2000) Org Lett 2:1843
68. Ogawa A, Sonoda N (1991) In: Trost BM, Fleming I (eds) Comprehensive organic synthesis. Pergamon, Oxford, p 461; Ogawa A, Sonoda N (1995) In: Katritzky AR, Meth-Cohn O, Rees CW (eds) Comprehensive organic functional group transformations. Pergamon, Oxford, p 231
69. Zhang Y, Yu Y, Lin R (1993) Synth Commun 23:189; Wang L, Zhang Y (1999) Synth Commun 29:3107; Zhou L-H, Zhang Y-M (1999) J Chem Res-S 28; Liu Y, Zhang Y (1999) Synth Commun 29:4043; Chen RC, Zhang Y (2000) Synth Commun 30:1331
70. Detty MR, Wood GP (1980) J Org Chem 45:80
71. Silveira CC, Braga AL, Larghi E (1999) Organometallics 18:5183
72. Nishiyama Y, Kawamatsu H, Funato S, Tokunaga K, Sonoda N (2003) J Org Chem 68:3599
73. Beletskaya IP, Sigeev AS, Peregudov AS, Petrovskii PV (2001) Russ J Org Chem 37:1703
74. Ranu BC, Mandal T, Samanta S (2003) Org Lett 5:1439
75. Nobrega JA, Goncalves SMC, Peppe C (2000) Tetrahedron Lett 41:5779
76. Kozikowski AP, Ames A (1978) J Org Chem 43:2735; Kozikowski AP, Ames A (1985) Tetrahedron 41:4821
77. Back TG, Collin S (1979) Tetrahedron Lett 2661; Back TG, Collin S, Kerr RG (1981) J Org Chem 46:1564
78. Segi M, Kato M, Nakajima T, Suga S, Sonoda N (1989) Chem Lett 1009
79. Sonoda N, Kondo K, Nagano K, Kambe N, Morimoto F (1980) Angew Chem Int Edit 19:308
80. Nishiyama Y, Katsuura A, Negoro A, Hamanaka S, Miyoshi N, Yamana Y, Ogawa A, Sonoda N (1991) J Org Chem 56:3776
81. Kato S, Kabuto H, Ishihara H, Murai T (1985) Synthesis 520
82. Ishihara H, Matsunami N, Yamada Y (1987) Synthesis 371
83. Huang X, Liang C-G (1999) J Chem Res-S 634
84. Harada S, Taguchi T, Tabuchi N, Narita K, Hanzawa Y (1998) Angew Chem Int Edit 37:1696; Liang C-G, Huang X (2000) J Chem Res-S 118
85. Zhong P, Xiong Z-X, Huang X (2000) Synth Commun 30:887; Liang C-G, Huang X (2000) J Chem Res-S 118
86. Huang X, Wang J-H (1999) Synlett 569
87. Weigert F (1903) Chem Ber 36:1007
88. Demuynck C, Thullier A (1969) B Soc Chim Fr 2434; Vedejs E, Nader B (1982) J Org Chem 47:3193

89. Bestmann HJ, Saalbaum H (1979) B Soc Chim Belg 88:951
90. Seyferth D, Hui RC (1984) Tetrahedron Lett 25:2623
91. Fujiwara S, Asai A, Shin-ike T, Kambe N, Sonoda N (1998) J Org Chem 63:1724
92. Takahashi H, Ohe K, Uemura S, Sugita N (1987) J Organomet Chem 334:c43; Uemura S, Takahashi H, Ohe K, Sugita N (1989) J Organomet Chem 361:63
93. Ryu I, Muraoka H, Kambe N, Komatsu M, Sonoda N (1996) J Org Chem 61:6396
94. Maeda H, Fujiwara S, Shin-ike T, Kambe N, Sonoda N (1996) J Am Chem Soc 118: 8160
95. Fujiwara S, Nishiyama A, Shin-ike T, Kambe N, Sonoda N (2004) Org Lett 6:453
96. Nishiyama Y, Tokunaga K, Kawamatsu H, Sonoda N (2002) Tetrahedron Lett 43: 1507
97. Pfenninger J, Heuberger C, Graf W (1980) Helv Chim Acta 63:2328
98. Boger DL, Mathvink RJ (1988) J Org Chem 53:3377; Boger DL, Mathvink RJ (1992) J Org Chem 57:1429
99. Boger DL, Mathvink RJ (1989) J Org Chem 54:1777
100. Boger DL, Hong J (2001) J Am Chem Soc 123:8515 (and references cited therein); Evans PA, Raina S, Ahsan K (2001) Chem Commun 2504 (and references cited therein); Boehm HM, Handa S, Pattenden G, Roberts L, Blake AJ, Li W-S (2000) J Chem Soc Perkin T 1 3522 (and references cited therein)
101. Ryu I, Sonoda N, Curran DP (1996) Chem Rev 96:177
102. Kozikowski AP, Ames A (1980) J Am Chem Soc 102:860
103. Back TG, Kerr RG (1982) Tetrahedron Lett 23:3241; Back TG, Kerr RG (1985) Tetrahedron 41:4759
104. Kuniyasu H, Ogawa A, Higaki K, Sonoda N (1992) Organometallics 11:3937
105. Zhao C-Q, Huang X (1996) Synth Comm 26:3607
106. Zhao C-Q, Huang X, Meng J-B (1998) Tetrahedron Lett 39:1933
107. Zhao C-Q, Li J-L, Meng J-B, Wang Y-M (1998) J Org Chem 63:4170
108. Byeon C-H, Chen C-Y, Ellis DA, Hart DJ, Li J (1998) Synlett 596
109. Byeon C-H, Hart DJ, Lai C-S, Unch J (2000) Synlett 119
110. Abe M, Tachibana K, Fujimoto K, Nojima M (2001) Synthesis 1243
111. Petasis NA, Lu S-P (1995) Tetrahedron Lett 36:2393
112. Petasis NA, Bzowej EI (1990) J Am Chem Soc 112: 6392
113. Piette J-L, Renson M (1970) B Soc Chem Belg 79:383; Piette J-L, Debergh D, Baiwit M, Llabres G (1980) Spectrochim Acta A 36:769
114. Kakigano T, Kanda T, Ishida M, Kato S (1987) Chem Lett 475
115. Kanda T, Nakaida S, Murai T, Kato S (1989) Tetrahedon Lett 30:1829
116. Suzuki H, Inamasu T, Ogawa T, Tani H (1990) J Chem Res S 56
117. Sasaki K, Aso Y, Otsubo T, Ogura F (1986) Chem Lett 977
118. Schiesser CH, Skidore MA (1997) J Chem Soc Perkin Trans 1: 2689
119. Zhang S, Zhang Y (1999) Heteroatom Chem 10:471
120. Inoue T, Takeda T, Kambe N, Ogawa A, Ryu I, Sonoda N (1994) Organometallics 13: 4543
121. Sasaki K, Aso Y, Otsubo T, Ogura F (1989) Chem Lett 607; Sasaki K, Mori T, Doi Y, Kawachi A, Aso Y, Otsubo T, Ogura F (1991) Chem Lett 415
122. Dabdoub MJ, Viana LH (1992) Synth Commun 22:1619
123. Hiiro T, Kambe N, Ogawa A, Miyoshi N, Murai S, Sonoda N (1987) Angew Chem Int Edit 26:1187; Hiiro T, Mogami T, Kambe N, Fujiwara S, Sonoda N (1990) Synth Commun 20:703
124. Hiiro T, Morita Y, Inoue T, Kambe N, Ogawa A, Ryu I, Sonoda N (1990) J Am Chem Soc 112:455
125. Inoue T, Kambe N, Ryu I, Sonoda N (1994) J Org Chem 59:8209

126. Chen C, Crich D, Papadatos A (1992) J Am Chem Soc 114:8313; Chen C, Crich D (1993) Tetrahedron Lett 34:1545; Crich D, Chen C, Hwang J-T, Yuan H, Papadatos A, Walter RI (1994) J Am Chem Soc 116:8937
127. Yamago S, Miyazoe H, Sawazki T, Goto R, Yoshida J-I Tetrahedron Lett (2000) 41:7517; Yamago S, Miyazoe H, Goto R, Hashidume M, Sawazaki T, Yoshida J-I (2001) J Am Chem Soc 123:3697
128. Wang L, Huang X (1994) Synth Commun 24:689
129. Zhao C-Q, Huang X (1997) Synth Commun 27:249

Top Curr Chem (2005) 251: 141–180
DOI 10.1007/b101008
© Springer-Verlag Berlin Heidelberg 2005

Dithiocarboxylic Acid Salts of Group 1–17 Elements (Except for Carbon)

Naokazu Kano · Takayuki Kawashima (⊠)

Department of Chemistry, Graduate School of Science, The University of Tokyo,
7-3-1 Hongo, Bunkyo-ku, Tokyo 113-0033, Japan
takayuki@chem.s.u-tokyo.ac.jp

Abstract In this chapter, we review dithiocarboxylic acid salts. Methods of synthesizing these salts are reviewed, such as insertion of carbon disulfide into the metal-carbon bond, reaction of a dithiocarboxylic acid with a transition metal salt in the presence or absence of a base, ligand exchange of dithiocarboxylato metal or ammonium salts with another metal complex bearing a leaving group, and so on. The dithiocarboxylic acid complexes exhibit a variety of coordination styles that have been characterized by X-ray crystallographic analysis. Spectroscopic (IR, ^{13}C, and UV-Vis) properties of the salts are summarized. The reactivities, structures, properties, characteristics and applications of each salt are discussed in detail for the salts of most elements (except for carbon).

Keywords Dithiocarboxylate salts · Dithiocarboxylato complex · Dithiocarboxylic acid · Thiocarbonyl group · Coordination

Abbreviations

acdc	2-amino-1-cyclopentene-1-dithiocarboxylato
ACP	alternative charge-polarization
AV	average valence
CDW	charge density wave
CP	charge-polarization
Cp*	1,2,3,4,5-pentamethylcyclopentadinenyl
1-D	one-dimensional
dmid	2-oxo-1,3-dithiole-4,5-dithiolato
dmit	1,3-dithiole-2-thione-4,5-dithiolato
dsit	2-thioxo-1,3-dithiole-4,5-diselenolato
DSC	differential scanning calorimeter
E	element
EXAFS	extended X-ray absorption fine structure
i-dto	1,1-dithiooxalato
i-dtoMe	O-methyl-1,1-dithiocarboxylato
IR	infrared
L	ligand
M	metal
MMX	halogen-bridged binuclear transition metal
mnt	1,2-dicyanoethene-1,2-dithiolato
NIR	near infrared
NMR	nuclear magnetic resonance
R	organic substituent
trto	trithiooxalato
tto	tetrathiooxalato
UV	ultraviolet
vis	visible
XPS	X-ray photoelectron spectroscopy

1
Introduction

Salts of dithiocarboxylic acids are compounds in which the thiol hydrogen atom of the dithiocarboxylic acid is replaced by another element. The terms "dithiocarboxylato" and "dithiocarboxylate" are used for the salts of transition metals and main group elements of dithiocarboxylic acids, respectively, in this chapter. Both dithiocarboxylate and dithiocarboxylato salts potentially have a variety of coordination styles and some interesting properties. Although salts that were crystallographically characterized in the initial studies on these salts are limited to transition metal complexes, many salts of main group elements have been studied and crystallographically characterized over the last two decades. This chapter describes the salts with the general form RCS_2EL_n, where R represents carbon substituents and E represents both transition metals and main group elements (except for hydrogen and carbon). There have already been some good reviews on studies in this area [1–3]. This chapter focuses on works published since 1990, in order to include new findings about these salts, although we also consider some older studies into their syntheses and basic properties. Since the delocalization of the C–S and C=S bonds in the dithiocarboxylato ligands of the transition metal complexes inhibits our ability to pin down their bond orders, both of them are drawn as a single bond in figures and schemes in this chapter. The coordination of thiocarbonyl sulfur to the transition metal is drawn in a similar way. On the other hand, the coordination between the thiocarbonyl sulfur and the main group element is drawn as a dotted line for the salts of main group elements. Starting with the syntheses of the salts, we cover the dithiocarboxylic acid salts of each group in the periodic table in turn (including some reactivities). Then we discuss the structures and spectroscopic properties of the salts of each group.

2
Synthetic Methods and Structures

2.1
General Synthetic Methods

Dithiocarboxylato salts of transition metals, RCS_2EL_n, are usually obtained by the following methods: (a) insertion of CS_2 into the C–E bond of the compound REL_n; (b) reaction of dithiocarboxylic acids with a transition metal salt in the presence or absence of a base; (c) ligand exchange of dithiocarboxylato metal or ammonium salts with another metal complex bearing a leaving group, like a halide, and; (d) conversion of a thiolate compound to a dithiocarboxylato compound (Scheme 1). There are several other miscellaneous synthetic methods [1, 2]. The dithiocarboxylate salts of main group elements are generally synthesized by methods (b), (c), and (d).

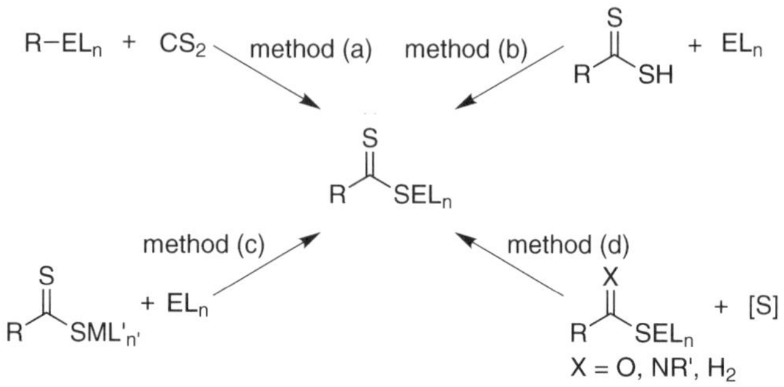

Scheme 1 Synthetic methods for dithiocarboxylic acid salts of Group 1–17 elements

Method (a) is the most atom-economical method. It is mostly used for synthesizing the salts of transition metals, alkali metals and alkaline earth metals. Some alkyllithiums and Grignard reagents are useful for synthesizing the corresponding lithium and magnesium dithiocarboxylate salts [4]. Method (b) is a straightforward reaction from the dithiocarboxylic acids to the corresponding salts. The alkali and ammonium salts are directly synthesized by deprotonation with metal hydrides such as LiH, NaH, and KH, rubidium and cesium acetates, or secondary and tertiary amines in aprotic nonpolar solvents [5, 6]. In method (c), a dithiocarboxylic acid salt forms other dithiocarboxylate salts by ligand exchange. Treatment of the dithiocarboxylic acid with secondary or tertiary amines for deprotonation and simultaneous or subsequent addition of chlorosilanes, chlorogermanes, chlorostannanes, and chloroplumbanes gives the corresponding group 14 element salt, respectively, via the intermediary ammonium salt [7–10]. Sodium salts, potassium salts, and cesium salts of dithiocarboxylic acids are also used [11–13]. Phosphorus and arsenic compounds are similarly synthesized from chlorophosphines and chloroarsines, respectively [14–16]. In method (d), the thioacyl group moiety is newly attached to the thiolate salts bearing an S–E bond, in contrast to the other three methods. Treatment of carboxylic chlorides with dithiophosphoric acids yields the corresponding anhydrides, which can be easily converted into thioacyl dithiophosphates upon treatment with an excess of dithiophosphoric acid [17, 18].

Many transition metal complexes bearing dithiocarboxylato ligands have been studied from various points of view. Their coordination modes depend on the metals and the other ligands on the metals. The complexes sometimes show interesting properties, such as electron conductivity. Synthetic methods are usually as same as those of the main-group-element complexes bearing dithiocarboxylato ligands. The insertion of carbon disulfide into the metal-carbon bond is especially used in the synthesis of transition metal complexes.

2.2
Structures and X-Ray Crystallographic Analysis

The most powerful tool for structural determination is X-ray crystallographic analysis, although good crystallinity of the salts is necessary. Several dithiocarboxylate salts have been characterized by X-ray crystallographic analysis. Dithiocarboxylate metal salts show various coordination types, which are mainly divided into six groups, as shown in Fig. 1. The ligand acts as an uncoordinated free anion (type I), a monodentate ligand (type II), a bidentate η^2-chelate ligand (type III), a monodentate ligand with additional interaction with another metal (type IV), a bridging ligand(s) to two metal atoms (type V), or a η^3-tridentate ligand (type VI). Such coordination types depend on the central metal, the R group of the dithiocarboxylato ligand (RCS_2), and the other ligand on the metal. In order to classify the ligands into coordination types, we need to know the locations of the metal atoms and the interatomic distances between each sulfur atom and metal atom. The bond angles around the thiocarbonyl carbon atoms give very little information about the structure and coordination types, because the S–C–S angles vary little from 120° in every salt. Types I–IV are not sharply separated, and it is difficult to distinguish the border between coordination and uncoordination; in other words, we cannot definitely state whether the ligand coordinates to the metal or not. In ambiguous and confusing cases, other spectral methods and theoretical calculations are used to classify the coordination styles.

In type I, the cation is too far from two sulfur atoms of the dithiocarboxylate ligand to coordinate and/or its Lewis acid properties are too weak to be coordinated. The dithiocarboxylate behaves as a free anion in this case. The inner salts of dithiocarboxylates are included in this category [19, 20]. Two C–S bonds exhibiting almost similar bond lengths (around 1.68 Å) are almost equivalent in these cases.

In type II, only the thiolate moiety of the dithiocarboxylate ligand coordinates to a metal atom. The thiocarbonyl sulfur is free from coordination and the lengths of the two C–S bonds are different in this case. The longer bond, which is usually shorter than the sum of the corresponding covalent radii, is considered to be a C–S single-bond, and the shorter one to be a C=S double-bond. This type of η^1-coordination is found in transition-metal complexes in which another donor in the organic substituent (R) of the dithiocarboxylato ligand chelates intramolecularly to the central metal forming a chelate ring (see Sects. 2.4.2, 2.4.3).

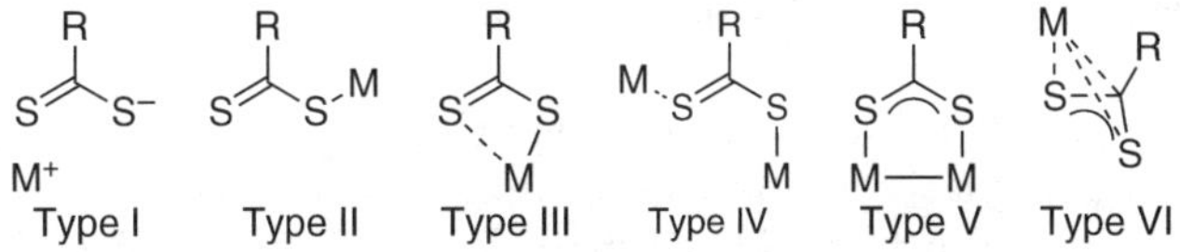

Fig. 1 Ways that dithiocarboxylate ligands can coordinate to metals

In many cases, however, another inter- or intramolecular coordination from the thiocarbonyl sulfur in the vicinity to the metal atom should be considered in order to discuss the coordination states of the dithiocarboxylic acid metal salts. In type III, the thiocarbonyl sulfur atom of the ligand coordinates to the atom to which the thiolate moiety is bound. This coordination style is often observed in main-group-element compounds bearing a dithiocarboxylate ligand, as well as in transition metal complexes [10, 15, 16]. Both the two C–S bonds and the two S–M bonds of a bidentate dithiocarboxylate ligand show different bond lengths in many cases, and the two sulfur atoms are different distances from the metal ion, giving unsymmetrical coordination to the metal. The C–S bond lengths are 1.59–1.78 Å. The difference in bond length between two C–S bonds is small if significant delocalization occurs. If the S–M bond lengths as well as the interactions between two sulfur atoms and the metal atom are the same, the C–S bond lengths are also equal and the double bond of the CS_2 moiety is completely delocalized.

In type IV, the thiolate sulfur atom coordinates to a metal atom and the thiocarbonyl sulfur coordinates to another metal atom. This type of coordination is seen in tetrathiooxalate complexes (see Sect. 2.7) [21–30]. This coordination style is also seen in polynuclear silver and copper complexes [31, 32].

In type V, the ligand bridges two metal atoms through two sulfur atoms. Each sulfur atom coordinates to one of two metal atoms in this case, respectively. The two C–S bonds are completely delocalized. This kind of coordination style is often observed in transition metal complexes, especially in the so-called MMX complexes (see Sect. 2.6) [16].

In type VI, the CS_2 moiety coordinates to one metal atom in a η^3-manner. The ligand works as a η^3-tridentate ligand, like a π-allyl ligand, in this type of compound, although such a coordination mode is rarely observed in dithiocarboxylic acid salts (see Sect. 2.4.4) [33–37].

Other types of coordination modes are also observed in several salts in which dithiocarboxylate ligands coordinate to three or more metal atoms. Such coordination modes are explained by combinations of the coordination types described above.

In the following sections, syntheses and structures of these salts are discussed in detail, as well as some of their properties and applications.

2.3
Alkali Metal and Alkaline Earth Metal Compounds

2.3.1
Group 1 Element Compounds (Li, Na, K, Rb, Cs)

Dithiocarboxylate salts of Group 1 elements are an important class of compounds for the synthesis of dithiocarboxylic acid derivatives. Anhydrous alkali salts of the dithiocarboxylic acid of 4-methyldithiobenzoic acid, such as **1–5**, are obtained in solid form by the reaction of the dithiocarboxylic acid **6** with alkali metal hydrides (LiH, NaH, and KH) and metal acetates (RbOAc and CsOAc), respectively, in aprotic solvents (Scheme 2) [6]. Lithium and sodium salts 1 and 2 are also obtained by treatment with butyllithium and sodium metal, respectively [5, 6]. Many useful preparation methods are provided by the reactions of trimethylsilyl dithiocarboxylate 7 with potassium, rubidium, and cesium fluorides [38]. Another synthetic method for sodium dithiobenzoate is the reaction of benzyl chloride with elemental sulfur in the presence of sodium methoxide [39].

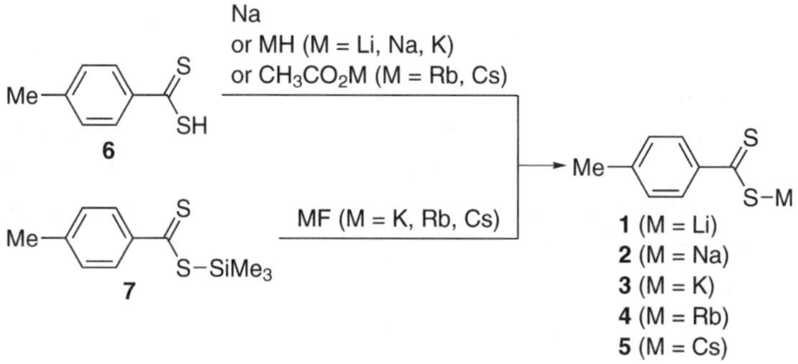

Scheme 2   Syntheses of alkali metal salts of 4-methyldithiobenzoic acid

The crystal structures of the lithium salts 8 with bulky aryl substituents exhibit monomeric structures and η^2-coordination of the dithiocarboxylate to the lithium atom (Fig. 2) [40]. Lithium complexes 9 and 10, bearing pyrazyl and pyridyl groups, respectively, also have monomeric structures, the former of which has η^2-coordination similar to that of 8. On the other hand, the lithium cation in 10 is chelated by one sulfur atom of the dithiocarboxylate and a nitrogen atom of the pyridyl group [4].

A series of 4-methyldithiobenzoate salts of potassium 3, rubidium 4 and cesium 5 show dimeric structures in which each dithiocarboxylate ligand is chelated to two metal cations (Fig. 3) [38]. The two metal atoms are located on the upper and lower sides of the plane involving four sulfur atoms of two dithiocarboxylate ligands. The metal cations further interact with the tolyl fragments in the neighboring molecules. The C–S bond lengths of the potassium

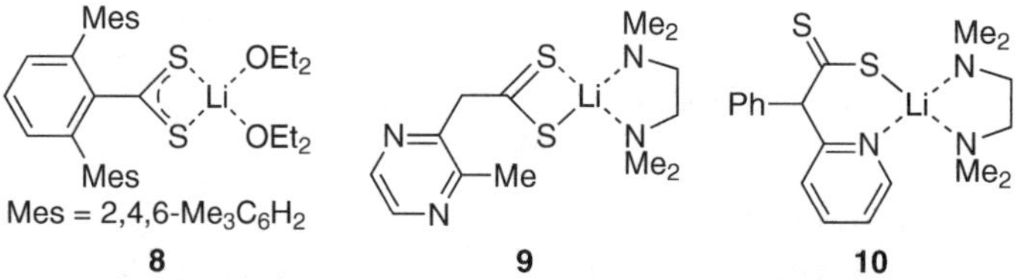

Fig. 2   Some lithium salts of dithiocarboxylic acids

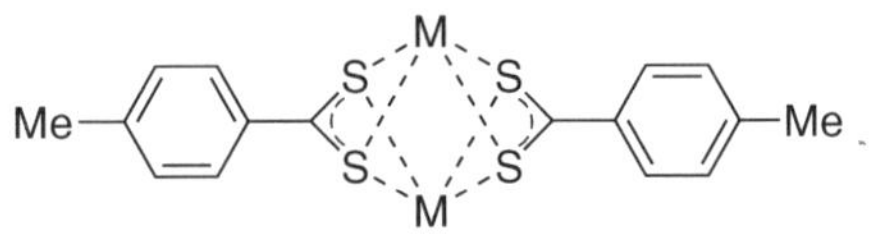

Fig. 3 Dimeric structures of the alkali salts of 4-methyldithiobenzoic acid

salt **3** are different from each other, while those of the rubidium and cesium salts, **4** and **5**, are almost equal. A similar situation is observed for M–S interatomic distances.

The dithiocarboxylate salts of alkali metals readily react with alkyl halides. The lithium, sodium, and potassium salts **1–3** readily react with methyl iodide to give the corresponding methyl ester **11** at room temperature in good yields, while similar reactions of rubidium and cesium salts **4** and **5** result in low yields (Scheme 3) [38].

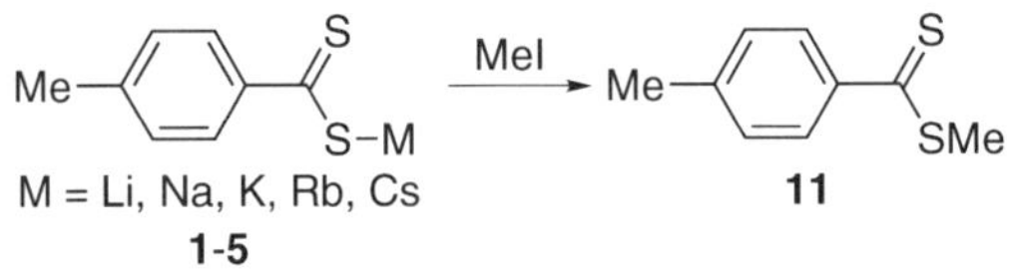

Scheme 3 Reaction of iodomethane with alkali metal salts of 4-methyldithiobenzoic acid

2.3.2
Group 2 Element Compounds (Mg)

Magnesium salts **12** are the most important among the dithiocarboxylate salts of alkaline earth metals. Magnesium salts **12** are prepared by insertion of CS_2 using Grignard reagents (Scheme 4) [41]. These salts **12** are treated with acid, alkyl halides and alkyl triflates to prepare dithiocarboxylic acids and esters, respectively [42–45]. A dinuclear aluminum-magnesium compound **13** is also synthesized by insertion of CS_2 into the Mg–C bond (Fig. 4) [46].

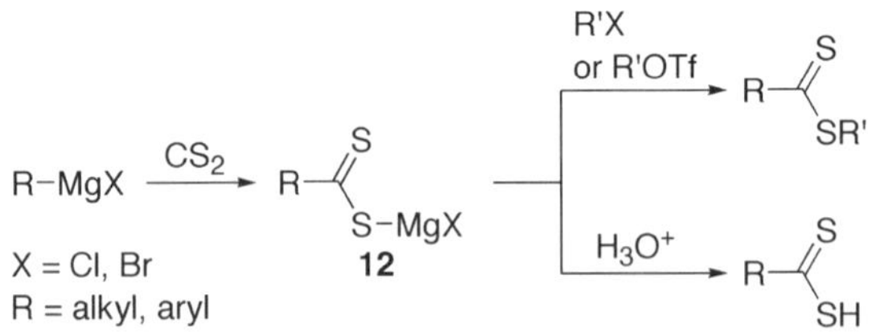

Scheme 4 Syntheses of magnesium salts of dithiocarboxylic acids and their reactions with electrophiles

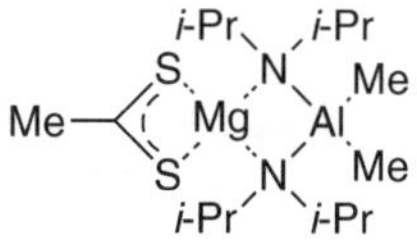

13

Fig. 4   The magnesium salt of dithioacetic acid

2.4
Transition Metal Compounds

2.4.1
Group 3 Element Compounds (Sc, Y)

Dithiocarboxylato complexes of the Group 3 elements scandium and yttrium have scarcely been studied. Only one example of the yttrium dithiocarboxylato complex [Y{η^2-(3,5-Me$_2$C$_6$H$_3$)CH$_2$CSS}Cp$_2^*$] **14** has been synthesized by insertion of CS$_2$ into the C–Y bond of the corresponding substituted benzyl complex [47]. IR spectra suggest that the coordination mode of the dithiocarboxylato ligand in **14** is η^2-bonding.

2.4.2
Group 4 Element Compounds (Ti, Zr, Hf)

Not many examples are known of zirconium and hafnium complexes bearing dithiocarboxylato ligands [48–50]. For titanium complexes, examples of the complexes **15** with new "scorpionate" ligands, [bis(3,5-dimethylpyrazol-1-yl)dithioacetato], have been reported (Fig. 5) [51]. The scorpionate ligand of these titanium complexes usually acts as a tridentate ligand, although introduction of two such ligands on the titanium atom results in the formation of **16**, in which the ligand acts as a bidentate ligand.

Another example of a titanium complex is the binuclear titanocene complex **18** which is obtained by the insertion of titanocene dicarbonyl into the S–S bond of a fused ring compound **17**, involving elimination of CO (Scheme 5) [52]. Both dithiocarboxylato and thiolato moieties are bound to each titanium

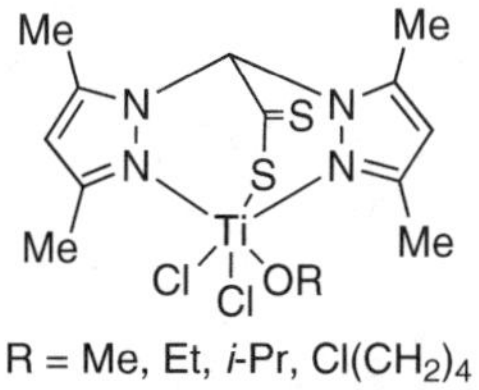

R = Me, Et, *i*-Pr, Cl(CH$_2$)$_4$

15 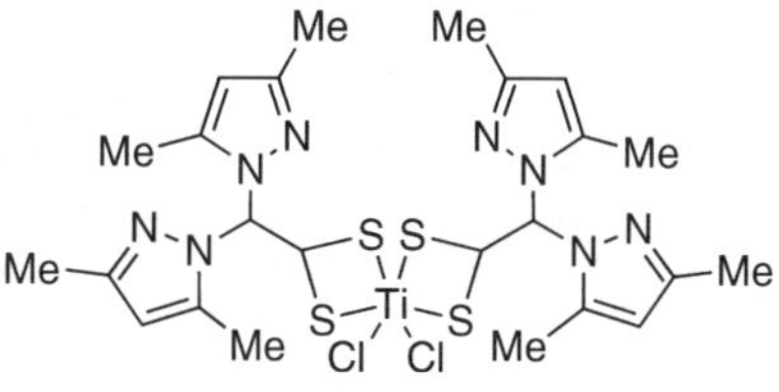   **16**

Fig. 5   Titanium complexes bearing one or two "scorpionate" ligands

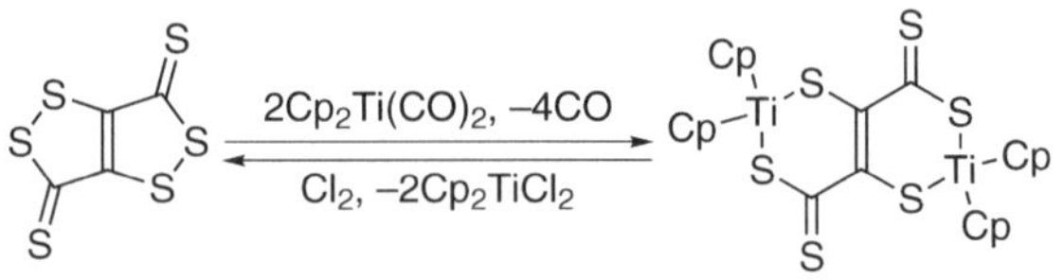

Scheme 5 Formation and a reaction of a binuclear titanocene complex

atom in η^1-coordination mode in **18**. The reaction of **18** with chlorine gives **17** and TiCp$_2$Cl$_2$.

2.4.3
Group 5 Element Compounds (V, Nb, Ta)

The oxovanadium(IV) complex **19** of 2-amino-1-cyclopentene-1-dithiocarboxylato (acdc) and the complex with *N*-methylated acdc ligands **20** have both been synthesized and characterized by various spectroscopic measurements in order to elucidate the coordination environment of the acdc ligand (Fig. 6) [53]. Simply introducing a methyl group onto the amino group to form **20** causes a change in the coordination around the oxovanadium(IV) center compared to **19**. Although **19** prefers bonding to two sulfur atoms of the dithiocarboxylato unit, **20** prefers bonding to one nitrogen and one sulfur atom, since the basicity of the amino group is enhanced.

There has been only one example of dithiocarboxylato complexes of tantalum – Ta(MeCSS)$_3$Cl$_2$ – which has only been characterized by IR and ^{1}H NMR spectroscopy and elemental analysis [54]. Among a few examples of niobium complexes, complex **21**, which was synthesized by the reaction of Nb(η^7-C$_7$H$_7$)-(η^5-C$_5$H$_5$) with dithioacetic acid, bears two dithiocarboxylato ligands and one η^2-S$_2$ ligand (Fig. 7) [55]. The coordination environment of the central Nb(V) atom is a severely distorted pentagonal bipyramid, where a η^5-C$_5$H$_5$ ring occupies one axial site, and a side-on bonded disulfide, η^2-S$_2$, occupies two equatorial sites. The remaining positions are occupied by two nonequivalent bidentate dithioacetato ligands. In the formation of **21**, desulfurization from dithioacetic acid and substitution with dithioacetic acid occur under mild reaction conditions. A similar reaction has been reported in the formation of divanadium complex **22** bearing two μ-η^2-S$_2$ bridges [56].

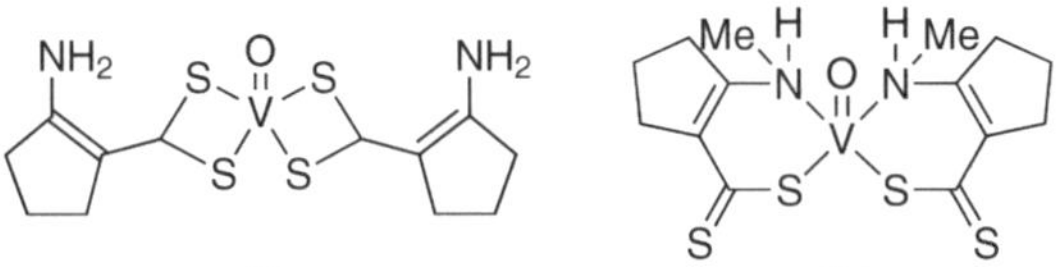

Fig. 6 Oxovanadium(IV) complexes bearing 2-amino-1-cyclopentene-1-dithiocarboxylato ligands

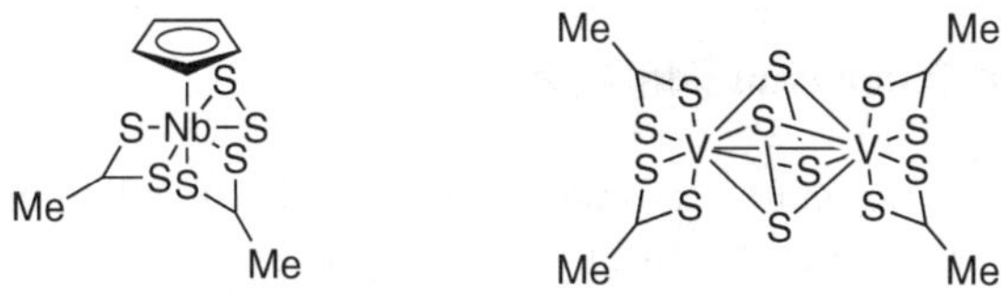

21   22

Fig. 7 Niobium and vanadium complexes bearing both dithiocarboxylato and η^2-S$_2$ ligands

2.4.4
Group 6 Element Compounds (Cr, Mo, W)

In contrast to the much less explored chromium complexes bearing dithiocarboxylato ligands [57–60], several examples of the molybdenum and tungsten complexes bearing dithiocarboxylato ligands have been reported. One synthetic method for dithiobenzoate complexes is to react the carbyne complexes of molybdenum and tungsten with elemental sulfur [61–64]. The former alkyne moiety of the alkylidyne complex **23** is sulfurized with elemental sulfur to form **24** in this reaction (Scheme 6). Sulfurization of **23** with propylene sulfide under careful control of reaction conditions provides the metallathiirene intermediate **25** which gives **24** after further sulfurization with elemental sulfur, indicating the intermediacy of the metallathiirene complexes in the formation of dithiocarboxylato complexes from alkylidyne complexes [65]. Furthermore, a rare example of a mononuclear tridentate dithiocarboxylato complex **26** was obtained by sulfurization from the corresponding alkylidyne complex (Fig. 8) [34].

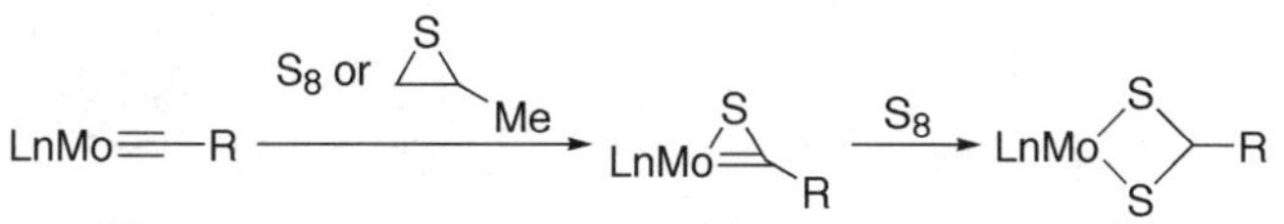

23   25   24

R = C$_6$H$_4$OMe-4; ML$_n$ = Mo(CO)$_2${HB(pyrazol-1-yl)$_3$}

Scheme 6 Formation of molybdenum complexes bearing dithiocarboxylato ligands by sulfurization of the alkylidyne complexes

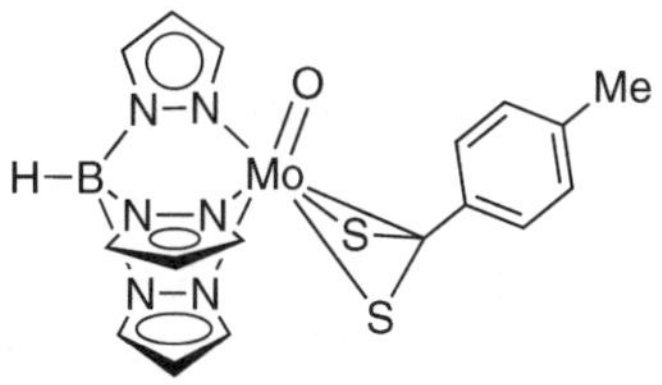

26

Fig. 8 Mononuclear molybdenum complex bearing a tridentate dithiocarboxylato ligand

2.4.5
Group 7 Element Compounds (Mn, Tc, Re)

Although there are not many examples of η^2-dithiocarboxylato manganese complexes [66], the nitrido technetium-99m complexes **27** bearing alkyldithiocarboxylato ligands have been studied in relation to their potential applications as radiopharmaceuticals, since they label leucocyte in whole blood cells with good efficiency and selectivity in vitro (Fig. 9) [44]. The design of the ligand was optimized and the best labeling result was obtained by **27** with aliphatic linear chains. Furthermore, **27** shows lymphocyte selectivity [67]. Such a property is helpful in revealing the subcellular distribution of the radiopharmaceutical in human blood cells via microautoradiographic analysis. The redox-active nitrido technetium-99m complex **28** bearing two ferrocenyldithiocarboxylato ligands has also been reported [68]. Both the neutral Fe(II) state and the cationic mixed-valence Fe(II)-Fe(III) state are synthesized at tracer level. The latter state can be converted easily into the former state by one-electron transfer with various reductants, such as triphenylphosphine and excess $SnCl_2$. Biodistribution studies in rats showed that both states are mostly retained in the lungs and liver without any significant uptake in organs such as the heart and brain [69].

Reported rhenium complexes possess one [70–74], two [75], and three [75] dithiocarboxylato ligands coordinated to the one rhenium atom to form a high coordination state, as in the case of the technetium complex [76]. One of these complexes **29** is synthesized from $Re(VII)S_4$ and $(PhCSS)_2$ (Fig. 10) [72]. An internal redox reaction occurs and Re(VII) is reduced to Re(III) in the synthetic process. Addition of sulfur-abstracting reagents, Ph_3P or Et_4NCN, produces the heptacoordinate complex **30** with the neutral capped octahedron structure or

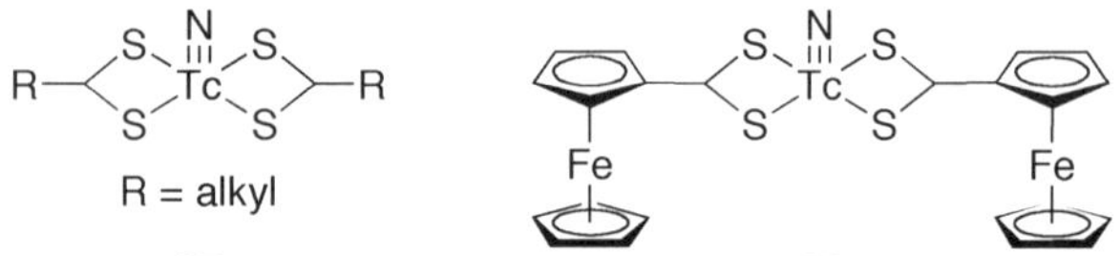

Fig. 9 Nitrido technetium complexes bearing dithiocarboxylato ligands

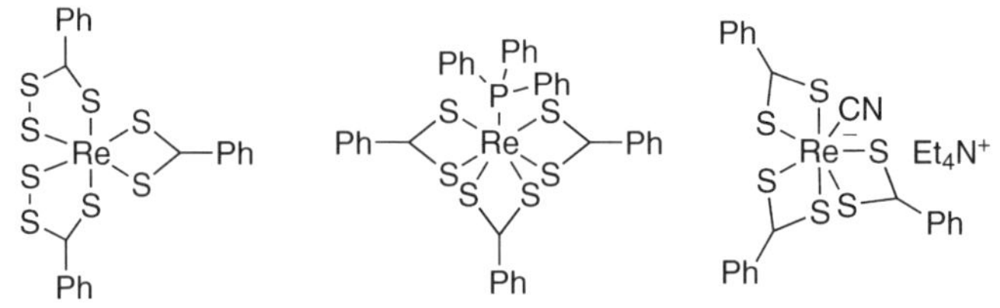

Fig. 10 Rhenium complexes possessing dithiobenzoato ligands

31 with the distorted pentagonal-bipyramidal structure, respectively [75]. These tris(dithiocarboxylato)rhenium complexes bear one more ligand, such as Ph_3P or CN^-, on the rhenium atom.

2.4.6
Group 8 Element Compounds (Fe, Ru, Os)

Whereas osmium dithiocarboxylato complexes are very rare, some dithiocarboxylato iron and ruthenium complexes have been synthesized by substituting into the metal-halogen complex with lithium dithiocarboxylate [77] and by inserting CS_2 into the carbon-metal bond [78, 79]. The reaction of a vinylidene complex 32 with CS_2 and NaOMe has also been reported to undergo a facile loss of HCl, followed by insertion of CS_2 to give the dithiocarboxylato complex 33 (Scheme 7) [80].

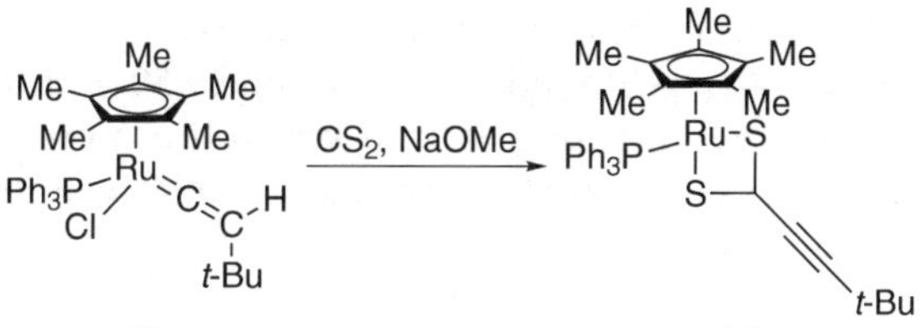

Scheme 7 Formation of a ruthenium complex bearing a dithiocarboxylato ligand from a vinylidene complex

The reaction of the iron complex 34 with CS_2 yields the corresponding complex 35 of bicyclic structures (Scheme 8) [81]. The reaction proceeds via a 1,3-dipolar cycloaddition of the C=S bond across the Fe–N=C unit, followed by insertion of an isocyanide to form the ferra[2.2.2]bicyclic complex 35.

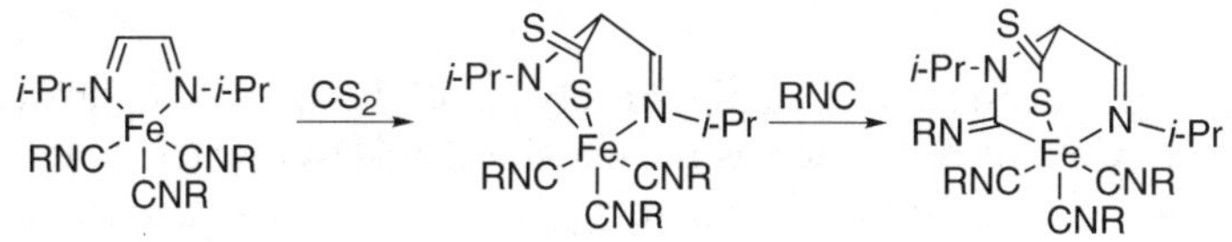

Scheme 8 Formation of bicyclic iron complexes bearing a dithiocarboxylato ligand

The insertion of CS_2 occurs at the Fe–C bond of the tetramethylfulvene-bridged diiron complex 36, producing the triiron complex 37 and the monoiron complex 38 (Scheme 9) [82]. A dithiocarboxylato ligand in 37 bridges three iron fragments in a novel coordination mode. One iron atom adopts the distorted-piano-stool geometry, while the other two iron atoms are distorted octahedral.

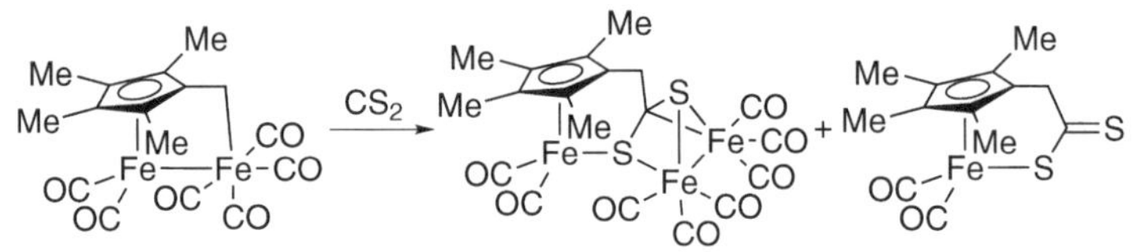

Scheme 9 Formation of a triiron complex and a monoiron complex bearing a dithiocarboxylato ligand

Each of the two pairs of iron atoms is bridged by sulfur atoms from the dithiocarboxylato ligand. The carbon atom of the dithiocarboxylato ligand interacts with an iron atom, adopting a pseudotetrahedral geometry in contrast to other dithiocarboxylato complexes. A reduced π-bonding interaction in the C–S bonds causes elongation of the C–S bonds in **37**.

The dithiocarboxylato complexes are also produced by reactions involving insertion of iron into the S–S bond. The reaction of **39** with $Fe_2(CO)_9$ gives the tetrairon complex **40**, in which each of two sulfur atoms of the dithiocarboxylato ligand are coordinated to two and three iron atoms, respectively (Scheme 10) [83–86].

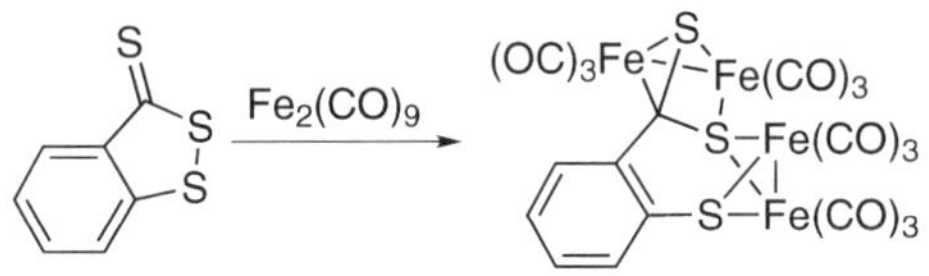

Scheme 10 Formation of a tetrairon complex

2.4.7
Group 9 Element Compounds (Co, Rh, Ir)

There have only been a few examples of cobalt and rhodium complexes bearing dithiocarboxylato ligands, such as the cobalt complex **41** which bears a bridging dithiobenzoate ligand, as shown in Fig. 11 [87–91], while there are a few (interesting) examples of iridium complexes. Treatment of a stable iridabenzene

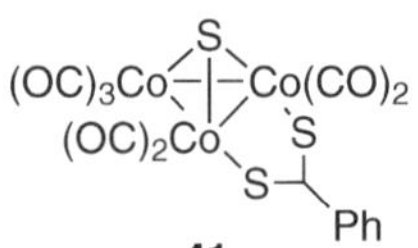

Fig. 11 Cobalt complex bearing a bridging dithiobenzoato ligand

complex **42** with CS_2 leads to [4+2]-cycloaddition reactions, giving the bicyclic complex **43** (Scheme 11) [92, 93]. The formal C=S double bond length of the adduct **43** is almost as same as the C–S single bond length, suggesting that a resonance structure (C^+–S^-) is an important contributor to the bonding. The CS_2 adduct **43** further undergoes cyclotrimerization reactions by linking subunits through thiocarbonyl sulfur centers after elimination of Et_3P [94]. The trimer **44** contains a novel organometallic trithia-12-crown-3 ring system.

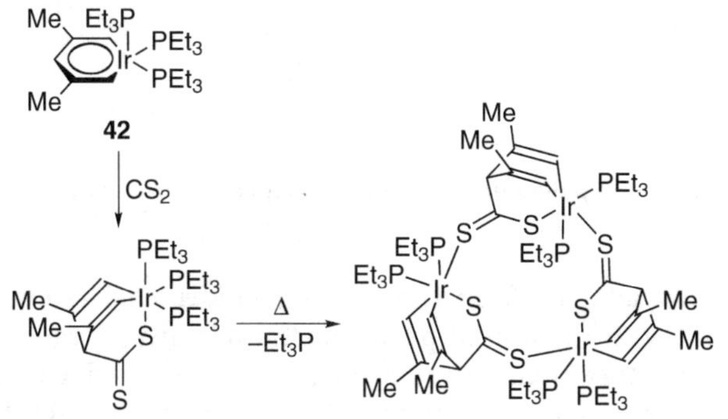

Scheme 11　Formation of cyclic iridium complexes bearing dithiocarboxylato ligands from an iridabenzene complex

2.4.8
Group 10 Element Compounds (Ni, Pd, Pt)

The insertion of CS_2 into a Ni–C coordination bond is proposed in the reaction of the cyclohexyne nickel(0) complex **45**, as in the case of the reaction with CO_2, although the product **46** could not be separated from the byproduct (Scheme 12) [95].

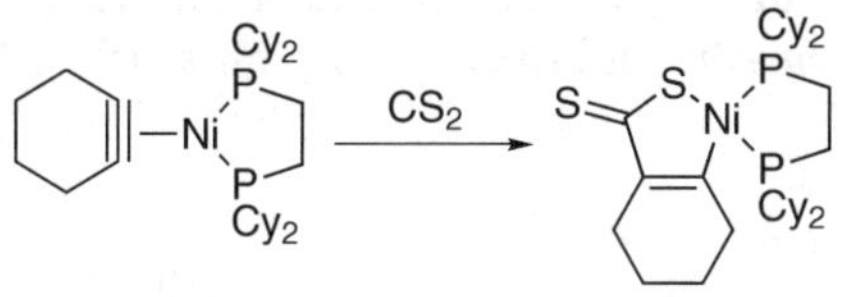

Scheme 12　Formation of a cyclic nickel complex bearing a dithiocarboxylic moiety from a cyclohexyne nickel(0) complex

Although few examples have been reported on the insertion of CS_2 into the Pd–C bond [96], the substitution of lithium dithiocarboxylate **47** for $PdCl_2$ and hydrolysis gives the palladium dithiocarboxylato complex **48** (Scheme 13) [97].

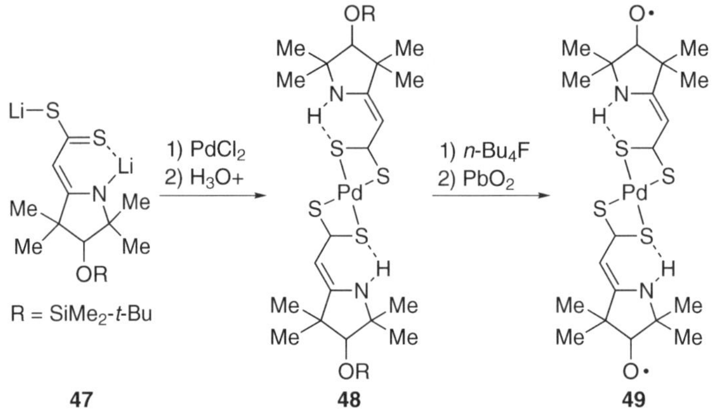

Scheme 13　Formation of palladium dithiocarboxylato complexes with a hydrogen bond

The dithiocarboxylato groups in **48** are not only bidentate ligands, but they also hydrogen-bond with the amino proton in the vicinity of each thiocarbonyl group. The three ring systems, the chelate and the six and five membered (imidazole) rings, are arranged in an almost planar fashion in **48**. The diradical derivative **49** shows strong antiferromagnetic coupling between the two spins at low temperatures. Other nickel, palladium, platinum complexes bearing dithiocarboxylato ligands have also been studied spectroscopically [98–103].

2.4.9
Group 11 Element Compounds (Cu, Ag, Au)

Synthesis of $M(Me_5C_5CSS)(R_3P)$ (M=Au, Ag, Cu) **50** is achieved from $Li(Me_5$-$C_5CSS)$ and $MCl(PR_3)_n$ (Fig. 12) [104]. Arylcopper(I) compounds ArCu (R=Ph, $3\text{-}MeC_6H_4$, $2\text{-}MeC_6H_4$) directly react with CS_2 in the presence of Ph_3P and undergo CS_2 insertion to form the corresponding dithiocarboxylato complexes $Cu(ArCSS)(Ph_3P)_2$ **51** [105]. Reaction of these complexes with MeI affords ArCSSMe. The dithiocarboxylato complexes of copper tend to form oligomers with various geometries around the copper atoms [106, 107]. Although $Cu(\eta^2$-

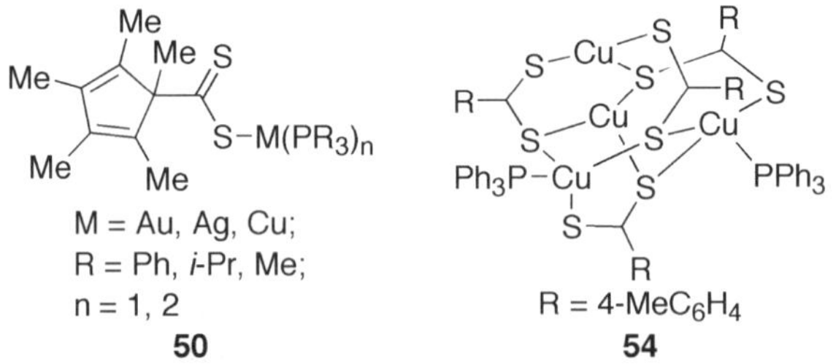

Fig. 12　Group 11 element compounds bearing dithiocarboxylato ligands

PhCSS)(Ph$_3$P)$_2$ **52** [105] and Cu(η^2-MeCSS)(Ph$_3$P)$_2$ **53** [108] have monomeric structures, the *p*-tolylcopper complex [{Cu(4-MeC$_6$H$_4$CSS)}$_4$(Ph$_3$P)$_2$] **54**, which is synthesized by the reaction of [{Cu(4-MeC$_6$H$_4$CSSS)}$_4$] with Ph$_3$P, has a tetrameric structure [31]. Four dithiocarboxylato ligands are bound to four copper atoms through an intricate bonding system, as each ligand is tridentate with one sulfur bound to one copper atom and the other to two atoms, so that each copper atom is bound to three sulfurs and is involved in three bridges. The reaction mechanism involves the desulfurization of the trithioperbenzoate ligand.

Silver dithiocarboxylato complexes also tend to form dimers, polymers and clusters with various geometries around the silver atoms [109–111]. For example, in the tetranuclear silver clusters **55** and **56** bearing dithiocarboxylato ligands, four silver atoms are in a roughly square-planar array and a highly distorted tetrahedral array, respectively (Fig. 13) [32]. In these complexes, each sulfur atom is attached to one or two silver atoms, respectively.

In the binuclear gold complex bearing a dithiobenzoato ligand **57**, the coordination geometries of the two gold centers contrast with those of the above silver and copper complexes (Fig. 14) [112]. The dithiobenzoato ligand coordinates to only one gold center while two dppm bridge the two gold centers. In the mononuclear gold complex **58**, two PPh$_2$C(PPh$_2$Me)CSS ligands act as bidentate *P,S*-chelates to form an almost square-planar geometry around the gold [113].

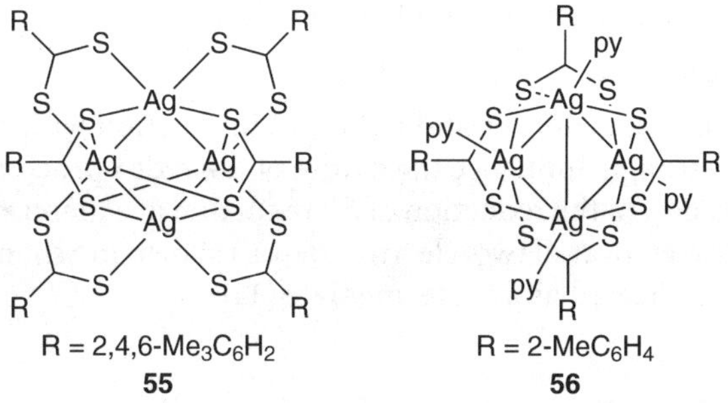

Fig. 13 Tetranuclear silver clusters bearing dithiocarboxylato ligands

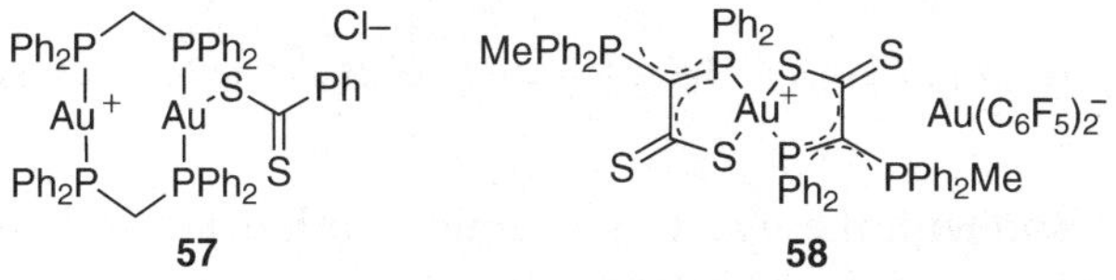

Fig. 14 Gold complexes bearing dithiocarboxylato ligands

2.4.10
Group 12 Element Compounds (Zn, Cd, Hg)

The zinc(II) complex **59** was obtained by the reaction of $Zn(MeCN)_4(BF_4)_2$ with the corresponding sodium salt of dithiocarboxylic acid and subsequent recrystallization from pyridine (Fig. 15) [114]. The crystal structure was shown to be a distorted trigonal bipyramidal geometry with one sulfur atom from each of the dithiocarboxylato ligands occupying the axial sites. In the absence of donor molecules such as pyridine, one sulfur atom from the dithiocarboxylato ligand coordinates to another zinc center to form a dimer-like Zn(II) complex **60** in the crystalline state [115]. EXAFS study indicates that **60** retains the dimeric nature in the mesophase.

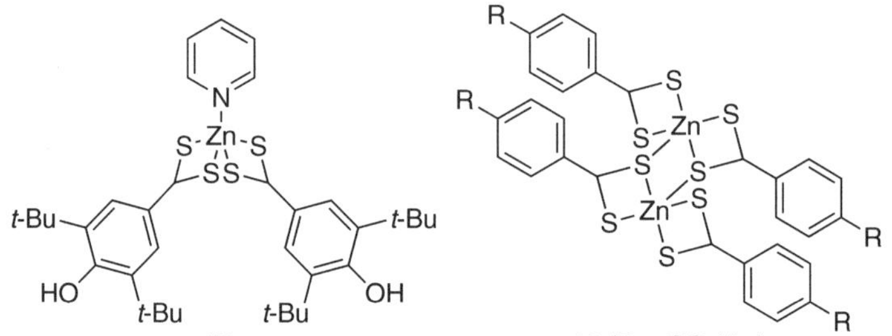

Fig. 15 Zinc complexes bearing dithiocarboxylato ligands

The lability of $Hg(acdc)_2$ **61** and the rapid interaction with elemental mercury electrode strongly influence the nature of the redox processes observed at mercury electrodes. In the reduction of **61**, reduction at a mercury electrode occurs according to an overall two-electron step, as shown in Scheme 14, although mercury(I) is implicated as an intermediate [116].

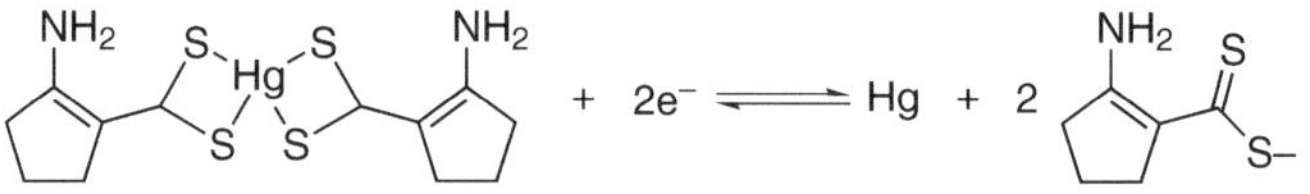

Scheme 14 The redox processes of $Hg(acdc)_2$ observed at a mercury electrode

In comparison with these results, cadmium dithiocarboxylato complexes have been explored much less extensively [48].

2.4.11
Lanthanoid and Actinoid Compounds (La, Ce, Sm, Th, U)

There have only been a few dithiocarboxylato complexes of lanthanoid and actinoid atoms. Only the samarium and uranium complexes have been crystallographically analyzed among such complexes, although the formations and spectral studies of a few lanthanum [117], cerium [48, 117] and thorium [117, 118] complexes have been reported.

The dithiocarboxylato samarium complex **62** was characterized crystallographically to have octacoordination and a distorted dodecahedral geometry where the two dithiocarboxylato ligands are located in the same plane (Fig. 16) [119]. The complex **62** was found to react readily with methyl iodide and benzoyl chloride to give the corresponding dithioesters, respectively, indicating that the complex **62** exhibits the same reaction mode with the corresponding alkali salts. Another samarium complex **63** was synthesized by treatment of CS_2 with $SmCp_2^*(\eta^3\text{-}CH_2CHCH_2)$ [120]. The allyldithiocarboxylato complex **63** slowly isomerizes to **64**. Both sulfur atoms coordinate to the samarium center.

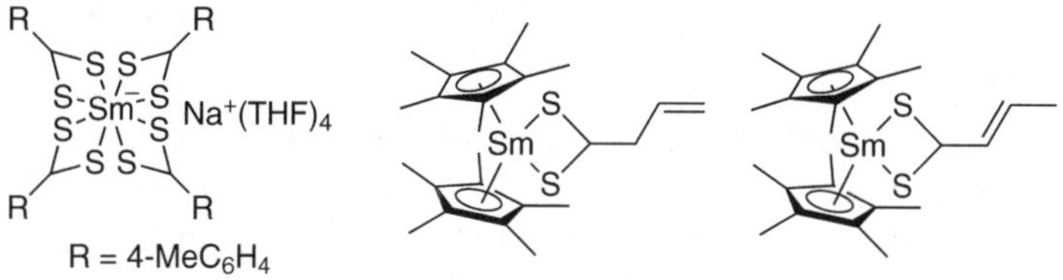

Fig. 16 Samarium complexes bearing dithiocarboxylato ligands

There have been a few uranium complexes **65, 66** in which the uranium(VI) atom is in a heptacoordinate and pentagonal-bipyramidal environment (Fig. 17) [121, 122]. The linear uranyl group is perpendicular to the equatorial plane in which four sulfur atoms of two dithioacetato groups and the oxygen atom of the ligand occupy the corners of an irregular pentagon in **65**.

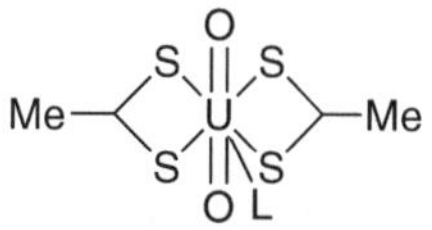

65 (L = Ph₃PO)
66 (L = Ph₃AsO)

Fig. 17 Uranium complexes bearing dithioacetato ligands

2.5 Main Group Element Compounds

2.5.1 Group 13 Element Compounds (B)

The reaction of amino-9-fluorenylideneborane **67** with $[FeCp(CO)_2]_2CS_2$ gives the corresponding [2+2] cycloadduct **68** with CS_2 (Scheme 15) [123]. Both the dithiocarboxy group and the boron atom are fixed in the four-membered ring of **68**.

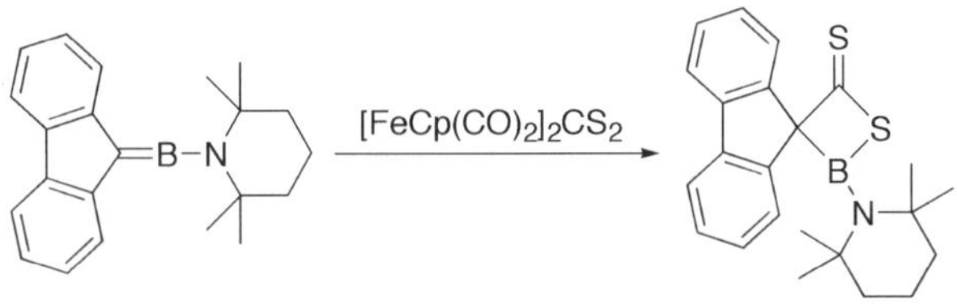

67 **68**

Scheme 15 Formation of the [2+2] cycloadduct of amino-9-fluorenylideneborane with CS_2

Other boryl esters of dithiocarboxylic acids **69** are obtained by sulfurization of the corresponding carboxylic acids, esters or amides with organoboron sulfides such as $(9\text{-BBN})_2S$ and bis(1,5-cyclooctanediylboryl) sulfide (Scheme 16) [124, 125]. Methanolysis of the boryl esters **69** gives the corresponding dithiocarboxylic acids.

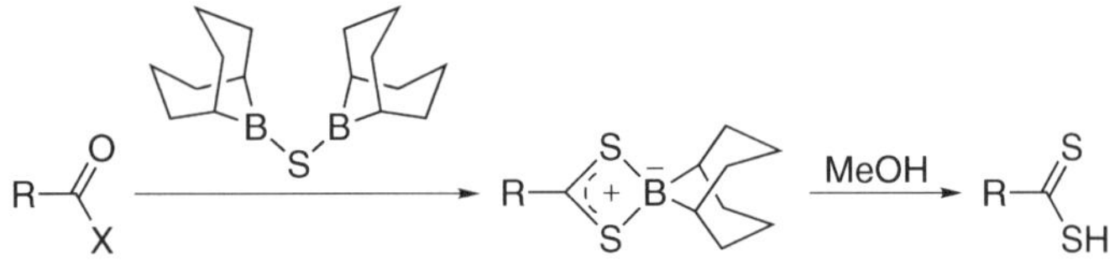

R = alkyl, aryl
X = OH, OR, NR'$_2$

69

Scheme 16 Formation of boryl esters of dithiocarboxylic acids by sulfurization of the corresponding carboxylic acids, esters or amides with an organoboron sulfide

The dithiocarboxylate salts of other Group 13 elements have not been studied extensively in recent years, although a few spectroscopic studies have been reported [126, 127].

2.5.2
Group 14 Element Compounds (Si, Ge, Sn, Pb)

Although CS_2 insertion into the metal-carbon bond is rarely used to synthesize dithiocarboxylate salts of Group 14 elements, divalent tin salts of the dithiocarboxylates **70** and **71** are synthesized by this method (Fig. 18). They are obtained by the stepwise insertion of CS_2 into two Sn–C bonds of a diarylstannylene, Sn-$[2,4,6-(t\text{-Bu})_3C_6H_2]_2$ [128]. Reaction of another stannylene Sn$\{2,4,6-[(Me_3Si)_2\text{-}CH]_3C_6H_2\}[2,4,6-(i\text{-Pr})_3C_6H_2]$ with CS_2 produces the stannylene-CS_2 ylide **72**, which is trapped by methyl acrylate to afford cyclic stannyl dithiocarboxylate **73** (Scheme 17) [129].

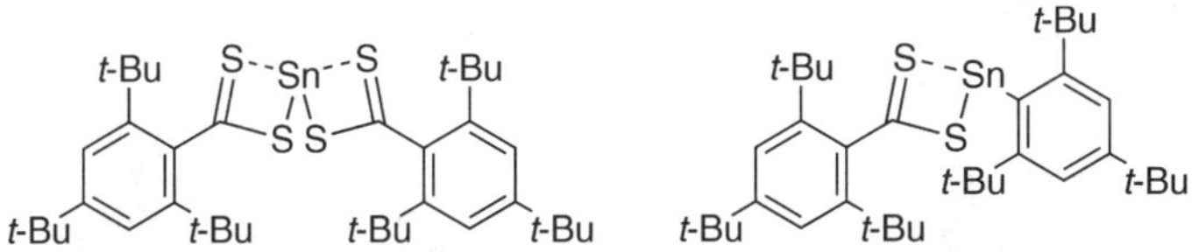

Fig. 18 Divalent tin compounds bearing bulky dithiocarboxylate ligands

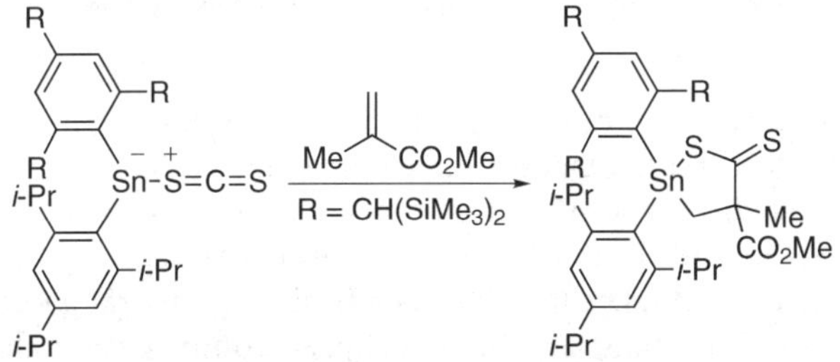

Scheme 17 Formation of a cyclic stannyl dithiocarboxylate from a stannylene-CS_2 ylide

A series of tetravalent Group 14 element compounds bearing one, two, or three dithiocarboxylate ligands **74** have been synthesized by the stoichiometric reactions of piperidinium or triethylammonium dithiocarboxylates with the corresponding chlorosilanes, chlorogermanes, chlorostannanes, and chloroplumbanes, respectively (Scheme 18) [7–10, 130]. The intramolecular weak nonbonded interactions between the thiocarbonyl sulfur and the central Group 14 elements (Ge, Sn, Pb) have been confirmed by X-ray crystallographic analysis and ^{13}C NMR, IR, and UV-Vis spectroscopy [10]. Based on the intramolecular distances between the thiocarbonyl sulfur and Group 14 elements, the affinities of the Group 14 elements are deduced to decrease in the order Sn>Pb≥Ge>Si>C.

A relatively strong Sn–S interaction is also found in tin compounds bearing two or three dithiocarboxylate ligands. The stannyl bis(dithiocarboxylates) **75** and **76** have been shown to take a distorted octahedral geometry, where the two

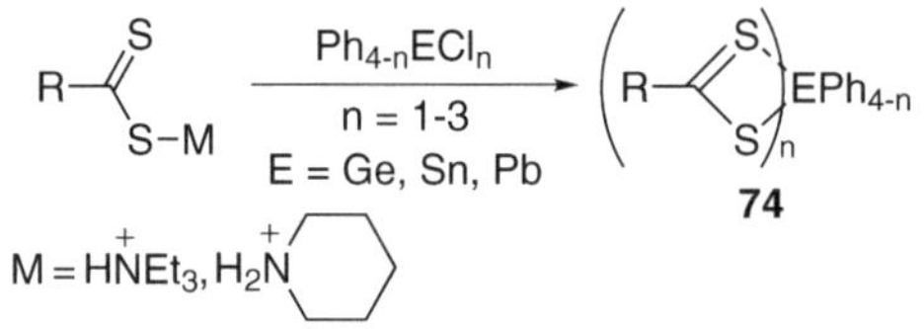

Scheme 18 Syntheses of a series of tetravalent Group 14 element compounds bearing one, two, or three dithiocarboxylate ligands

dithiocarboxylate ligands are bound to the tin atom as a bidentate ligand, and the stannyl tris(dithiocarboxylate) **77** exhibits a heptacoordinate pentagonal bipyramidal structure (Fig. 19) [10, 11, 131, 132].

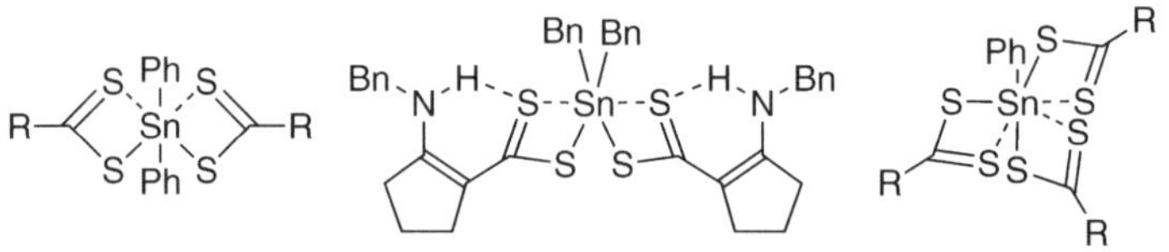

75 (R = 4-MeC$_6$H$_4$)   **76**   **77** (R = 2-MeC$_6$H$_4$)

Fig. 19 Tetravalent tin compounds bearing dithiocarboxylate ligands

The disilane bearing four mesityldithiocarboxylate ligands **78** has been reported to be the first heptacoordinate disilane (Fig. 20) [12]. For one silicon atom of **78**, three thiocarbonyl sulfur atoms are closely located on the opposite sides of two Si–S bonds or one Si–Si bond, respectively, giving heptacoordination to the silicon atom. Disilane **78** has a basic tetrahedral geometry around each silicon, and maintains an almost eclipsed conformation along the Si–Si bond axis. Such a conformation and the high coordination states of the silicon atoms in **78** are attributable to the interaction of the lone pair of each thiocarbonyl sulfur atom with the corresponding σ* orbitals of the Si–S or Si–Si bonds on their opposite sides [133].

Among the salts of Group 14 elements, silicon compounds are especially sensitive to moisture and they easily react with water and ethanol at room temperature to give the corresponding dithiocarboxylic acids [7, 12, 13]. In contrast,

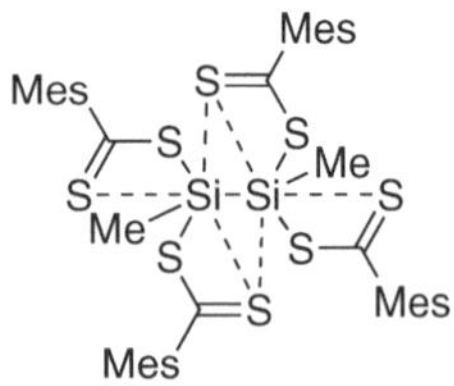

78 (Mes = 2,4,6-Me$_3$C$_6$H$_2$)

Fig. 20 Disilane bearing four dithiocarboxylate ligands

stannyl dithiocarboxylate **79** is not reactive toward methanol, even at reflux temperature, whereas it readily reacts with primary and secondary amines at room temperature to give the corresponding thioamides and bis(triphenyl-stannyl) sulfide (Scheme 19) [8]. The reaction of **79** with sodium methoxide gives the sodium dithiobenzoate **2**, indicating an attack on the tin atom, while an attack on the thiocarbonyl carbon atom is suggested in the similar reaction of plumbyl dithiocarboxylate **81** [9].

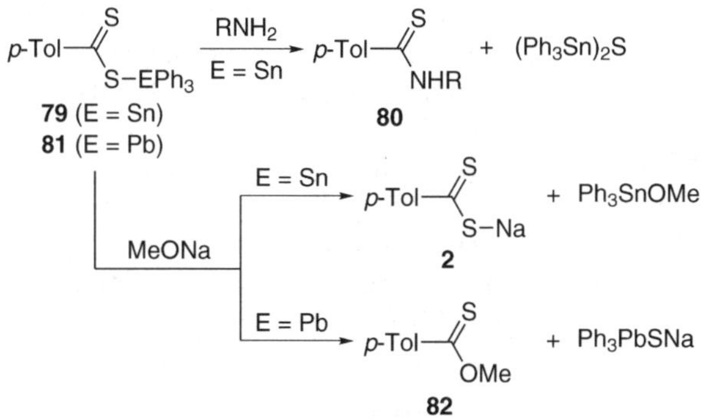

Scheme 19 Reactions of stannyl and plumbyl dithiocarboxylates with amines and sodium methoxide

2.5.3
Group 15 Element Compounds (N, P, As, Sb)

Ammonium salts of the dithiocarboxylate **83**, prepared by treating the corresponding dithiocarboxylic acid with amines, have been utilized to synthesize both dithiocarboxylic esters [134] and the salts of other elements [14, 15, 59, 135, 136]. Tertiary and secondary amines such as triethylamine and piperidine are usually used to prepare these ammonium salts (Scheme 20) [137]. The quaternary tetramethylammonium salts of dithiobenzoic acids **84** and **85** are synthesized by treatment of the corresponding sodium salts, **2** and **86**, respectively, with tetramethylammonium chloride (Scheme 21). Their crystal structures have been elucidated to show monomeric structures in which the tetramethylammonium cation is located outside the plane of the dithiocarboxylate group [38]. The almost equal C–S bond lengths indicate the delocalization of negative charge on the dithiocarboxyl group.

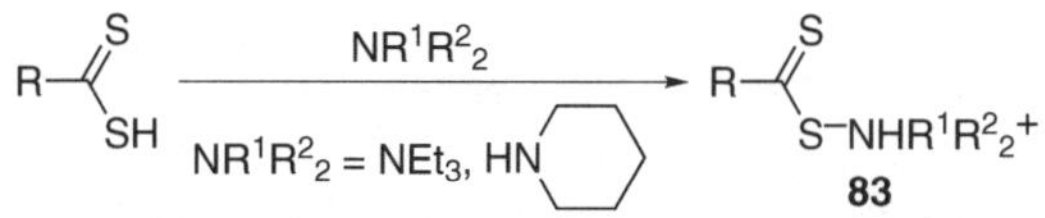

Scheme 20 Syntheses of ammonium dithiocarboxylates from the corresponding dithiocarboxylic acids

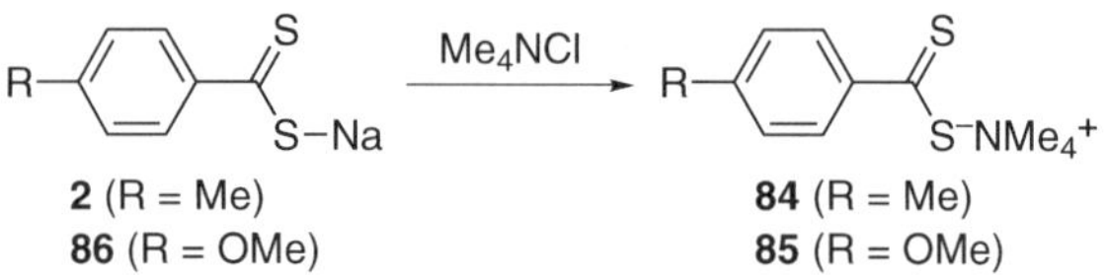

Scheme 21 Syntheses of tetramethylammonium salts of dithiobenzoic acids

There are some reports of syntheses and reactivities of inner salts of dithio-carboxylate, such as the bis(*N,N*-substituted amino)carbenium dithiocarboxylates **87** [138–143]. The carbenium and dithiocarboxylate moieties are nearly perpendicular to each other in a quasi-plane [19, 20, 144]. Its sulfur atom reacts not only with methyl iodide to give carbenium iodide **88** [145], but also with organometallic reagents RM (M=Li, MgX) to give thiolates **89** [146], revealing a unique ability to serve both as a nucleophile and as an electrophile (Scheme 22). Furthermore, **87** is *S*-iminated by [*N*-(*p*-tolylsulfonyl)imino]phenyliodinane to provide a novel inner salt **90** that formally possesses a thione-*S*-imide structure **90'** as one of its canonical structures [143].

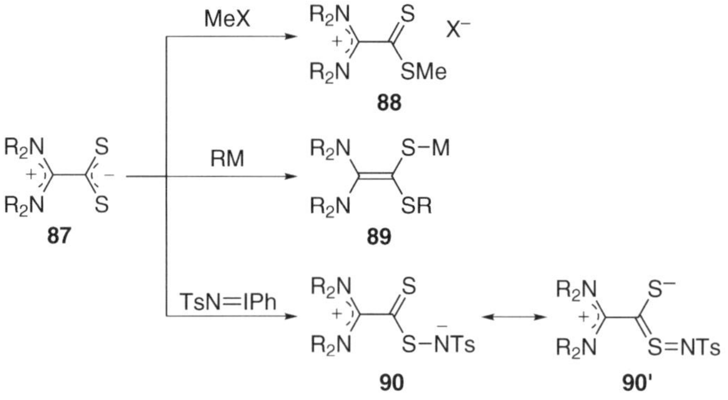

Scheme 22 Some reactions of bis(*N,N*-substituted amino)carbenium dithiocarboxylates

Treatment of acyl chlorides with a 2-sulfenyl-2-thioxo-1,3,2-dioxaphos-phinane **91** gives the corresponding thioacyl dithiophosphates **92** (Scheme 23) [17, 18]. The dithiophosphates **92** are useful thioacylating reagents [14, 147–149].

A series of thioacylsulfenylphosphines **93** [15], -arsines **94** [16], -stibines **95** [150], and -bismuthines **96** have been synthesized by treatment of piperidinium

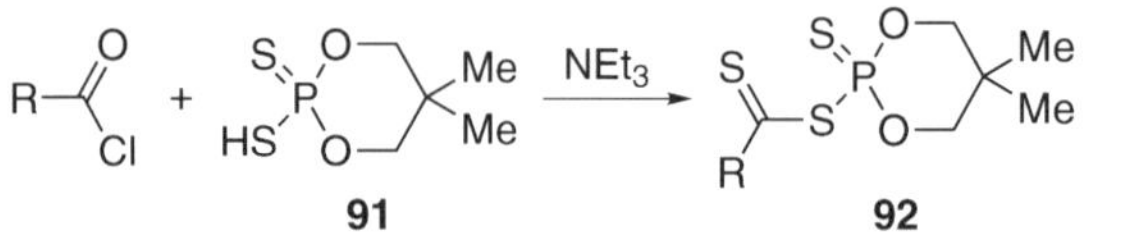

Scheme 23 Formation of thioacyl dithiophosphates

salts or sodium salts of dithiocarboxylates with halophosphines and haloarsines, respectively (Scheme 24). The ^{31}P NMR spectra of **93** show upfield shifts as the number of dithiocarboxylate ligands is increased. The arsenic compounds **94** display intramolecular interactions between the thiocarbonyl sulfur and the arsenic atoms. Such interactions are supported by natural bond order (NBO) analyses performed on the model compounds. Reaction of the phenylbis-(dithiocarboxy)arsine **97** with piperidine gives the novel compounds **98** and **99** (Scheme 25).

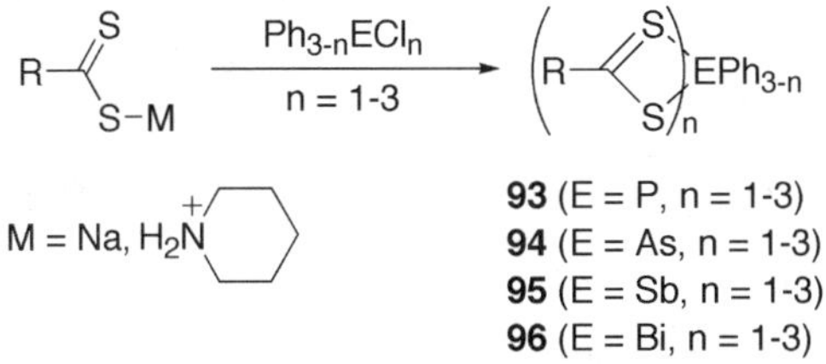

R
S
S–M
$Ph_{3-n}ECl_n$
n = 1-3
EPh_{3-n}
M = Na, H_2N^+
93 (E = P, n = 1-3)
94 (E = As, n = 1-3)
95 (E = Sb, n = 1-3)
96 (E = Bi, n = 1-3)

Scheme 24 Syntheses of a series of thioacylsulfenylphosphines, -arsines, -stibines, and -bismuthines

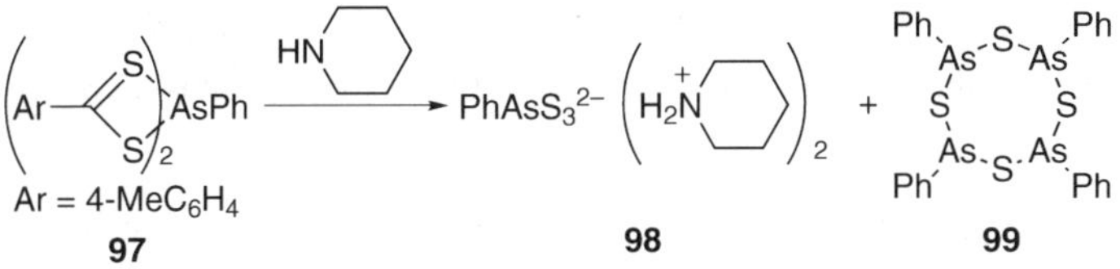

Ar
AsPh
HN
$PhAsS_3^{2-}$
H_2N^+
Ph
As
S
Ar = 4-MeC_6H_4
97
98
99

Scheme 25 Reaction of a phenylbis(dithiocarboxy)arsine with piperidine

2.5.4
Group 16 Element Compounds (O, S, Se, Te)

Many dithiocarboxylate compounds of sulfur have been reported. Bis(thioacyl) disulfides **100**, dimers of dithiocarboxylate ligands, are easily synthesized by oxidative dimerization of the corresponding dithiocarboxylic acids (Scheme 26) [151–154]. Their reactivities toward cycloaddition reactions and S–S bond cleavage have been investigated [155–157]. Bis(thioacyl) trisulfides **101** and tetrasulfides **102** are also prepared by treating the dithiocarboxylic acid with SCl_2 and S_2Cl_2, respectively [135].

For cyclic compounds, cyclic polysulfides **103** which contain six sulfur atoms in a ring have been reported (Fig. 21) [158]. Other heterocyclic compounds bearing dithiocarboxylate units in the ring such as 1,2-dithiole-3-thiones **104** [159, 160] and 5*H*-1,2,3-dithiazole-5-thiones **105** have been studied [161–165]. Some 3*H*-1,2-dithiol-3-thiones are found in cruciferous vegetables, and a structurally similar synthetic dithiolthione **106** called oltipraz has attracted wide interest [166–168]. Oltipraz **106** has been shown to inhibit chemically-induced

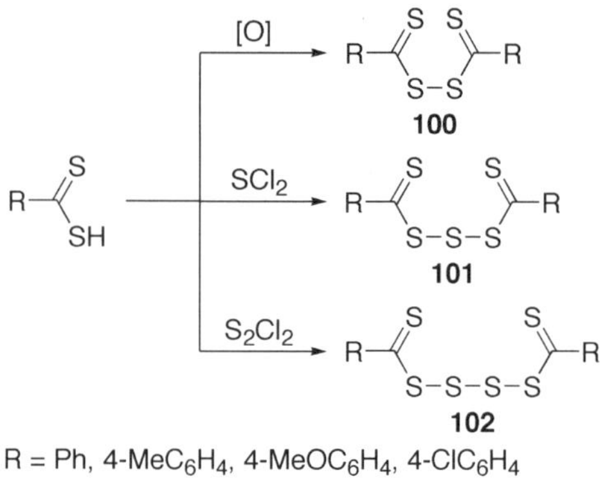
[O]
SCl2
S2Cl2
R = Ph, 4-MeC6H4, 4-MeOC6H4, 4-ClC6H4

100

101

102

Scheme 26 Syntheses of bis(thioacyl)disulfides, -trisulfides, and tetrasulfides from dithiocarboxylic acids

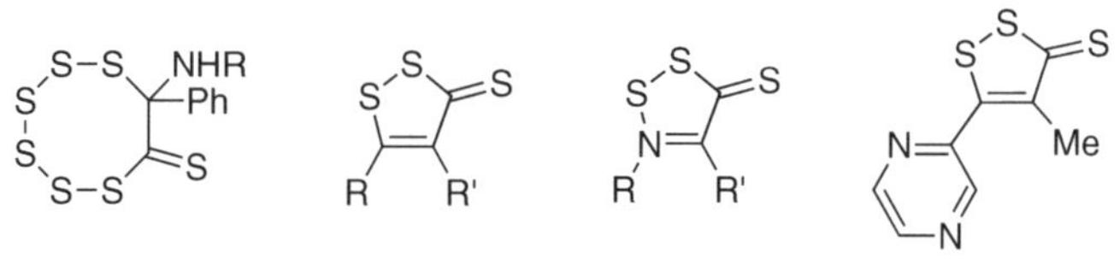

103 (R = alkyl) **104** **105** **106**

Fig. 21 Some heterocyclic compounds bearing dithiocarboxylate units in the ring

carcinogenesis in a variety of animal models. It has a unique ability to protect several target organs from structurally diverse carcinogens. Preclinical and clinical data continue to support the development of oltipraz as a chemopreventive reagent for clinical usage.

There are a few reports of syntheses of thioacylsulfenyl chalcogen compounds of oxygen, selenium and tellurium, such as the thioacylsulfenate ester **107** [169] and its heavier chalcogen derivatives **108** [169], the haloselenium- and halotellurium dithiocarboxylates **109** [170, 171], and the selenium- and tellurium bis(dithiocarboxylates) **110** (Fig. 22) [171, 172]. The chemistries of oxygen, selenium, and tellurium compounds bearing dithiocarboxylate ligands have not been studied as closely as those of the sulfur compounds.

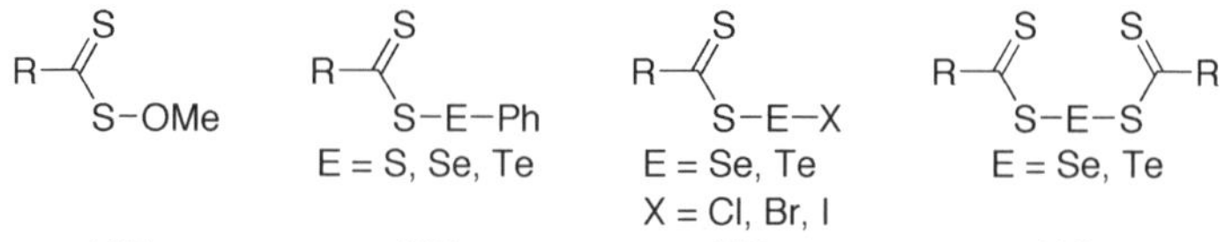
E = S, Se, Te
E = Se, Te
X = Cl, Br, I
E = Se, Te

107 **108** **109** **110**

Fig. 22 Some thioacylsulfenyl chalcogen compounds

2.5.5
Group 17 Element Compounds (Br, I)

On the one hand, the reaction of imidazolium-2-dithiocarboxylate **111** with bromine gives the corresponding cationic thioacylsulfenyl bromide **112** [173]. On the other hand, treatment of **111** with an equimolar amount of iodine gives the charge transfer complex **113**. Two iodine atoms and a thiolate sulfur atom are arranged in almost a linear fashion in **113**. Further treatment of **113** with an excess amount of iodine causes oxidative coupling, affording the dicationic disulfide **114** (Scheme 27).

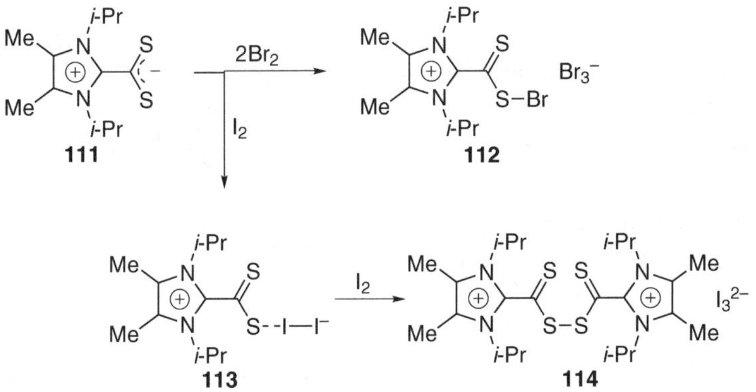

Scheme 27   Reactions of an imidazolium-2-dithiocarboxylate with bromine and iodine

2.6
MMX-Type Transition Metal Complexes

Halogen-bridged (X) one-dimensional (1-D) binuclear transition-metal (MM) complexes – so-called "1-D MMX chains" – have attracted much attention because of their interesting physical properties, such as metallic conduction, metal-insulator transition with 1-D charge ordering, and so on [174]. One of the MMX complexes **115** bears four bridging dithiocarboxylato ligands on two metal atoms (Fig. 23). The electronic ground states of the MMX complexes **115** can be controlled by varying the compositions of the substituents, the metal ions and the bridging halogen ions, and also the light, heat, and pressure. There have been several reports on MMX complexes **115** with platinum and alkyldithiocarboxylato ligands, where the alkyl groups are methyl [174–176], ethyl [177–179], propyl [180], and butyl [179, 181] groups with counterions such as bromide, iodide, and perchlorate [182]. The complex $Pt_2(MeCSS)_4I$ **116**, which is synthesized by oxidizing the binuclear platinum complex **117** with iodine [176, 183], has attracted particular attention among these 1-D MMX complexes. Complex **116** displays metallic conduction around room temperature; the first observation of this for a 1-D MMX complex.

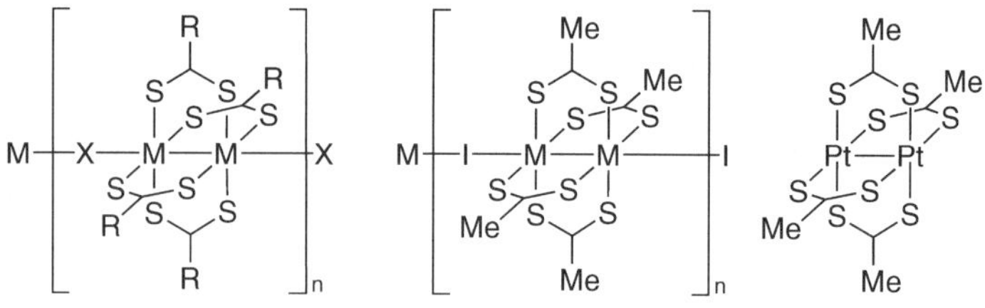

115   **116** (M = Pt)   **117**
118 (M = Ni)

Fig. 23 1-D MMX complexes and a dinuclear platinum complex bearing dithiocarboxylato ligands

An important feature of MMX compounds is an increase in the internal degrees of freedom upon introducing a binuclear metal unit into the mixed-valence linear chain system [174, 179]. This property enables a variety of electronic structures, represented by the extreme valence-ordering states shown below:

a. average valence (AV) state: $-M^{2.5+}-M^{2.5+}-X^--M^{2.5+}-M^{2.5+}-X^--$
b. charge density wave (CDW) state: $-M^{2+}-M^{2+}-X^--M^{3+}+M^{3+}-X^--$
c. charge-polarization (CP) state: $-M^{2+}-M^{3+}-X^--M^{2+}-M^{3+}-X^--$
d. alternate charge-polarization (ACP) state: $-M^{2+}-M^{3+}-X^--M^{3+}-M^{2+}-X^--$

In the case of the CP state, where an unpaired electron exists at the M^{3+} site per dimer unit, no cell doubling occurs and so either a Mott-Hubbard insulator or a one-dimensional metal is expected. Conversely, the electronic structures of the CDW state and the ACP state are regarded as Peierls and spin-Peierls states, respectively [178, 184, 185]. The four states of the MMX complexes are distinguished by the valence states of the metal and the bond distances in the crystal structure.

In **116**, the dithioacetato ligands and the iodines bridge two platinum atoms [176, 186, 187]. Each hexacoordinate platinum atom is surrounded by four sulfur atoms in an approximately square-planar arrangement, and by an iodine and another platinum atom at apical positions, respectively, to form an almost octahedral geometry. Therefore the Pt–Pt–I– unit constitutes the neutral 1-D chain structure. This complex has a helical arrangement of four dithiocarboxylato ligand planes around the central Pt–Pt axis. The C–S bonds are completely delocalized. The lattice parameters and bond lengths of Pt–Pt and Pt–I differ depending on the temperature, especially at low temperatures [187]. The electrical resistivity parallel to the chain axis (*b*) also displays a dependence on temperature [185].

The conductivity at room temperature ($\sigma=13$ S cm^{-1}) is much higher than that of other MMX chain compounds. The MMX system of **116** is found to exhibit metallic conduction between 300 and 340 K. The metal-semiconductor transition is observed at 300 K, and then below 300 K the resistivity increases slowly with temperature down to 150 K. Above 340 K, the resistivity decreases with temperature up to 350 K, which is correlated with the first-order phase transition observed at 340 K in X-ray analysis and DSC measurements.

Various spectroscopic data reveal that the complex **116** is in a one-dimensional metallic phase in the AV state above the metal-semiconductor transition temperature, 300 K. Variable temperature measurements of **116** show that it is in a semiconducting phase with the CP state between 90 and 300 K, and in an insulating phase with the ACP state below 90 K with the bridging I^- ions at the midpoints of two $Pt_2(MeCSS)_4$ dimers [176, 184, 185, 187, 188]. The metal-insulator transition that occurs at 300 K originates from the Mott transition. Low-temperature X-ray single-crystal analysis [187] suggests that a slight increase in the crystal values below 90 K might relate to the expected spin-Peierls distortion associated with MeCSS ligand twisting. Quantum chemical calculations [189, 190] suggest that Peierls distortion in the $Pt_2(MeCSS)_4I$ chain can be relayed to a twisting distortion of MeCSS ligand, resulting in the CDW state. Other spectroscopic methods, such as IR, Raman, polarized Raman, UV-Vis-NIR, polarized reflectance, Mössbauer, and XPS, as well as electrical conductivity (or resistivity), magnetic susceptibility, thermoelectric power, molar differential scanning calorimetry, and heat capacity measurements also show that the valence-ordering models of **116** change depending on the temperature [175, 185, 191–193].

The nickel complex $Ni_2(MeCSS)_4I$ **118** has also been studied [194–198].

2.7
Metal Complexes Bearing Di-, Tri-, or Tetra-Thiooxalato Ligands

2.7.1
Metal Complexes Bearing a Teterathiooxalato Ligand

When there is a donor group such as an amino, oxo and phosphino group in the dithiocarboxylato ligand, the metal center often forms a chelate to the metal center with intramolecular coordination of the donor group instead of the thiocarbonyl sulfur in the dithiocarboxylato complexes as described in the previous sections. Since a tetrathiooxalato (tto, $C_2S_4^{2-}$) ligand has two dithiocarboxylato moieties, it potentially exhibits some coordination modes. Almost all tto complexes, however, show the side-on/side-on coordination mode depicted in **119**, where each of two metals is bound to one thiolato moiety together with intramolecular coordination to a thiocarbonyl sulfur atom of the other dithiocarboxylato moiety of the same tto ligand (Fig. 24). Since tto^{2-} and divalent

119   **120**

Fig. 24 Transition metal complexes bearing a tetrathiooxalato ligand metal ions such as Pd^{2+}, Cu^{2+}, Co^{2+}, and Ni^{2+} easily form polymers **120** through the bridging tto ligand, end-capping (by coordinating other ligands to the metal) is usually used to obtain monomeric tto complexes [199, 200].

A major synthetic route to a tto complex is salt metathesis using tto salts such as $(Et_4N)_2$tto [24, 28]. For example, the nickel complex **121** bearing the dmit ligand (dmit=1,3-dithiole-2-thione-4,5-dithiolato, $C_3S_5^{2-}$) is synthesized by reaction of a stoichiometric ratio of tetrabutylammonium bromide, Cs_2(dmit) and $(Et_4N)_2$tto, followed by addition of $NiCl_2(H_2O)_6$ [25]. Treatment of the Ni(dmit)$_2$ complex **122** with $Na_2S_2O_4$ also gave **121** (Scheme 28) [21]. The conversion of $dmit^{2-}$ to tto^{2-} is realized for the first time in the latter method.

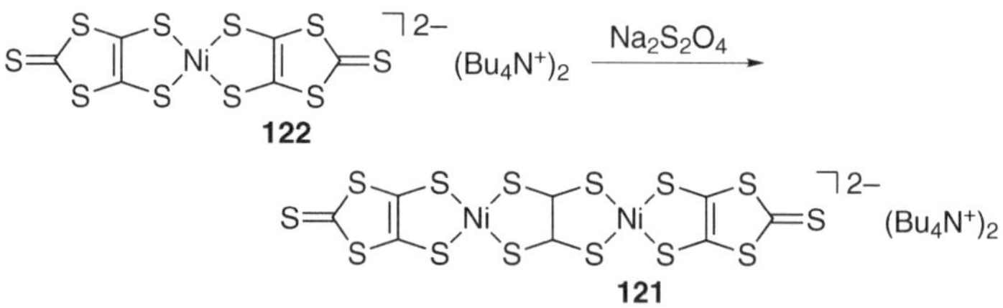

Scheme 28 Synthesis of a nickel complex bearing both dmit and tto ligands

The tto unit in complex **119** usually exists in a planar conformation. All the C–S and C=S bond lengths of **119** are shorter than the sum of the covalent radii in the C–S single bond and longer than those of the C=S double bond. These structural features indicate delocalization of the π-system. Comparing the structures of several tto complexes with those of oxalato complexes suggests that tto is more capable of delocalizing electrons from the metal than related C_2O_4 bridges are. As a result of delocalization, many complexes bearing a tto ligand have been reported to show short C–C bond lengths. When the C–C bond length is closer to that of the C=C double-bond length rather than that of a C–C single-bond in such complexes, the bridging ligand should be regarded not as a tto ligand, but an ethylenetetrathiolato ligand, unless other spectroscopic data clearly indicate the oxidation state of the metal.

Chemical oxidation of the ethylenetetrathiolato bridged complexes gives the corresponding tto complexes, as indicated by electrochemical oxidation. For example, the bimetallic ethylenetetrathiolato complex **123** is oxidized by $AgBF_4$ in the presence of $P(OMe)_3$ to give the corresponding tto-bridged complex **124** (Scheme 29) [23].

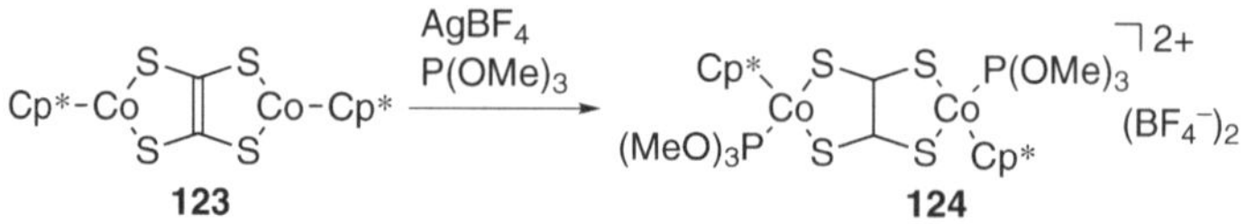

Scheme 29 Formation of a dicationic cobalt tto complex

Since transition metal complexes **122** with dmit ligands have received significant attention due to their metal-like electronic properties, incorporation of more sulfur atoms into the periphery of the structure is expected to stabilize the interactions via direct S···S overlap on the adjacent inter- and intrastack ions, and to extend the delocalization. Extended conjugated π-systems are shown in **125** where chalcogen-rich 1,2-dithiolato or 1,2-diselenolato ligands such as dmit are used as the capping ligands, and tto is used as the bridging ligand surrounding the square-planar coordinating metals [201]. The π-system of the tto bridging ligand is highly delocalized, resulting in a planar structure for most of these complexes, as judged by the C–C and C–S bond lengths of the tto fragment. Some copper complexes such as **126–128** exhibit exceptionally slight tetrahedral distortion from square planar coordination around the coordination spheres of the metal centers [22]. The planar complexes **125** have close-packed and segregated arrangements in the crystal lattice. The intrastack interactions with adjacent stacking units in **125** allow delocalized electronic systems [22, 29, 30]. These structural features suggest that these compounds may find use as conducting materials. Some chalcogen-rich ligands other than dmit have also been used as capping ligands when constructing electrically conducting molecules, such as the 1,2-dicyanoethene-1,2-dithiolato (mnt) ligand, 2-thioxo-1,3-dithiole-4,5-diselenolato (dsit), the 2-oxo-1,3-dithiole-4,5-dithiolato (dmid) ligand, and the cyclic dithiolato ligand $(MeOCO)_2C_2S_2C_2S_2^{2-}$ [22, 26, 202]. The counterions most commonly used for such dianionic tto complexes **125** are ammonium cations, such as tetrabutylammonium.

The insoluble salt formed by the reaction of **121** with (tetrathiafulvalene)$_3$-$(BF_4)_2$ displays a conductivity of $0.4\ S\ cm^{-1}$ at room temperature. A black material obtained by the electrocrystallization of **121** in the presence of Bu_4NBr under a constant current yields an even higher conductivity: $0.5\ S\ cm^{-1}$ [25, 27].

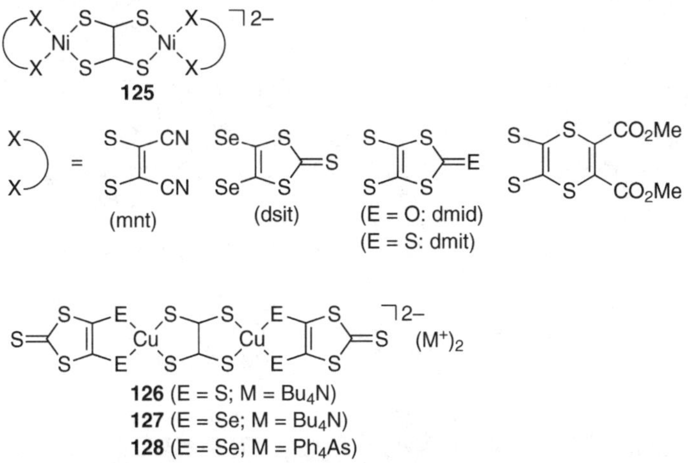

Fig. 25 Several dinuclear nickel and copper complexes bridged by a tetrathiooxalato ligand

2.7.2
Metal Complexes Bearing a Trithiooxalato or a Dithiooxalato Ligand

The multidentate, chalcogen-rich ligands trithiooxalato (trto) and 1,1-dithiooxalato (i-dto) make it possible to construct low-strain five-membered (side-on) or four-membered (end-on) chelate rings for topological reasons. For example, the silver complex **129** bearing a trto ligand represents a side-on/end-on coordination type [203]. Both silver atoms in **129** are bound to sulfur atoms of the bridging ligand. The silver atoms and the trto ligand form a five-membered chelate ring as well as a four-membered chelate ring simultaneously, whereas the oxygen atom is not involved in any coordination. Conversely, the corresponding copper complex **130** shows side-on/side-on coordination similar to the most tto complexes [204]. The silver and copper complexes **131** and **132**, bearing another chalcogen-rich ligand, i-dto, also show the side-on/side-on coordination mode of the bridging ligand [204]. The bridging trto ligand in **130** and the i-dto ligand in **132** are far from planar, in contrast to the most of tto complexes. The coordination modes in **129–133** are explained by MO calculations.

Binuclear bridged copper complexes with capping triphenylphosphine, such as **130** and **132**, are light sensitive in solution [205]. Photolysis of the i-dto complex **132** in dichloromethane gives the binuclear copper complex $(Ph_3P)_2Cu(\mu\text{-}Cl)_2Cu(PPh_3)$, with the total loss of the bridging i-dto ligand, whereas that of the trto complex **130** in pyridine affords the mononuclear copper complex, $(Ph_3P)_2Cu(SH)(py)$. In contrast, the corresponding tto complex **133** remains unchanged.

The *O*-methyl-1,1-dithiocarboxylato (i-dtoMe) ligand which corresponds to the methylated derivative of the i-dto ligand does not behave as a tetradentate ligand, in contrast to the corresponding i-dto complexes. Both the silver and copper complexes **134** and **135** bearing the i-dtoMe ligand form a mononuclear

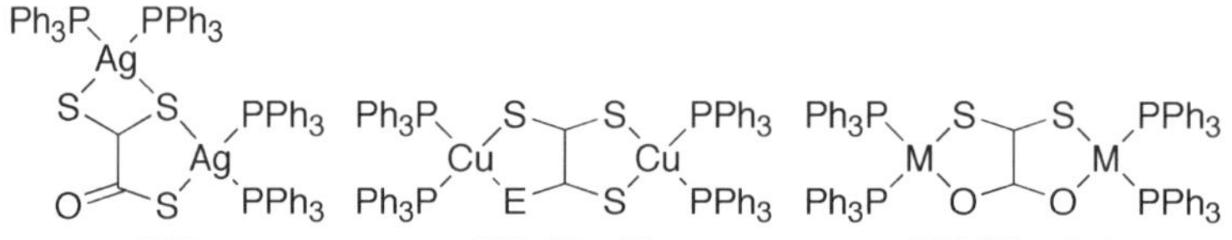

129 **130** (E = O) **133** (E = S) **131** (M = Ag) **132** (M = Cu)

Fig. 26 Some dinuclear silver and copper complexes bridged by a di- or trithiooxalato ligand

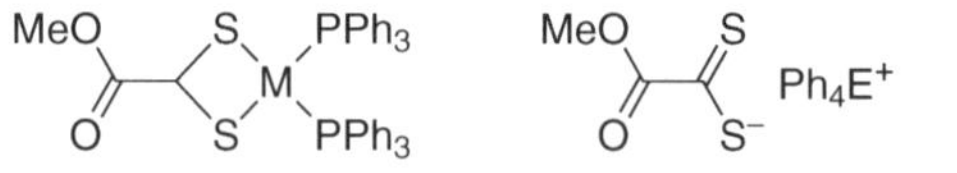

134 (M = Ag) **135** (M = Cu) **136** (E = P) **137** (E = As)

Fig. 27 Some compounds and salts bearing an *O*-methyl-1,1-dithiocarboxylato ligand structure with the end-on (S,S′) coordination mode [206]. ^{13}C NMR spectroscopic studies show that the coordination in the complexes **134** and **135** causes shifts of the ^{13}C signals of the dithiocarboxyl-carbon to higher field than those encountered for the tetraphenylphosphonium and -arsonium salts, **136** and **137**, due to the π-acceptor behavior of this type of ligand [207].

3
Spectroscopic Properties

3.1
^{13}C NMR Spectroscopic Properties

^{13}C NMR data are lacking for many transition metal complexes. In the ^{13}C NMR spectra that we do have for such compounds, the thiocarbonyl carbon nuclei of the dithiocarboxylic acid salts of transition metal elements show diverse chemical shifts, ranging from δ_C=191.8–259.6, except for the case of one example of the η^3-coordinated iron complex **37** (δ_C=129.7) [80, 82]. The chemical shifts from main-group-element compounds similarly range from δ_C=194.8–277.2 [38, 123]. Although the chemical shifts are influenced by the coordination styles of the ligand and the element that the ligand is bound to, the organic substituent (R) on the CS_2 unit of the RCS_2 ligand affects the chemical shift dramatically. For example, while 4-MeOC$_6$H$_4$CS$_2$NMe$_4$ **85** gives δ_C=253.4, piperidinium adamantyldithiocarboxylate and Pr^{i}CS$_2$NMe$_4$ give δ_C=220.2 and 275.4, respectively [38, 124]. According to systematic studies of the ^{13}C=S chemical shifts of germanium, tin, and lead derivatives of mono-, bis-, and tris(4-methoxybenzene-dithiocarboxylate), the ^{13}C=S chemical shifts for aliphatic dithiocarboxylates are smaller than those of aromatic ones (δ_C=225.5–265.3) [10, 130]. Furthermore, the double-bond character of the C=S unit is an important factors of the ^{13}C NMR chemical shift. If the dithiocarboxylate unit takes part in the conjugated system and thiocarbonyl sulfur forms a hydrogen bond, like in **76**, the thiocarbonyl carbon shows an upfield shift (see Sect. 2.5.2) [132].

3.2
IR Spectroscopy Properties

IR spectroscopy provides some information about the coordination state of the dithiocarboxylate ligand to the metal and the contribution of the C=S double-bond character. Whereas the C=S stretch usually ranges from 1050 to 1200 cm^{-1}, the bands due to the symmetric and anti-symmetric stretching bands of the CS_2 moiety are observed at lower frequencies for the dithiocarboxylic acid salts. They are usually observed in the range 800–1250 cm^{-1}, and the $\nu_{as}CS_2$ vibration is higher than the $\nu_{sym}CS_2$ vibration. Assigning them, however, is difficult due to overlaps with other stretching bands in the fingerprint region. Indeed, stretching bands are sometimes assigned incorrectly in some reports.

The difference ($\Delta = \nu_{as}CS_2 - \nu_{sym}CS_2$) between anti-symmetric and symmetric CS_2 stretching bands for dithiocarboxylate ligands can be used as a criterion to distinguish the coordination mode of the ligand. Monodentate ligands give large Δ values, while bidentate ligands give small Δ values [60]. In some studies, however, only one of the bands has been reported as the $\nu_{as}CS_2$ or νC=S vibration. For example, νC=S vibrations at about 1200 cm^{-1} have been reported for Group 16 element compounds [14].

In compounds of main group elements, the band ranges from 870 to 1285 cm^{-1} [6, 102]. In transition metal compounds, it is observed from 810–1385 cm^{-1} [50, 53]. If the C=S double-bond character is low, as seen for a free dithiocarboxylate anion, the stretching shifts to lower wavenumbers [38, 124]. This is also true for complexes where the dithiocarboxylate ligand is conjugated with an α,β-unsaturated double-bond, and/or complexes with a C=S$\cdots$H hydrogen bond [50]. If the C=S double-bond character is high in the dithiocarboxylic acid salt, the νCS vibrations appear at 1100–1200 cm^{-1} [82, 95].

It is worth mentioning that the $\nu_{as}CS_2$ vibrational bands of aliphatic alkali salts appear at a lower frequency than those of aromatic ones, at 900–1050 cm^{-1} [4, 6, 38]. This is in contrast to the $\nu_{as}CO_2$ or $\nu_{as}COS$ vibrations of alkali metal carboxylates and thiocarboxylates.

3.3
UV-Visible Spectroscopic Properties

Dithiocarboxylic acid salts are usually colored. The colors derive from the n-π^* and π-π^* transitions of the dithiocarboxylate ligand. In UV-Visible spectroscopy, characteristic absorption maxima at 250–550 nm are observed in the visible regions of the electronic spectra of compounds of main group elements [14, 169]. The absorptions at longer and shorter wavelengths are apparently due to π-π^* and n-π^* transitions, respectively, of the thiocarbonyl group. Some studies describe the π-π^* transition alone. In some transition metal complexes, such as the MMX-type complexes, metal-to-ligand charge transfer bands are observed in addition to the π-π^* and n-π^* transitions [177, 184, 192].

The alkali salts show two or three band maxima between 250 and 500 nm, indicating that the bands below 350 nm are π-π^* and that the other band over 350 nm is due to n-π^* transitions [4, 6, 171]. These absorption maxima are rather insensitive to the effects of heavy metal atoms, although they are sensitive to the geometry around the metal. Similarly, Group 14 element compounds have two characteristic absorption maxima at 300–320 and 520–530 (aromatic) or 470–490 nm (aliphatic), respectively [7]. The latter n-π^* transitions show a tendency towards a blue shift, while the former π-π^* transitions show a tendency towards a red shift on increasing the atomic number of the element. The order (Si>Ge>Sn>C) of the n-π^* transition may reflect the capacity for $d\pi$-$p\pi$ bonding between the Group 14 element and the sulfur atom.

Finally, in terms of the intensities of molar absorption coefficients, we should mention the n-π^* transitions of the thiocarbonyl group of bis(thioacyl) di- and tri-sulfides. They are observed at 480–490 nm (aliphatic) and 500–530 nm (aromatic), which are unusually high ($\log\varepsilon>3$) for dithiocarboxylate compounds [136].

References

1. Kato S, Murai T (1992) In: Patai S (ed) The chemistry of acid derivatives, supplement B, vol 2, part 1. Wiley, New York, p 803
2. Couconvanis D (1969) In: Lippard SJ (ed) Progress in inorganic chemistry, vol 11. Wiley, New York, p 233
3. Murai T, Kato S (1995) In: Katritzky AR, Meth-Cohn O, Rees CW (eds) Comprehensive organic functional group transformations, vol 5. Pergamon, Oxford, UK, p 545
4. Ball SC, Cragg-Hine I, Davidson MG, Davies RP, Raithby PR, Snaith R (1996) Chem Commun 1581
5. Kato S, Ito K, Hattori R, Mizuta M, Katada T (1978) Z Naturforsch B 33:976
6. Kato S, Yamada S, Goto H, Terashima K, Mizuta M, Katada T (1980) Z Naturforsch B 35:458
7. Kato S, Mizuta M, Ishii Y (1973) J Organomet Chem 55:121
8. Kato S, Kato T, Yamauchi T, Shibahashi Y, Kakuda Y, Mizuta M, Ishii Y (1974) J Organomet Chem 76:215
9. Katada T, Kato S, Mizuta M (1975) Chem Lett 1037
10. Kato S, Tani T, Kitaoka N, Yamada K, Mifune H (2000) J Organomet Chem 611:190
11. Sharma R, Singh Y, Rai AK (2000) Phosphorus Sulfur 166:221
12. Kano N, Nakagawa N, Kawashima T (2001) Angew Chem Int Edit 40:3450
13. Kato S, Chiba S, Goto M, Mizuta M, Ishida M (1982) Synthesis 457
14. Kato S, Goto M, Hattori R, Nishiwaki K, Mizuta M, Ishida M (1985) Chem Ber 118:1668
15. Tani K, Matsuyama K, Kato S, Yamada K, Mifune H (2000) B Chem Soc Jpn 73:1243
16. Tani K, Hanabusa S-i, Kato S, Mutoh S-y, Suzuki S-i, Ishida M (2001) J Chem Soc Dalton 518
17. Doszczak L, Rachon J (2000) Chem Commun 2093
18. Doszczak L, Rachon J (2002) J Chem Soc Perkin T 1 1271
19. Nakayama J, Kitahara T, Sugihara Y, Sakamoto A, Ishii A (2000) J Am Chem Soc 122:9120
20. Fujihara T, Ohba T, Nagasawa A, Nakayama J, Yoza K (2002) Acta Crystallogr C 58:558
21. Bai J-F, Zuo J-L, Shen Z, You X-Z, Fun H-K, Chinnakali K (2000) Inorg Chem 39:1322
22. Pullen AE, Zeltner S, Hoyer E, Abboud KA, Reynolds JR (1996) Inorg Chem 35:4420
23. Guyon F, Jourdain IV, Knorr M, Lucas D, Monzon T, Mugnier Y, Avarvari N, Fourmigué M (2002) Eur J Inorg Chem 2026
24. Holloway GA, Klausmeyer KK, Wilson SR, Rauchfuss TB (2000) Organometallics 19:5370
25. Pullen AE, Olk R-M, Zeltner S, Hoyer E, Abboud KA, Reynolds JR (1997) Inorg Chem 36:958
26. Yang X, Doxsee DD, Rauchfuss TB, Wilson SR (1994) J Chem Soc Chem Commun 821
27. Pullen AE, Zeltner S, Olk R-M, Hoyer E, Abboud KA, Reynolds JR (1997) Inorg Chem 36:4163
28. Holloway GA, Rauchfuss TB (1999) Inorg Chem 38:3018
29. Piotraschke J, Pullen AE, Abboud KA, Reynolds JR (1995) Inorg Chem 34:4011
30. Teschmit G, Strauch P, Barthel A, Reinhold J, Kirmse R (1999) Z Naturforsch B 54:832
31. Manotti Lanfredi AM, Tiripicchio A, Camus A, Marsich N (1989) J Chem Soc Dalton 753
32. Schuerman JA, Fonczek FR, Selbin J (1989) Inorg Chim Acta 160:43
33. Scott F, Kruger GJ, Cronje S, Lombard A, Raubenjeimer HG, Behn R, Rufinska A (1990) Organometallics 9:1071

34. Hughes AK, Malget JM, Goeta AE (2001) J Chem Soc Dalton 1927
35. Coons DE, Laurie JCV, Haltiwanger RC, Rakowski Dubois M (1996) J Am Chem Soc 109:28
36. Coons DE, Haltiwanger RC, Rakowski Dubois M (1987) Organometallics 6:2417
37. Bruce MI, Hambley TW, Snow MR, Swincer AG (1984) J Organomet Chem 273:361
38. Kato S, Kitaoka N, Niyomura O, Kitoh Y, Kanda T, Ebihara M (1999) Inorg Chem 38:496
39. Yordanov ND, Ganeva E, Gochev G (1995) Z Anorg Allg Chem 621:699
40. Ives C, Fillis EL, Hagadorn JR (2003) Dalton Trans 527
41. Houben J, Pohl H (1907) Ber Deut Chem Ges 40:1303
42. Radamas SR, Srinivasan PS, Ramachandran J, Sastry VVS (1983) Synthesis 605
43. Mavellec F, Roucoux A, Noiret N, Moisan A, Patin H, Duatti A (2003) J Labelled Compd Rad 46:319
44. Demaimay F, Noiret N, Roucoux A, Patin H, Bellande E, Pasqualini R, Moisan A (1997) Nucl Med Biol 24:439
45. Hagiwara T, Ishizuka M, Fuchikami T (1998) Nippon Kagaku Kaishi 750
46. Chang C-C, Chen J-H, Srinivas B, Chiang MY, Lee G-H, Peng S-M (1997) Organometallics 16:4980
47. den Haan K, Luinstra, GA, Meetsma A, Teuben JH (1987) Organometallics 6:1509
48. Kato S, Wakamatsu M, Mizuta M (1974) J Organomet Chem 78:405
49. Tyagi SB, Singh B, Kapoor RN (1986) Transit Met Chem 11:74
50. Tyagi SB, Singh B, Kapoor RN (1985) Transit Met Chem 10:73
51. Otero A, Fernández-Baeza J, Antiñolo A, Carrillo-Hermossila F, Tejeda J, Lara-Sánchez A, Sánchez-Barba L, Fernández-Lopez M, Rodriguez AM, López-Solera I (2002) Inorg Chem 41:5193
52. Steudel R, Kustos M, Münchow V, Westphal U (1997) Chem Ber/Recueil 130:757
53. Kim Y-J, Kim Y-I, Choi S-N (2000) Polyhedron 19:2155
54. Santini-Scampucci C, Wilkinson G (1976) J Chem Soc Dalton 807
55. Andras MT, Duraj SA (1993) Inorg Chem 32:2874
56. Duraj SA, Andras MT, Kibala PA (1990) Inorg Chem 29:1232
57. Herberhold M, Haumaier L (1983) Chem Ber 116:2896
58. Cauquis G, Deronzier A (1980) J Inorg Nucl Chem 42:1447
59. Herberhold M, Ott J, Haumaier L (1985) Chem Ber 118:3143
60. Tsipis CA, SigalasMP, Hadjikostas CC (1983) Z Anorg Allg Chem 505:53
61. Anderson S, Hill AF, Nasir BA (1995) Organometallics 14:2987
62. Hill AF, Malget JM (1997) J Chem Soc Dalton 2003
63. Kläui W, Hardt T, Schulte H-J, Hamers H (1995) J Organomet Chem 498:63
64. Cook DJ, Hill AF (2003) Organometallics 22:3502
65. Cook DJ, Hill AF (1997) Chem Commun 955
66. Mace WJ, Main L, Nicholson BK, Hagyard M (2002) J Organomet Chem 664:288
67. Demaimay F, Roucoux A, Dazord L, Noiret N, Moisan A, Patin H (1999) Nucl Med Biol 26:225
68. Bolzati C, Uccelli L, Duatti A, Venturini M, Morin C, Chéradame S, Refosco F, Ossola F, Tisato F (1997) Inorg Chem 36:3582
69. Uccelli L, Bolzati C, Boschi A, Duatti A, Morin C, Pasqualini R, Giganti M, Piffanelli A (1999) Nucl Med Biol 26:63
70. Adams RD, Chen L, Wu W (1994) Angew Chem Int Edit 33:568
71. Adams RD, Chen L, Huang M (1995) J Chin Chem Soc 42:11
72. Mévellec F, Roucoux A, Noiret N, Patin H, Tisato F, Bandoli G (1999) Inorg Chem Commun 2 230
73. Gerber TIA, du Preez JGH, Kemp HJ (1994) J Coord Chem 33:245

74. McConnachie CA, Stiefel EI (1999) Inorg Chem 38:964
75. McConnachie CA, Stiefel EI (1997) Inorg Chem 36:6144
76. Mévellec F, Tisato F, Refosco F, Roucoux A, Noiret N, Patin H, Bandoli G (2002) Inorg Chem 41:598
77. Adams H, McHugh PE, Morris MJ, Spey SE, Wright PJ (2001) J Organomet Chem 619:209
78. Cadierno V, Gamasa MP, Gimeno J, Lastra E (1996) J Organomet Chem 510:207
79. Bruce MI, Liddel MJ, Snow MR, Tiekink ERT (1988) J Organomet Chem 352:199
80. Bruce MI, Hall BC, Zaitseva NN, Skelton BW, White AH (1998) J Chem Soc Dalton 1793
81. De Lange PPM, Alberts E, van Wijnkoop M, Frühauf H-W, Vrieze K, Kooijman H, Spek AL (1994) J Organomet Chem 465:241
82. Hashimoto H, Tobita H, Ogino H (1994) Organometallics 13:1055
83. Bird PH, Siriwardaneli U, Shaver A, Lopez O, Harpp DN (1981) J Chem Soc Chem Commun 513
84. Laifa EA, Benali-Cherif N, Merazig H (2003) Acta Crystallogr E 59:m108
85. Laifa EA, Benali-Cherif N, Berrah F (2003) Acta Crystallogr E 59:m80
86. Laifa EA, Benali-Cherif N (2003) Acta Crystallogr E 59:m283
87. Vastag S, Gervasio G, Marabello D, Szalontai G, Marko L (1998) Organometallics 17:4218
88. Bellitto C, Flamini A, Piovesana O (1979) J Inorg Nucl Chem 41:1677
89. Shrivastav A, Singh NK, Srivastava G (2002) Bioorg Med Chem 10:2693
90. Prasad R, Kumar A (2002) Thermochim Acta 383:59
91. Belay M, Edelmann, FT (1994) J Organomet Chem 479:C21
92. Bleeke JR, Behm R, Xie Y-F, Clayton TW Jr, Robinson KD (1994) J Am Chem Soc 116:4093
93. Bleeke JR, Behm R, Xie Y-F, Chiang MY, Robinson KD, Beatty AM (1997) Organometallics 16:606
94. Bleeke JR, Behm R, Beatty AM (1997) Organometallics 16:1103
95. Bennett MA, Johnson JA, Willis AC (1996) Organometallics 15:68
96. Vicente J, Abad J-A, Bergs R, de Arellano MCR, Martinéz-Viviente E (2000) Organometallics 19:5597
97. Hintermaier F, Mihan S, Dialer H, Volodarsky LB, Beck W (2001) Z Anorg Allg Chem 627:1901
98. Piotraschke J, Strauch P, Kirmse R (1995) Z Anorg Allg Chem 621:159
99. BorerLL, Kong JV, Keihl PA, Forkey DM (1987) Inorg Chim Acta 129:223
100. Usón R, Forniés J, Usón MA, Carranza MM (1989) Inorg Chim Acta 155:71
101. de Oliveira LFC, Santos PS (1995) Vib Spectrosc 10:57
102. Safavi A, Fotouhi L (1997) Can J Chem 75:1023
103. Safavi A, Fotouhi L (1997) J Electroanal Chem 434:93
104. Otto H, Werner H (1987) Chem Ber 120:97
105. Camus A, Marsich N, Nardin G (1980) J Organomet Chem 188:389
106. Biagini-Cingi M, Manotti-Lanfredi A-M, Ugozzoli F, Camus A, Marsich N (1997) Inorg Chim Acta 262:69
107. Kondo M, Shimizu E, Horiba T, Tanaka H, Fuwa Y, Nabari K, Unoura, K, Naito T, Maeda K, Uchida F (2003) Chem Lett 32:944
108. Deivaraj TC, Vittal JJ (2001) Acta Crystallogr E 57:m566
109. Manotti Lanfredi A-M, Ugozzoli F, Asaro F, Pellizer G, Marsich N, Camus A (1991) Inorg Chim Acta 190:71
110. Brunner H, Hollman A, Zabel M (2001) J Organomet Chem 630:169
111. Manotti Lanfredi AM, Ugozzoli F, Camus A, Marsich N (1995) J Chem Crystallogr 25:37
112. Wang J-C (1994) Acta Crystallogr C 50:704
113. Bardají M, Gimeno C, Jones PG, Laguna A, Laguna M, Merchán F, Romeo I (1997) Organometallics 16:1083

114. Darensbourg DJ, Adams MJ, Yarbrough JC (2002) Inorg Chem Commun 5:38
115. Guillon D, Bruce DW, Maldivi P, Ibn-Elha M, Dhillon R (1994) Chem Mater 6:182
116. Safavi A, Gholivand MB (1996) Can J Chem 74:95
117. El-Gyar SA, El-Gahami MA, Kandeel MM (1991) Phosphorus Sulfur 62:119
118. Trnka TM, Bonanno JB, Bridgewater BM, Parkin G (2001) Organometallics 20:3255
119. Kanda T, Ibi M, Mochizuki K-i, Kato S (1998) Chem Lett 957
120. Evans WJ, Seibel CA, Ziller JW, Doedens RJ (1998) Organometallics 17:2103
121. Bombieri G, Croatto U, Forsellini E, Zarli B, Graziani R (1972) J Chem Soc Dalton 560
122. Bandoli G, Bortolozzo G, Clemente DA, Croatto U, Panattoni C (1971) Inorg Nucl Chem Lett 7:401
123. Channareddy S, Glaser B, Mayer EP, Nöth H, Helm SW (1993) Chem Ber 126:1119
124. Koester R, Kucznierz R, Seide G, Betz P (1992) Chem Ber 125:1023
125. Koester R, Kucznierz R (1992) Liebigs Ann Chem 1081
126. Singh SK, Singh Y, Rai AK, Mehrotra RC (1992) Phosphorus Sulfur 68:211
127. Singh SK, Singh Y, Rai AK, Mehrotra RC (1989) Polyhedron 8:633
128. Weidenbruch M, Grobecker U, Saak W, Peters E-M, Peters K (1998) Organometallics 17:5206
129. Saito M, Tokitoh N, Okazaki R (1995) Organometallics 14:3620
130. Hartke K, Wagner U (1996) Chem Ber 129:663
131. Trassoli A, Sedaghat T, Neumüller B, Ghassemzadeh M (2001) Inorg Chim Acta 318:15
132. Trassoli A, Asadi A, Hitchcock PB (2002) J Organomet Chem 645:105
133. Naruse Y, Inagaki S, Kano N, Nakagawa N, Kawashima T (2002) Tetrahedron Lett 43:5759
134. Gotthardt H, Pflaumbaum W (1987) Chem Ber 120:411
135. Nakamoto M, Tanaka K, Tanaka T (1985) B Chem Soc Jpn 58:1816
136. Kato S, Nishiwaki M, Inagaki S, Ohshima S, Ohno Y, Mizuta M, Murai T (1985) Chem Ber 118:1684
137. Kato S, Mizuta M (1972) B Chem Soc Jpn 45:3492
138. Nakayama J (1993) Sulfur Lett 15:239
139. Nakayama J, Akiyama I (1992) Chem Commun 1522
140. Akimoto K, Nakayama J (1997) Heteroatom Chem 8:505
141. Otani T, Sugihara Y, Ishii A, Nakayama J (1998) Heteroatom Chem 9:703
142. Miyashita I, Matsumoto K, Kobayashi M, Nagasawa A, Nakayama J (1998) Inorg Chim Acta 283:256
143. Nakayama J, Zhao L, Otani T, Sugihara Y, Ishii A (2000) Chem Lett 1248
144. Nagasawa A, Akiyama I, Mashima S, Nakayama J (1995) Heteroatom Chem 6:45
145. Nakayama J, Otani T, Tadokoro T, Sugihara Y, Ishii A (2001) Chem Lett 768
146. Nakayama J, Otani T, Sugihara Y, Ishii A (1998) Chem Lett 321
147. Doszczak L, Przychodzen W, Witt D, Rachon J (2002) J Chem Soc Perkin T 2 1747
148. Doszczak L, Rachon J (2002) Synthesis 1047
149. Doszczak L, Przychodzen W, Witt D, Rachon J (2002) Phosphorus Sulfur 177:1851
150. Kimura M, Iwata A, Itoh M, Yamada K, Kimura T, Sugiura N, Kanda T, Ishida M, Kato S (2004) Dalton Trans (submitted)
151. Cea-Olivares R, Garcia-Montalvo V, Hernández-Ortega S, Rodríguez-Narváez C, García y García P, López-Cardoso M, de March P, González L, Elias L, Figeredo M, Font J (1999) Tetrahedron Asym 10:3337
152. Niyomura O, Kitoh Y, Nagayama K-i, Kato S (1999) Sulfur Lett 22:195
153. Akimoto K, Masaki K, Nakayama J (1997) B Chem Soc Jpn 70:471
154. Awad IMA (1993) Phosphorus Sulfur 78:237
155. Thang SH, Chong YK, Mayadunne RTA, Moad G, Rizzardo E (1999) Tetrahedron Lett 40:2435

156. Bouhadir G, Legrand N, Quiclet-Sire B, Zard SZ (1999) Tetrahedron Lett 40:277

157. Akimoto K, Masaki K, Nakayama J (1996) B Chem Soc Jpn 69:2091

158. Asinger F, Neuray D, Witte EC, Hartig J (1972) Monatsh Chem 103:1661

159. McKinnon DM (1996) In: Katritzky AR, Rees CW, Scriven EFV (eds) Comprehensive heterocyclic chemistry II. Pergamon, Oxford, UK, Ch 3.11, p 569

160. Pedersen CT (1995) Sulfur Reports 16:173

161. Christoforou IC, Koutentis PA, Rees CW (2002) J Chem Soc Perkin T 1 1236

162. Chang Y-G, Kim K, Park YJ (2002) J Heterocyclic Chem 39:29

163. Koutentis PA, Rees CW (1998) J Chem Soc Perkin T 1 2505

164. Emayan K, English RF, Koutentis PA, Rees CW (1997) J Chem Soc Perkin T 1 3345

165. Lee H-S, Kim K (1996) Tetrahedron Lett 37:3709

166. Clapper ML (1998) Pharmacol Ther 78:17

167. Kensler TW, Groopman JD, Sutter TR, Curphey TJ, Roebuck BD (1999) Chem Res Toxicol 12:113

168. Kensler TW, Helzlsouer KJ (1995) J Cell Biochem Suppl 22:101

169. Kato S, Ono Y, Miyagawa K, Murai T, Ishida M (1986) Tetrahedron Lett 27:4595

170. Schnabel W, von Deuten K, Klar G (1982) Phosphorus Sulfur 13:345

171. Kato S, Ito Y, Ohta Y, Goto K, Kimura M, Mizuta M, Murai T (1985) Chem Ber 118:1696

172. Kato S, Ono Y, Miyazaki T, Itoh Y, Aoyama T, Ishida M, Murai T (1987) Nippon Kagaku Kaishi 1457

173. Kuhn N, Bohnen H, Henkel G (1994) Z Naturforsch B 49:1473

174. Kitagawa H, Mitani T (1999) Coord Chem Rev 190–192:1169

175. Kitagawa H, Onodera N, Sonoyama T, Yamamoto M, Fukawa T, Mitani T, Seto M, Maeda Y (1999) J Am Chem Soc 121:10068

176. Bellitto C, Flamini A, Gastaldi L, Scaramuzza L (1983) Inorg Chem 22:444

177. Kobayashi A, Kitagawa H, Ikeda R (2002) Mol Cryst Liq Cryst 379:315

178. Mitsumi M, Murase T, Kishida H, Yoshinari T, Ozawa Y, Toriumi K, Sonoyama T, Kitagawa H, Mitani T (2001) J Am Chem Soc 123:11179

179. Mitsumi M, Kitamura K, Morinaga A, Ozawa Y, Kobayashi M, Toriumi K, Iso Y, Kitagawa H, Mitani, T (2001) Angew Chem Int Edit 41:2767

180. Mitsumi M, Umebayashi S, Ozawa Y, Toriumi K, Kitagawa H, Mitani T (2002) Chem Lett 258

181. Ikeuchi S, Saito K, Nakazawa Y, Sato A, Mitsumi M, Toriumi K, Sorai M (2002) Phys Rev B 66:115110

182. Okura H, Oyama M, Okazaki S (2002) Electrochem Commun 2:367

183. Bellitto C, Flamini A, Piovesana O, Zanazzi PF (1980) Inorg Chem 19:3632

184. Kitagawa H, Onodera N, Ahn J-S, Mitani T, Kim M, Ozawa Y, Toriumi K, Yasui K, Manabe T, Yamashita M (1996) Mol Cryst Liq Cryst 285:311

185. Kitagawa H, Onodera N, Mitani T, Toriumi K, Yamashita M (1997) Synth Met 86:1931

186. Yamashita M, Wada Y, Toriumi K, Mitani T (1992) Mol Cryst Liq Cryst 216:207

187. Kitagawa H, Nakagami S, Mitani T (2001) Synth Met 116:401

188. Shirotani I, Kawamura A, Yamashita M, Toriumi K, Kawamura H, Yagi T (1994) Synth Met 64:265

189. Borsch SA, Passides K, Robert V, Solonenko AO (1998) J Chem Phys 109:4562

190. Robert V, Petit S, Borsch SA (1999) Inorg Chem 38:1573

191. Kitagawa H, Yamamoto M, Onodera N, Mitani T (1999) Synth Met 103:2151

192. Kitagawa H, Onodera N, Mitani T (2000) Mol Cryst Liq Cryst 342:111

193. Miyazaki Y, Wang Q, Sato A, Saito K, Yamamoto M, Kitagawa H, Mitani T, Sorai M (2002) J Phys Chem B 106:197

194. Kitagawa H, Sonoyama T, Mitani T, Seto M, Maeda Y (1999) Synth Met 103:2159
195. Bellitto C, Dessy G, Fares V (1985) Inorg Chem 24:2815
196. Makiura R, Kitagawa H, Ikeda R (2002) Mol Cryst Liq Cryst 379:309
197. Ikeda R, Kimura N, Ohki H, Furuta T, Yamashita M (1995) Synth Met 71:1907
198. Bellitto C, Dessy G, Fares V (1985) Mol Cryst Liq Cryst 120:381
199. Pyrasch M, Amirbeyki D, Tieke B (2002) Colloids Surface A 198–200:425
200. Pyrasch M, Amirbeyki D, Tieke B (2002) Adv Mater 13:1188
201. Kobayashi H (1997) Curr Opin Solid State Mater Sci 2: 440
202. Teschmit G, Strauch P, Barthel A, Reinhold J, Kirmse R (2000) Inorg Chim Acta 298:94
203. Strauch P, Dietzsch W, Golic L, Sieler J, Franke A, Münzberg I, Trübenbach K, Kirmse R, Reinhold J, Hoyer E (1995) Inorg Chem 34:763
204. Strauch P, Golic L (1996) Z Anorg Allg Chem 622:1236
205. Strauch P, Dietzsch W, Golic L (1997) Z Anorg Allg Chem 623:129
206. Strauch P, Dempe B, Kempe R, Dietzch W, Sieler J, Hoyer E (1994) Z Anorg Allg Chem 620:498
207. Piotraschke J, Strauch P (1995) Inorg Chim Acta 236:177

Top Curr Chem (2005) 251: 181–225
DOI 10.1007/b101009
© Springer-Verlag Berlin Heidelberg 2005

Carbodithioic Acid Esters

Akihiko Ishii (✉) · Juzo Nakayama

Department of Chemistry, Faculty of Science, Saitama University, Sakura-ku,
Saitama 338–8570, Japan
ishiiaki@chem.saitama-u.ac.jp

Abstract Recent works (published over the last decade) on the syntheses and reactions of dithioic acid esters (dithioates) are reviewed. Dithioates have been synthesized by various means, including: (i) reaction of dithioic acids and their salts with electrophiles; (ii) thionation of S-substituted thioates; (iii) [3,3]-sigmatropic (thio-Claisen) rearrangements of ketene alkyl allyl dithioacetals and [2,3]-sigmatropic rearrangement; (iv) elimination of one of the S-substituents of ketene dithioacetal; (v) sulfuration of iminium salts. Dithioates are useful substrates for aldol reactions, conjugate additions and cycloadditions. Dithioates undergo thiophilic and carbophilic additions with nucleophiles, the nature of which depends on the character of the nucleophile. Reactivites of α-carbenium dithioates derived from inner salts of dithioic acids are discussed. Five-, six-, and seven-membered heterocyclic compounds with -C(S)S- groups in their ring systems, cyclic anhydrides of dithioic acids, S-M dithioates (M=Group 14, 15, and 16 elements), and transition metal complexes of dithioic acids are also briefly described.

Keywords Dithioates · Dithioic acid · Thionation · Thiophilic addition · Carbophilic addition

Abbreviations

Ac	acetyl
ACDA	2-amino-1-cyclopentene-1-carbodithioic acid
1-Ad	1-adamantyl
Ans	4-methoxyphenyl (4-anisyl)
Boc	*tert*-butoxycarbonyl
Bu	butyl
s-Bu	*sec*-butyl
t-Bu	*tert*-butyl
cat	catalyst
Cp	cyclopentadienyl
DABCO	1,4-diazabicyclo[2.2.2]octane
DBH	1,3-dibromo-5,5-dimethylhydantoin
de	diastereomer excess
DMAD	dimethyl acetylenedicarboxylate
DMF	*N,N*-dimethylformamide
DMPU	1,3-dimethyl-3,4,5,6-tetrahydro-2(1H)-pyrimidinone (*N,N'*-dimethylpropyleneurea)
DMSO	dimethyl sulfoxide
dr	diastereomer ratio
E^+	electrophiles
ee	enantiomer excess
Et	ethyl
h	hour(s)
HMPA	hexamethylphosphoric triamide
LDA	lithium diisopropylamide
LHMDS	lithium hexamethyldisilazide
MBT	2-mercaptobenzothiazole
LR	Lawesson's reagent
m-CPBA	*m*-chloroperoxybenzoic acid
Me	methyl
Mes	mesityl, 2,4,6-trimethylphenyl
Ms	methanesulfonyl
NBS	*N*-bromosuccinimide
NCS	*N*-chlorosuccinimide

NIS	*N*-iodosuccinimide
Ph	phenyl
Pr	propyl
i-Pr	isopropyl
py	pyridine
rt	room temperature
TBDMS	*tert*-butyldimethylsilyl
Tf	trifluoromethanesulfonyl (triflyl)
THF	tetrahydrofuran
TMS	trimethylsilyl
Tol	4-methylphenyl
Ts	4-toluenesulfonyl

1
Introduction

The chemistry of dithioic acid esters (dithioates) has a very wide scope, from materials science to biological science. Their relatively straightforward syntheses and high reactivities have enabled us to investigate their structures, biological activities, and their use in organic and inorganic syntheses [1]. In this article, we survey papers published on the syntheses and reactions of dithioic acid alkyl and aryl esters (*S*-alkyl and *S*-aryl dithioates). We focus mainly on recent works (published during 1993–2003), although we sometimes also refer to other historically important research. There are several types of heterocyclic compounds that contain -C(=S)S- groups in their ring systems, and we also review some of these cyclic dithioates. While free dithioic acids and their alkali and alkaline-earth metal salts are beyond the scope of this article, we briefly discuss *S*-M dithioates [M=Group 4–16 (except carbon) elements] too.

2
Synthesis of Carbodithioic Acid Esters

2.1
From Dithioic Acids and Their Salts

Alkylations of dithioic acids or their salts comprise the most straightforward route to alkyl dithioates. Dithioic acids react with alkyl halides under basic conditions to yield the dithioates, as illustrated in Eq. 1 [2, 3].

$$(1)$$

Dithioate salts are readily prepared in situ by reaction of a carbanion with CS_2, and are alkylated using various electrophiles (Table 1). Compounds that have acidic hydrogens activated by -CN (Entries 1–3) [4–6], -CO_2Et (Entry 4) [7], $PhS(O)_2$- (Entry 5) [8], -S(O)- (Entry 6) [9], and -C(O)- (Entries 7–9) [10, 11] groups are converted to the corresponding alkyl carbodithioates by treating them with CS_2 under basic conditions, and then an alkyl iodide. Dithioate groups in α-cyano dithioates **1a–c** prepared in this way can be used as building blocks in syntheses of heterocyclic compounds, as shown in Eqs. 2 to 4 [4–6]. In the case of sulfoxides, the best results are accomplished using 4-fluorophenyl methyl trithiocarbonate (**2**) instead of CS_2 and methyl iodide (see Entry 6) (Scheme 1) [9, 21]. The chirality on the sulfoxide sulfur atom of (*R*)-cyclohexyl methyl sulfoxide (**3**) is mostly retained in the reaction to give the chiral β-sulfinyl dithioic acid ester **4** (Eq. 5) [9, 22].

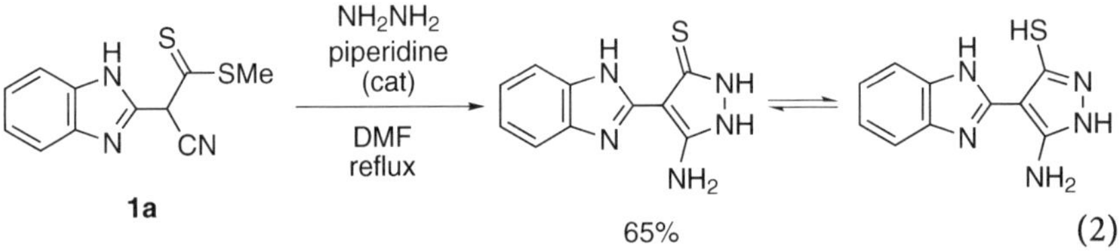

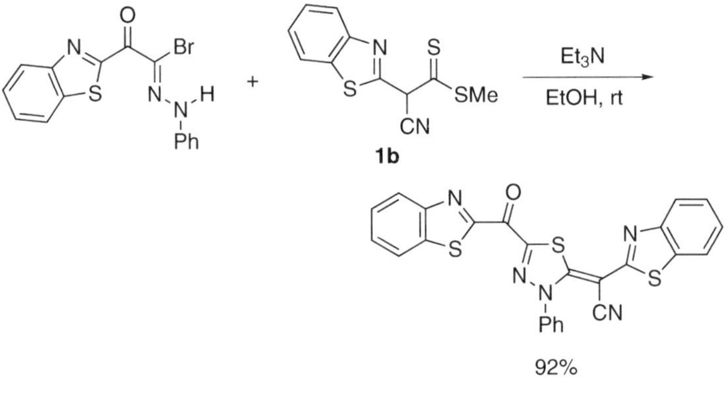

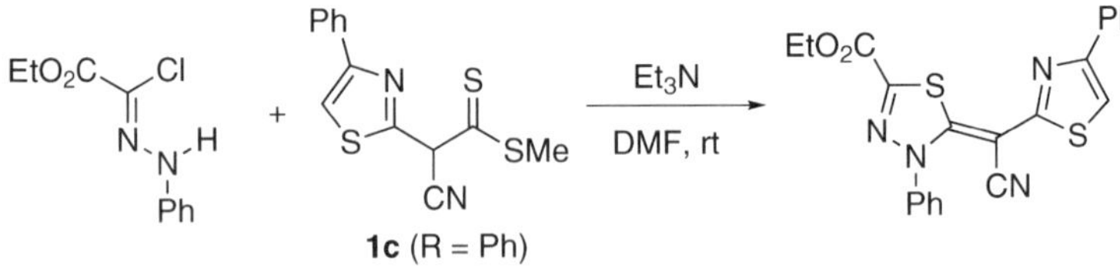

Table 1 Syntheses of carbodithioates using CS_2

Entry	Starting compound	Reagents and conditions	Product	Reference
1	benzimidazol-2-yl-CH$_2$CN	1. EtONa, EtOH, 40-50 °C 2. CS$_2$, reflux * 3. MeI, EtOH * The Na salt was isolated in 58% yield. **From the Na salt.	**1a** 78%**	[4]
2	benzothiazol-2-yl-CH$_2$CN	1. CS$_2$, KOH, DMF, rt 2. MeI	**1b** 82%	[5]
3	4-R-thiazol-2-yl-CH$_2$CN	1. CS$_2$, KOH, DMF, rt 2. MeI	**1c** R = Ph (74%), (74%)	[6]
4	6-Me-7-oxo-[1,3,4]thiadiazolo-pyrimidin-2-yl-CH$_2$CO$_2$Et	1. CS$_2$, NaH, DMF 2. MeI	41%	[7]
5	PhS(O)$_2$CH$_2$R	1. BuLi (1 equiv), THF 2. CS$_2$ (1.5 equiv) 3. BuLi (1 equiv) 4. MeI (1 equiv)	R = H (89%), n-C$_7$H$_{15}$ (88%), –CH$_2$CH$_2$- (79%)	[8]
6	t-Bu–S(O)–CH$_3$	1. BuLi (2 equiv) 2. CS$_2$ 3. MeI	33%	[9]

Table 1 (continued)

Entry	Starting compound	Reagents and conditions	Product	Reference
7	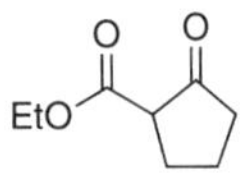 (ethyl 2-oxocyclopentanecarboxylate)	1. K_2CO_3, DMF 2. CS_2 3. RI	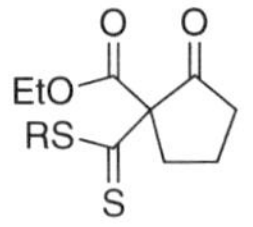 (1-ethoxycarbonyl-2-oxocyclopentyl dithioester) $R = Me, Bu$	[10]
8	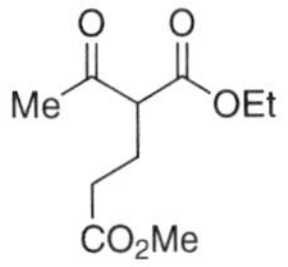 (Me–CO–CH(CO$_2$Et)–CH$_2$CH$_2$–CO$_2$Me)	1. K_2CO_3, DMF 2. CS_2 3. BuI	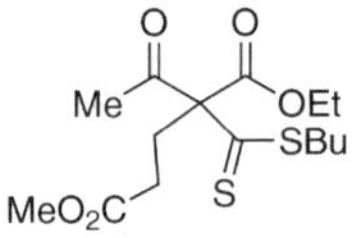 (corresponding SBu dithioester product)	[10]
9	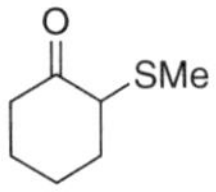 (2-(methylthio)cyclohexanone)	1. NaH, THF, 0 °C 2. CS_2 3. MeI	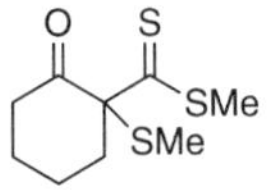 (product with C(=S)SMe and SMe) 90%	[11]
10	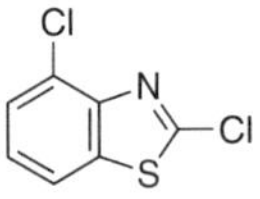 (4-chloro-2-chlorobenzothiazole)	CS_2, NaOH, DMSO	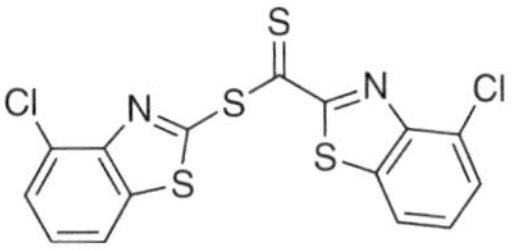 (bis(4-chlorobenzothiazol-2-yl) dithioester)	[12]
11	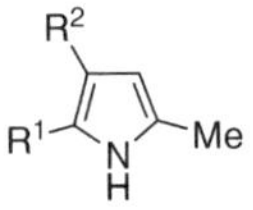 (pyrrole: R^2 at 4-position, R^1 at N-C, 2-Me) $R^1 = Me, Ph, 2\text{-furyl}, 2\text{-thienyl}; R^2 = H, Me$ $R^1\text{-}R^2 = (CH_2)_4$	1. CS_2, KOH, DMSO 2. EtI	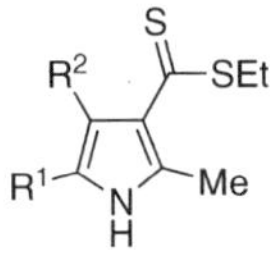 (3-(ethylthiocarbonothioyl)pyrrole) 36–61%	[13]
12	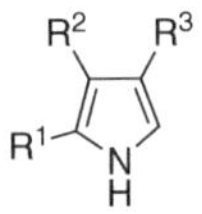 (pyrrole: R^2, R^3, R^1) $R^1 = Me, t\text{-Bu}; R^2 = H, Me; R^3 = H, Me$	1. CS_2, KOH, DMSO 2. EtI	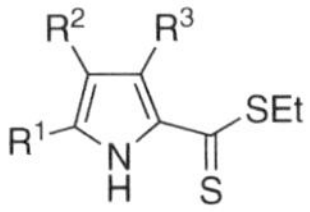 (2-(ethylthiocarbonothioyl)pyrrole) 46–62%	[13,14]

Table 1 (continued)

Entry	Starting compound	Reagents and conditions	Product	Reference
13	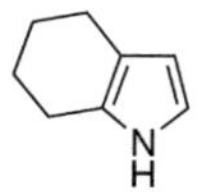	1. CS_2, KOH, DMSO 2. 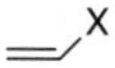	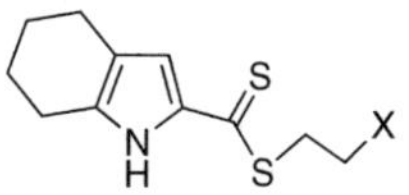 $X = CN$ (62%), CO_2Me (36%), $CONH_2$ (52%)	[15]
14	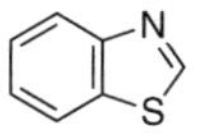	1. BuLi, THF 2. CS_2 3. CF_3SO_3Me	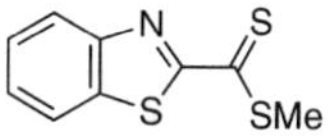 67%	[16]
15	ArMgBr	1. CS_2 2. MeI	Ar = 1-naphthyl (96%), 2-naphthyl (82%), 4-PhC_6H_4 (89%), 4-n-$C_8H_{17}OC_6H_4$ (84%), 6-MeO-2-naphthyl (91%)	[17]
16	PhMgBr	1. CS_2 2. 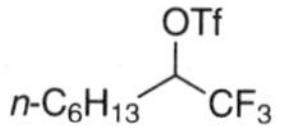 **5**	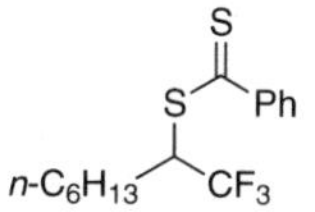 **6** 63%	[18]
17	RMgX	1. CuI 2. CS_2 3. **8**	**7** 31–75% R = 3-pyridyl, Ph, Bu, s-Bu, Pr, i-Pr, Me, $TMSCH_2$, Mes, $PhCMe_2CH_2$; X = Cl, Br	[19]
18	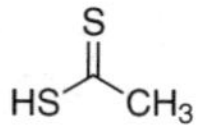	1. CS_2, LHMDS, THF 2. MeI	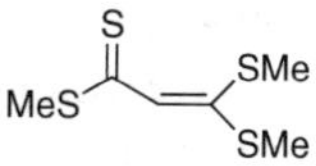 69%	[20]

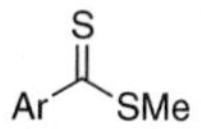

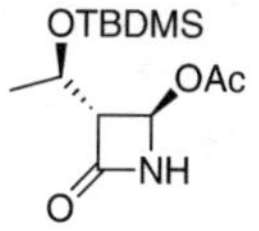

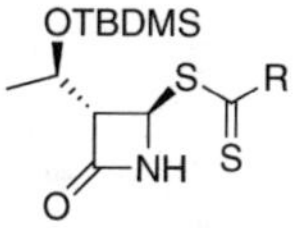

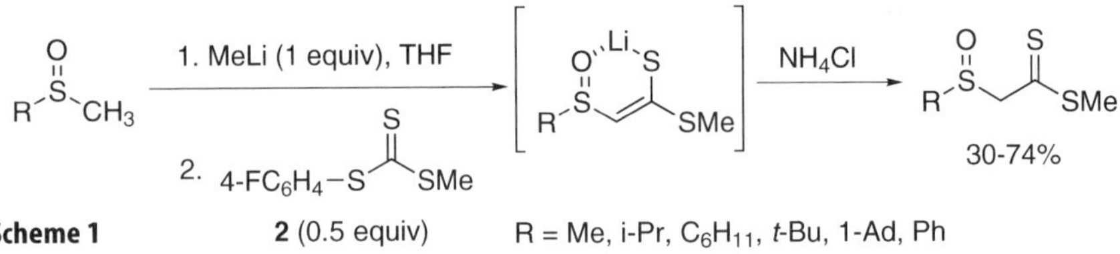

Scheme 1

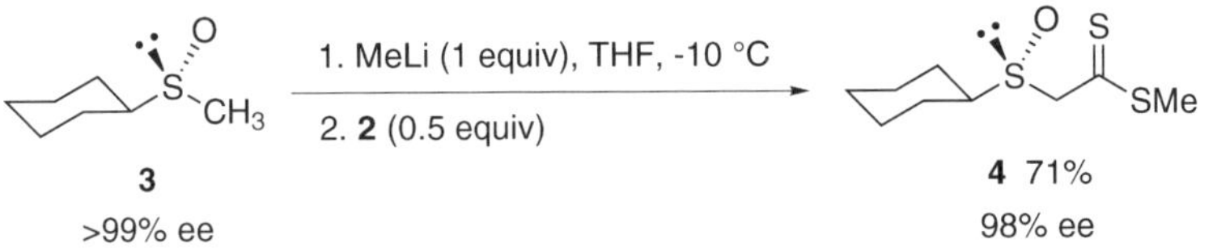

$$R = Me, \text{ i-Pr}, C_6H_{11}, \text{ t-Bu}, \text{1-Ad}, Ph$$

3 >99% ee

4 71% 98% ee   (5)

Pyrroles are deprotonated with KOH in DMSO, and the resulting anions react with CS_2 at the unsubstituted carbon atom in the ring, not the nitrogen atom (Entries 11–13) [13–15]. The resulting potassium carbodithioates can be trapped with electron-deficient alkenes using conjugate addition (Entry 13) [15]. Triflates (Entries 14 and 16) [16, 18] and acetates (Entry 17) [19] also serve as electrophiles. α-Trifluoromethyl triflate **5** (Entry 16) can be allowed to react with benzenecarbodithioate to give the α-trifluoromethyl dithioate **6** [18]. The azetidinone dithioic acid esters **7** (Entry 17) have been prepared in 31–75% yields using organocuprates, CS_2, and acetoxy azetidinone **8** [19]. The azetidinone dithioic acid esters **7** have been converted to 2-substituted penems **9** in good yields (Scheme 2) [19].

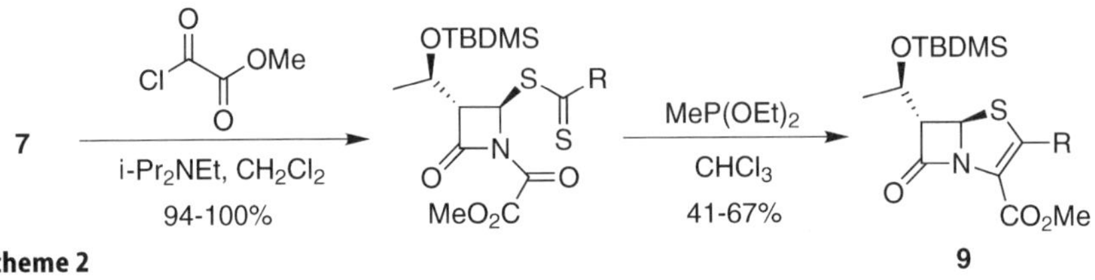

Scheme 2

9

3-Silyloxyalk-2-enylphosphonium salts undergo nucleophilic substitution with various nucleophiles [23]. The phosphonium salt **10**, prepared in situ from cyclohexenone on treatment with TfOTBDMS and triphenylphosphine in THF, reacted with lithium dithioate **11** to yield dithioate **12** with a 78% yield (Eq. 6).

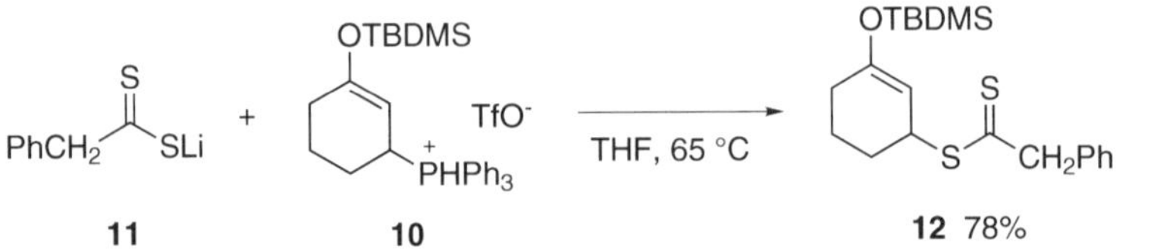

11   **10**   **12** 78%   (6)

Treatment of dithioacetic acid with 2 equiv of BuLi generates the dilithium dithioenolate **13**, which reacts at the terminal carbon with the carbonyl carbon of enone **14**. The resulting lithium dithioate is methylated with methyl iodide to provide the methyl β-hydroxy dithioate **15** (Scheme 3) [24].

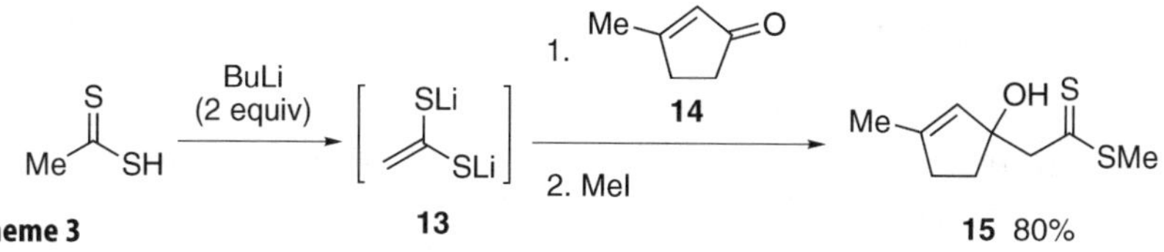

Scheme 3

Phenyl dithioate **16** was prepared by treating the sodium salt of **17** with thiophenol in the presence of boron trichloride, as shown in Eq. 7, where the intervention of the *S*-boryl dithioate is considered [25].

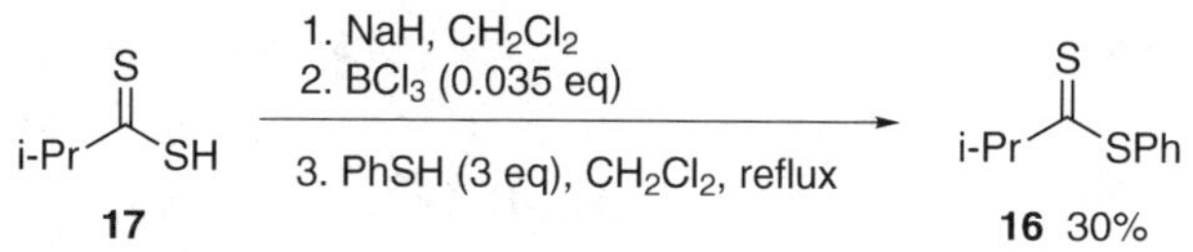

$$(7)$$

2.2
From *S*-Substituted Thioates or Carboxylic Acids

When heated with Lawesson's reagent (LR) or P_4S_{10}, *S*-alkyl and *S*-aryl thioates are thionated to give dithioates in low to high yields (Eq. 8 and Table 2). The chirality at the β-position is retained upon thionation with LR (Entry 1) [26]. The reaction of *S*-ethyl 3-(*N*-*tert*-butoxycarbonyl)aminopropanethioate (Entry 2) with LR gives the dithioate **18** in low yield, probably due to the lack of selectivity of LR toward the thioester carbonyl versus the carbamate carbonyl [27]. The dithioate **18** was prepared by reacting the thioacyl-*N*-phthalimide **19** with ethanethiol in higher yield (Eq. 9) [27, 32].

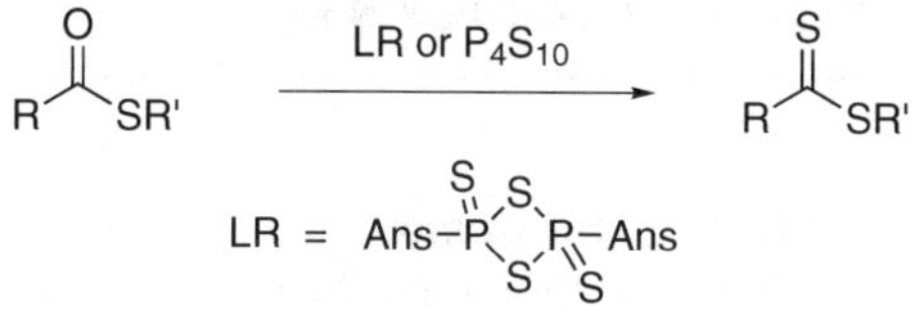

$$(8)$$

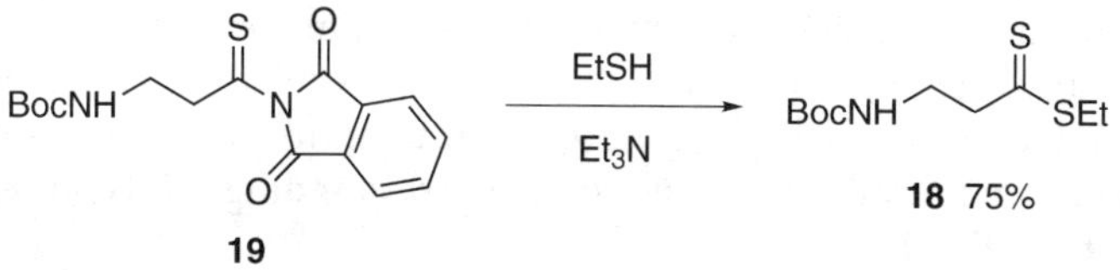

$$(9)$$

Table 2 Syntheses of dithioates with Lawesson's reagent (LR) or P_4S_{10}

Entry	Starting compound	Reagents and conditions	Product	Reference
1	Ph–CH(Me)–CH$_2$–C(=O)–SEt	LR, PhMe	Ph–CH(Me)–CH$_2$–C(=S)–SEt, 94%	[26]
2	BocNH–CH$_2$CH$_2$–C(=O)–SEt	LR, PhMe, 100 °C, 12 h	BocNH–CH$_2$CH$_2$–C(=S)–SEt, **18** 23%	[27]
3	Et–C(=O)–S–C$_6$H$_4$–R	LR, PhMe, reflux	Et–C(=S)–S–C$_6$H$_4$–R, R = H, Me, Cl, Br, SH	[28]
4	R–C$_6$H$_4$–C(=O)–S–CH(Me)–Ph	LR, PhMe, reflux, 4 h	R–C$_6$H$_4$–C(=S)–S–CH(Me)–Ph, R = H (88%), Me (91%), MeO (82%), Br (90%)	[29]
5	7-Cl-benzothiophene-2-C(=O)–SEt	P_4S_{10}, PhMe	7-Cl-benzothiophene-2-C(=S)–SEt	[30]
6	R^1–C(=O)–OH	R^2SH, P_4S_{10} (20 mol%), PhMe, 110 °C, 4 h	R^1–C(=S)–SR^2, 45-96%	[31]

R^1 = Pr, Ph, p-XC$_6$H$_4$ (X = Me, t-Bu, MeO, F, Cl, Br, I, CF$_3$, NO$_2$, CN), o-MeC$_6$H$_4$, m-FC$_6$H$_4$, m-NO$_2$C$_6$H$_4$, 3,4,5-(MeO)$_3$C$_6$H$_2$, 2-thienyl
R^2 = PhCH$_2$, Pr, Bu, t-Bu, Ph, o-MeOC$_6$H$_4$, p-ClC$_6$H$_4$, 2-thienyl

Carboxylic acids can be converted directly to the dithioic acid esters on treatment with a thiol and 20 mol% of P_4S_{10} (Entry 6) [31]. It has been proposed that this reaction involves the formation of the O-aroyl (or acyl) trithiophosphate derivative **20**, the reaction of these intermediates with thiols, and thionation of the resulting S-substituted thioates with P_4S_{10}. Alcohols such as benzyl alcohol, t-butyl alcohol, 1-phenethyl alcohol were used instead of thiols; it was claimed that these alcohols are initially converted to the corresponding thiols on reaction with P_4S_{10} (Eq. 10) [31].

20

$$Ar\text{-}COOH \xrightarrow[\text{solvent}]{\text{ROH, P}_4\text{S}_{10}\ (1\ \text{equiv})} Ar\text{-}C(=S)\text{-}SR \qquad (10)$$

30-75%

Ar = Ph, p-XC$_6$H$_4$ (X = MeO, F, NO$_2$)

R = PhCH$_2$, t-Bu, 1-phenethyl

solvent (temp./°C): PhH (80), PhMe (110), PhCl (110),
(ClCH$_2$)$_2$ (80), CCl$_4$ (70)

2.3
By [3,3]-(Thio-Claisen) and [2,3]-Sigmatropic Rearrangements

The α-sulfinyl ketene dithioacetals **21** rearrange readily at room temperature (5–45 h) to give γ,δ-unsaturated α-sulfinyl dithioates **22** in good isolated yields with high diastereoselectivity (93:7->99:1), as shown in Eq. 11 [12, 33]. To rationalize the diastereoselectivity, **I** was proposed, based on the Felkin-Anh model, as the conformation that led to the transition state of the rearrangement (Eq. 12). Similarly, β-trimethylsiloxy (**23**) [34] or β-hydroxy [35] ketene dithioacetals underwent thio-Claisen rearrangements, yielding the corresponding α-allyl β-hydroxy dithioates **24** (Eq. 13). In the case of S-crotylic ketene dithioacetal **23** (R^1=Et, R^2=Me), four diastereoisomers were formed in the ratio of 1:6:90:3, where the *syn-syn* diastereomer was predominant. The conformation **II** was proposed for the formation of the major isomer (Eq. 14).

$$(11)$$

21

R^1 = Me, t-Bu, i-Pr, cyclohexyl
R^2 = H, Me

22 42-63%
>86% de

$$\xrightarrow{[3,3]} \qquad (12)$$

I   **22**

$$\text{(13)}$$

R^1 = i-Pr, R^2 = H; dr = 86:14
R^1 = Et, R^2 = Me; dr = 1:6:90:3

$$\text{(14)}$$

The thio-Claisen rearrangement of **25** was examined under Zeolite-catalyzed and uncatalyzed conditions (Eq. 15) [36]. Under Zeolite-catalyzed conditions, only *anti* α-allyl β-hydroxy dithioate **26** was obtained in high yield. Under uncatalyzed conditions, the *syn* isomer was the major product (8:1–19:1). This diastereoselectivity switch has been explained by transition model **III**.

$$\text{(15)}$$

R = Me, Et, Pr, *t*-Bu, Ph

A [2,3]-sigmatropic rearrangement of γ,δ-unsaturated carbene (**27**) can be employed as the key step in the synthesis of naturally-occurring halogenated monoterpenes. The carbene **27** is generated when tosylhydrazone **28** is treated

Scheme 4

with NaH, and rearrangement of **27** yields the dithioate **29**. Further elaboration of the dithioate **29** leads to the racemic antitumor agent **30** in four steps (Scheme 4) [37].

2.4
From Ketene Dithioacetal

β-Oxo ketene dithioacetal **31** has been converted to β-oxo dithioate **32** by treating it with hydrogen sulfide in the presence of boron trifluoride etherate in refluxing dioxane (Eq. 16) [38]. Dimsyl sodium, generated from DMSO and NaH, acts as a base toward 2-ylidene-1,3-dithiolane **33**, prompting the fragmentation of the dithiolane ring, giving β-hydroxy α,β-unsaturated vinyl dithioate **34** in high yields (Scheme 5) [39]. Dimsyl sodium also induces S-demethylation of ketene dimethyl dithioacetal, giving β-hydroxy α,β-unsaturated methyl dithioate [39].

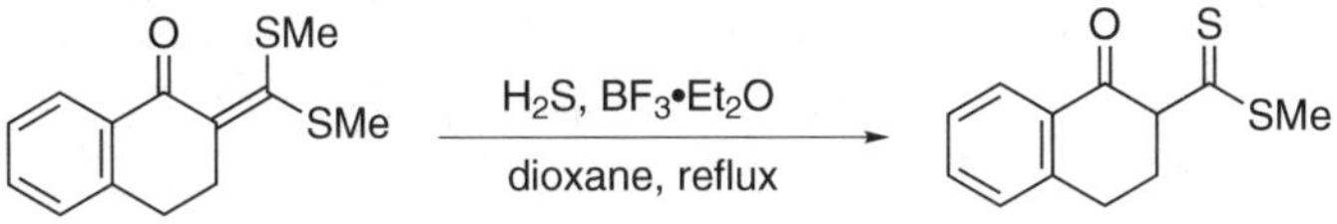

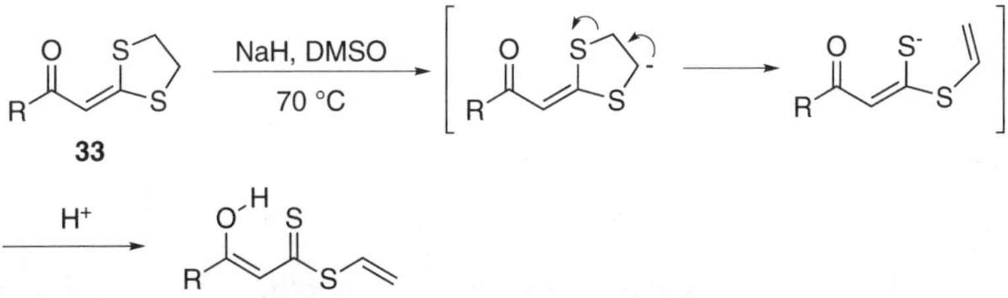

$$\text{(16)}$$

Scheme 5   R = Ph (92%), Me (82%)

The α-phosphonodithioacrylic acid esters **35** were prepared in moderate yields by reacting phosphonoketene dithioacetal **36** with trifluoromethanesulfonic acid and tetrabutylammonium bromide (or iodide) (Scheme 6) [40]. This

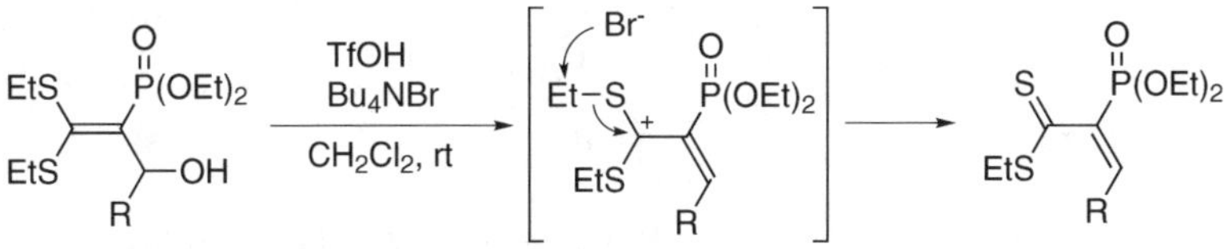

Scheme 6

transformation was also achieved by treating with triphenylphosphine-CBr$_4$ in dichloromethane [40].

2.5
From Iminium Salts

Iminium salts, prepared by *S*-methylating thioamides with methyl iodide (Scheme 7) [26], TfOMe (Scheme 8) [41], or dimethyl sulfate (Scheme 8) [41], can be reacted with H$_2$S to yield methyl dithioates. Trifluorothioacetamide **37** is first converted to the dichloride **38**, and then **38** is allowed to react with thiols to provide the iminium chlorides **39**, which was treated with H$_2$S to give the trifluorodithioacetates **40** in moderate to good yields (Scheme 9) [41].

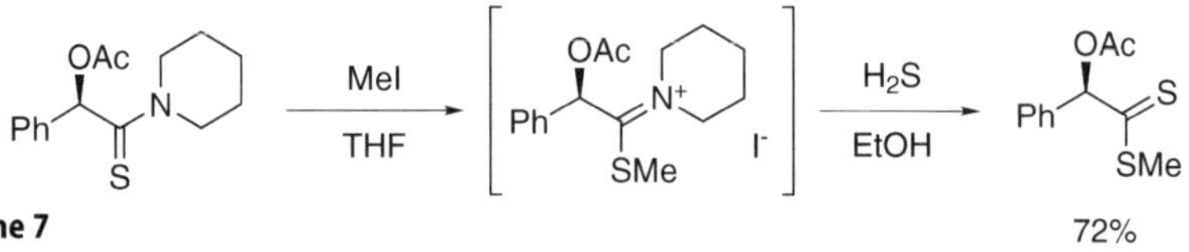

Scheme 7   72%

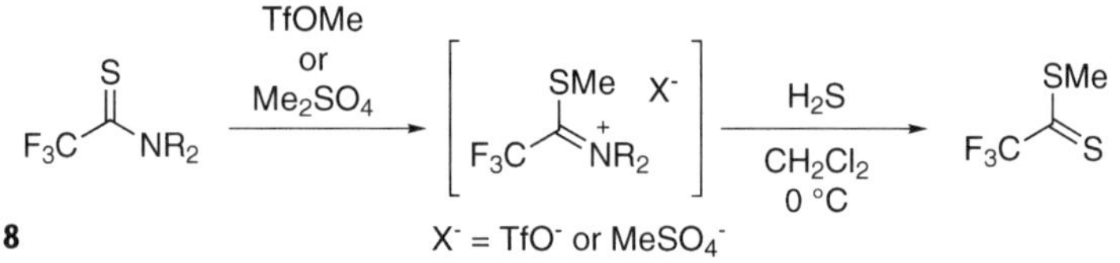

Scheme 8   X⁻ = TfO⁻ or MeSO₄⁻

In a modified Pinner reaction, the cyano group of the *N*-acyl amino nitrile **41** was converted to the ethylthio iminium salt **42** on treatment with ethanethiol in liquid HF at 77 K. The iminium salt **42** was treated with H$_2$S gas in pyridine at 0 °C to give a dithioester derivative of the *N*-acyl amino acid **43** (Scheme 10) [42]. This method has also been used to synthesize dithioesters of dipeptides.

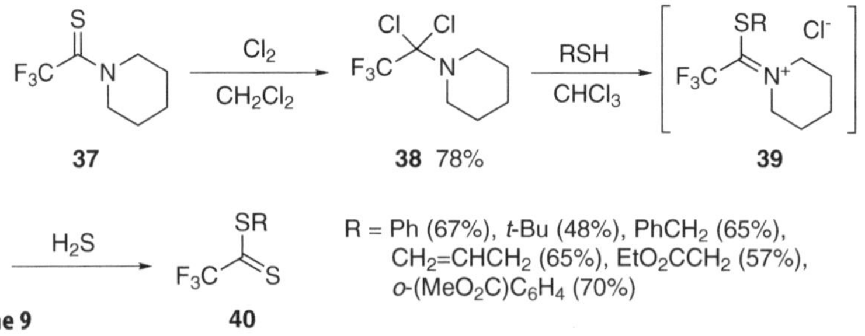

37   **38** 78%   **39**

R = Ph (67%), *t*-Bu (48%), PhCH$_2$ (65%), CH$_2$=CHCH$_2$ (65%), EtO$_2$CCH$_2$ (57%), *o*-(MeO$_2$C)C$_6$H$_4$ (70%)

Scheme 9   **40**

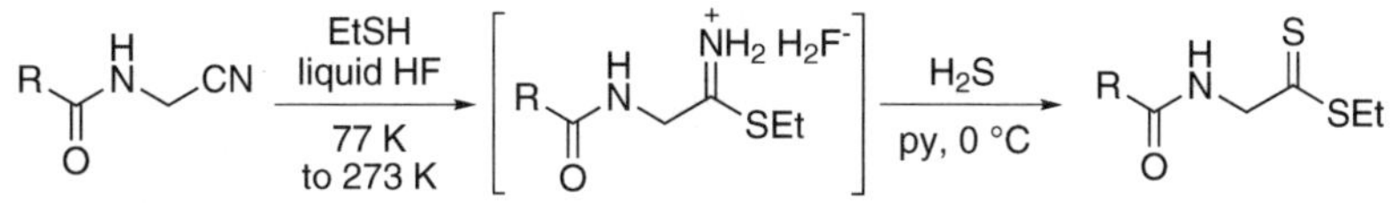

Scheme 10

R = Me (42%), Ph (87%), p-XC$_6$H$_4$ [X = Cl (68%), Br (48%), F (68%), I (48%), MeO (51%)]

2.6
Miscellaneous

Aryl methyl ketones can be converted to α-keto dithioates, as shown in Scheme 11 [43]. Therefore, on treatment with iodine and pyridine, the ketone **44** yields pyridinium salt **45** , and **45** can then be treated first with elemental sulfur in the presence of triethylamine in DMSO and DMF and then with methyl iodide to give S-methyl α-keto dithioate **46**. The reaction of 1-(diethylamino)-2-(phenylthio)acetylene (**47**) with elemental sulfur in refluxing chlorobenzene gives compound **48** in which the dithioester and thioamide groups directly connected to each other (Eq. 17) [44].

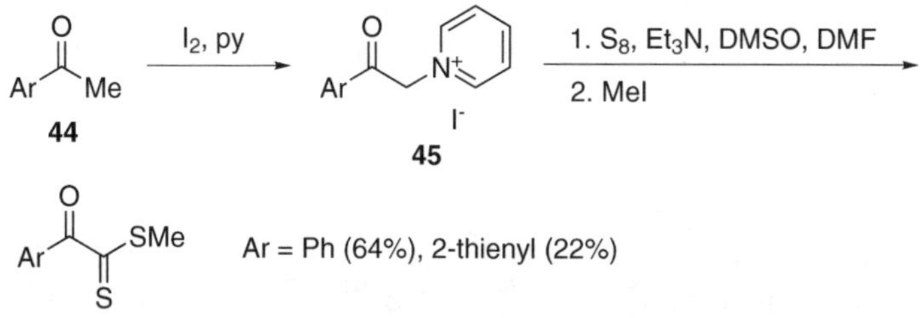

Ar = Ph (64%), 2-thienyl (22%)

Scheme 11   **46**

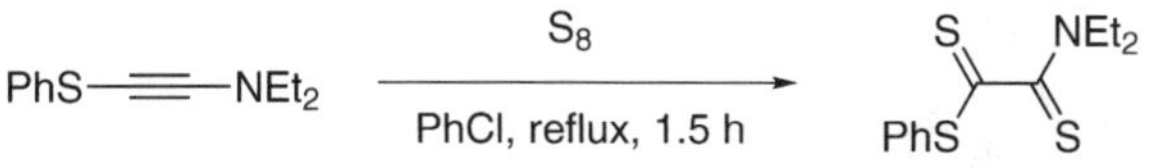

$$\tag{17}$$

47   **48** 74%

Perfluoroalkanedithioic acid ester **49** is synthesized by reacting α,α-dichloro sulfide **50** with zinc sulfide in acetonitrile (Scheme 12) [45], where the zinc sulfide is prepared from ZnCl$_2$, NH$_4$OH, and Na$_2$S. Interestingly, this synthesis does not work with pure ZnS. This dithioester serves as an excellent dienophile toward butadienes.

The reaction of 3-(methylthio)-3H-1,2-dithiolylium iodide (**51**) with 1,3-diaminopropan-2-ol gives bis(enaminodithioester) isomers **52** in an isomeric mixture with 60% combined yield (Eq. 18); these isomers are potential tetradentate ligands [46].

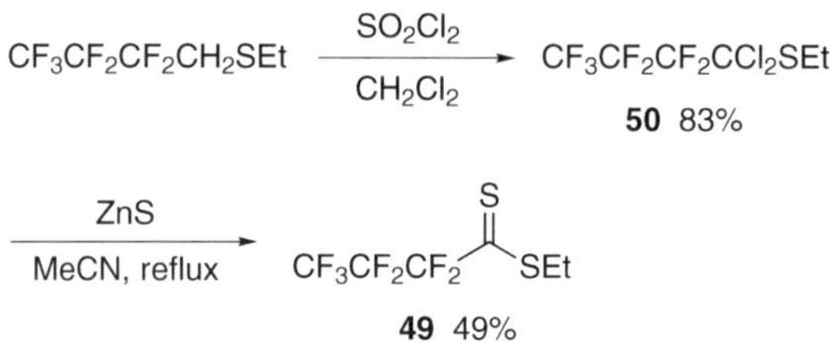

$$CF_3CF_2CF_2CH_2SEt \xrightarrow[CH_2Cl_2]{SO_2Cl_2} CF_3CF_2CF_2CCl_2SEt$$

50 83%

$$\xrightarrow[MeCN,\ reflux]{ZnS} CF_3CF_2CF_2-C(=S)-SEt$$

Scheme 12   **49** 49%

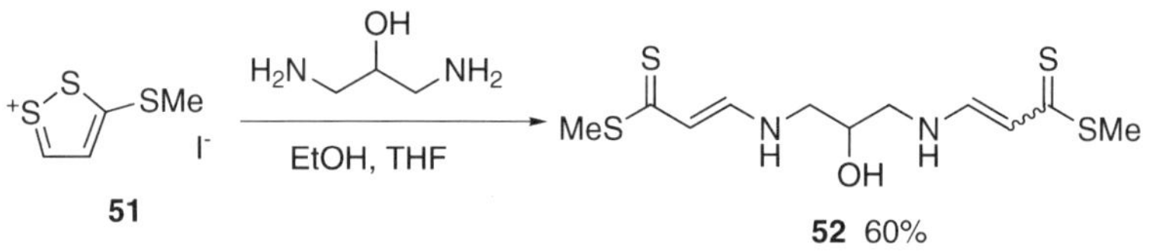

51 + [reagent] $\xrightarrow[EtOH,\ THF]{H_2N-CH_2-CH(OH)-CH_2-NH_2}$ **52**   (18)

52 60%

β-Oxo dithioester **53** was prepared from ketone **54** on treatment with dimethyl trithiocarbonate in the presence of *t*-BuOK in a mixture of benzene and DMF (Scheme 13) [47]. Spectroscopic observations revealed that the dithioester **53** exists mostly in the enolic form **55**. The reaction of **53/55** with hydrazine gives the tricyclic compound **56**. A Diels-Alder reaction of 9-methylanthracene with **57** forms the adduct **58** in 93% yield (Scheme 14) [48]. The thermolysis of **58** in refluxing chloroform resulted in the elimination of $PhSO_2^-$, yielding anthracenecarbodithioate **59** with a yield of 82% through the carbocation intermediate **60**.

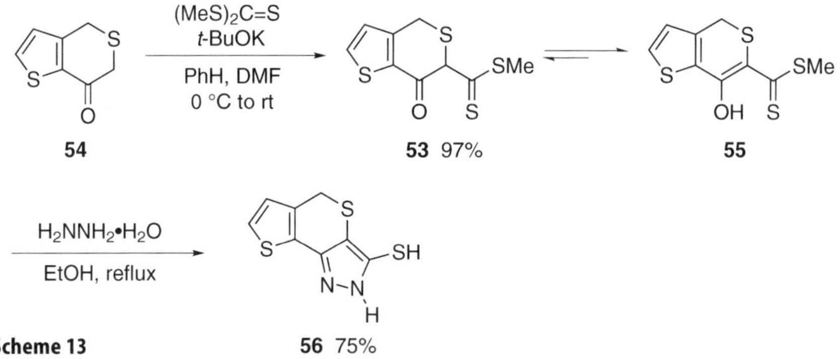

54 $\xrightarrow[\substack{PhH,\ DMF \\ 0\ °C\ to\ rt}]{\substack{(MeS)_2C=S \\ t-BuOK}}$ **53** 97% ⇌ **55**

54   **53** 97%   **55**

$\xrightarrow[EtOH,\ reflux]{H_2NNH_2 \cdot H_2O}$ **56** 75%

Scheme 13   **56** 75%

tert-Alkyl radicals, generated by thermolysis of the corresponding azo compounds, react with disulfides of the type ZC(S)SSC(S)Z (Z=RO, RS, R_2N, and Ph) to furnish the respective ZC(S)SR (R=*tert*-alkyl) [49, 50]. In the case of bis(thiobenzoyl) disulfide (**61**), *t*-alkyl dithiobenzoates **62** are obtained in good yields (Eq. 19). The reaction is carried out in refluxing cyclohexane [49] or refluxing ethyl acetate [50] under inert atmosphere.

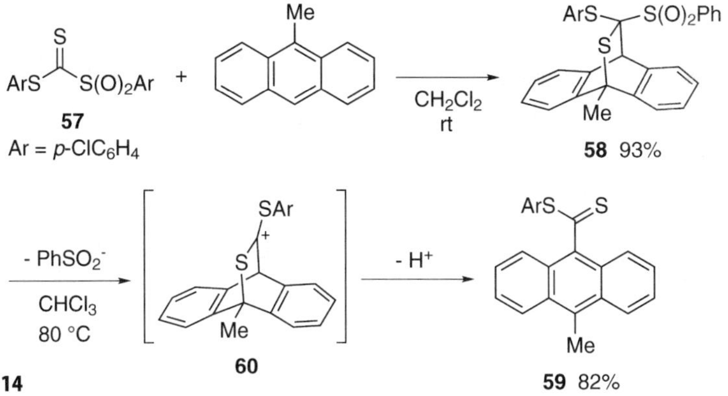

Scheme 14

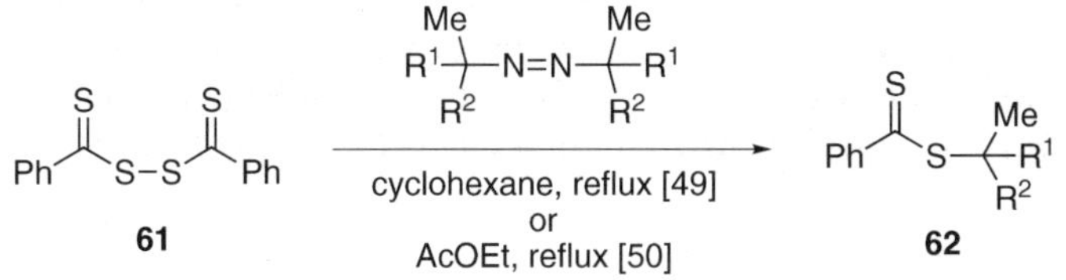

$$R^1 = Me;\ R^2 = CN\ (56\%)\ [49]$$
$$R^1 = Me;\ R^2 = CO_2Me\ (65\%)\ [49]$$
$$R^1 = CH_2CH_2CO_2H;\ R^2 = CN\ (68\%)\ [50]$$

(19)

3
Reactions

3.1
Reaction at α-Position: Aldol Reaction and Conjugate Addition

Anions generated at the α-position of a dithioester react with aromatic aldehydes to give the β-hydroxy dithioates **63** (Scheme 15) [17, 51]. A diastereomeric mixture of **63** (R=PhCH₂) was obtained in low to moderate yields when 3-phenylpropanedithioate was employed [17]. The dithioates **63** lead to (*E*)-α,β-unsaturated dithioates **64** on treatment with MsCl in pyridine.

A spirocyclic dithiolactone **65** was synthesized in high yield by sulfenyl group-assisted dehydration followed by intramolecular cyclization of β-hydroxy γ-(phenylthio) dithioate **66** (Scheme 16) [52, 53].

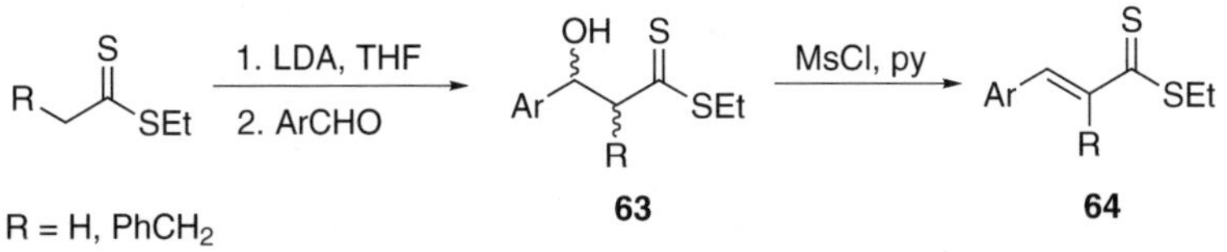

R = H, PhCH₂

Scheme 15

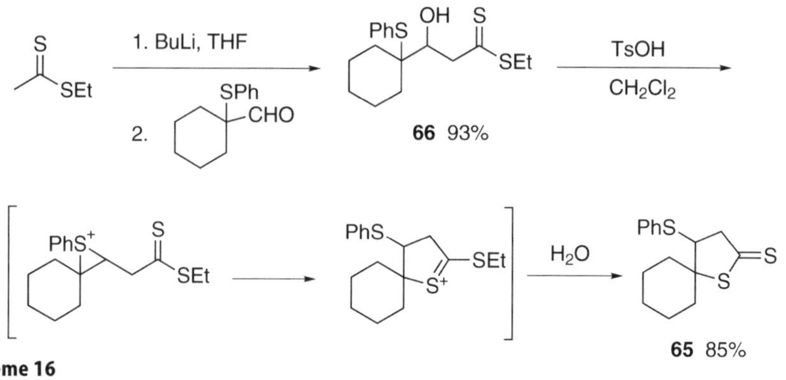

Scheme 16

Acid-catalyzed condensation of α-phosphonyl dithioate **67** with aromatic bis-morpholino aminal **68** yielded α-phosphonyl α,β-unsaturated dithioate **69** in high yield, which underwent hetero-Diels-Alder reactions with alkenes to give the 3,4-dihydro-2H-thiopyrans **70** in high yields (Scheme 17) [54].

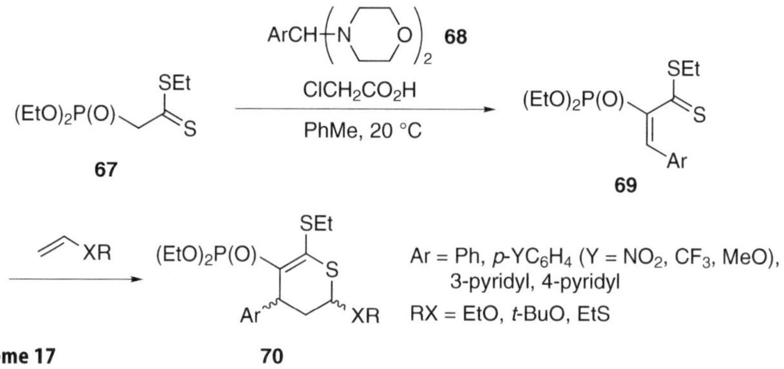

$$Ar = Ph, p\text{-}YC_6H_4 \ (Y = NO_2, CF_3, MeO),$$
3-pyridyl, 4-pyridyl
$$RX = EtO, t\text{-}BuO, EtS$$

Scheme 17   **70**

A conjugate addition of α-sulfonyl dithioate **71** to acrylaldehyde took place quantitatively in the presence of t-butyltetramethylguanidine (**72**) as the base (Eq. 20) [8].

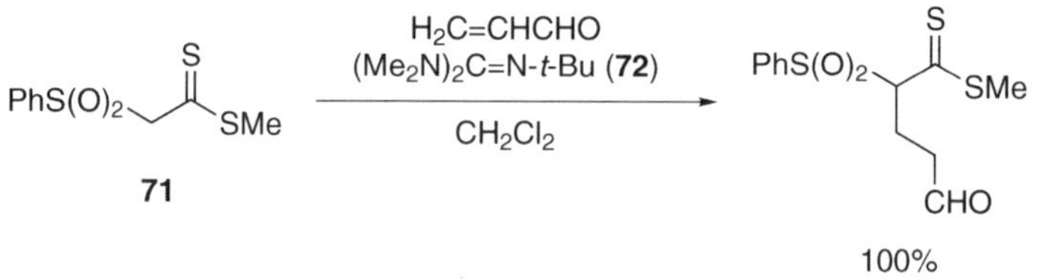

(20)

100%

3.2
Cycloaddition

Perfluoroalkanedithioates readily react with 1,3-butadienes to give thiopyran derivatives [41, 45, 55]. The reaction of butyl trifluorodithioacetate (**73**) with 1-(trimethylsilyloxy)buta-1,3-diene (**74**) produces a mixture of two regioisomers **75** and **76**, each of which consists of two diastereomers present in the ratios of 2.5:1 and 2.3:3.2, respectively [45]. The reaction of methyl trifluorodithioacetate (**77**) with diphenyldiazomathane (**78**) in petroleum ether at –20 °C produces thiirane **79** almost quantitatively [41]. On the other hand, the reaction with ethyl diazoacetate (**80**) in ether at temperatures between 0 °C and room temperature gives thiirane **79** stereoselectively with an 86% yield, along with a 3% yield of 1,3-dithiolane **81**. The formation of **79** and **81** is explained by the intervention of the thiocarbonyl ylide **82** which undergoes a conrotatory ring closure to give thiirane **79**, or a further reaction with **77** to give 1,3-dithiolane **81**.

$$\text{(21)}$$

$$
\begin{array}{c}
\text{73} \quad + \quad \text{74} \quad \xrightarrow{\text{rt}} \quad \text{75} \quad + \quad \text{76}
\end{array}
$$

$$\text{(22)}$$

$$
\text{77} \quad + \quad N_2CR^1R^2 \quad \longrightarrow \quad \text{79} \quad + \quad \text{81}
\end{array}
$$

78: $R^1 = R^2 = Ph$ 98%   -

80: $R^1 = CO_2Et$: $R^2 = H$ 86%   3%

82

A Diels-Alder cycloaddition of methyl pentamethylcylopentadiene-5-carbodithioate (**83**: X=S) with *N*-phenylmaleimide takes place with complete anti-π-facial selectivity to give *anti*-**84**, while that of the corresponding ester **83** (X=O) shows syn-π-facial selectivity to give *syn*-**84** predominantly, as shown in Eq. 23 [56]. This selectivity was interpreted in terms of the deformation of the frontier molecular orbitals of **83**, which is dependent on the orbital energy difference between π_{HOMO} of the diene moiety and n_S [C(S)SMe] or n_O [C(O)OMe] of the 5-substituent.

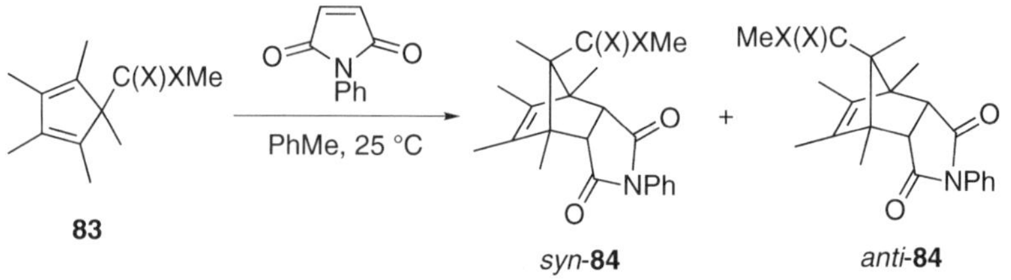

$$(23)$$

	syn-**84**		*anti*-**84**
X = S	0	:	100
X = O	84	:	16

3.3
Thiophilic and Carbophilic Additions

3.3.1
Thiophilic Addition

A characteristic feature of thiocarbonyl compounds is thiophilic addition to the thiocarbonyl sulfur atom by nucleophiles [1]. Reactions of dithioates and thiocarbonyl *S*-oxides (sulfines) with organolithium reagents have been investigated. Addition of organolithium reagents to methyl 2-pyridinecarbothioate (**85**) takes place at the thiocarbonyl sulfur atom, and the resulting carbanion **86** is trapped with an electrophile (E$^+$) to give the dithioacetal **87** in moderate to high yields, as shown in Scheme 18 [57]. The reaction of methyl 3-pyridinecarbothioate with organolithium reagents occurs in the same manner to give the corresponding adducts (21–85%). The reaction of perfluoroalkanedithioates **88** with Grignard reagents and BuLi yields the perfluoroketene dithioacetals **89** by elimination of the fluoride ion from the α-position (Eq. 24) [58].

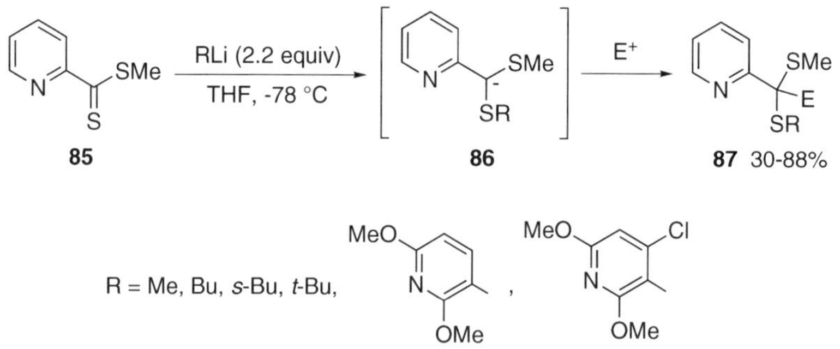

Scheme 18   Electrophile (E$^+$) = H$_2$O, MeI, CH$_2$=CHCH$_2$Br, PhCH$_2$Br, MeCHO

Oxidation of methyl 3-pyridinecarbodithioate (**90**) with *m*-CPBA at 0 °C gives an (*E*) and (*Z*) mixture of the *S*-oxides (sulfines) **91** with a 79% yield

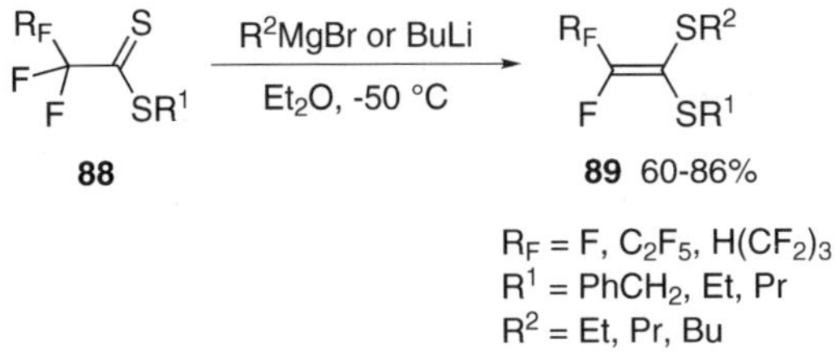

$$R_F = F, C_2F_5, H(CF_2)_3$$
$$R^1 = PhCH_2, Et, Pr$$
$$R^2 = Et, Pr, Bu$$

(24)

89 60-86%

(Scheme 19) [57]. The reaction of **91** with MeLi or BuLi produces a diastereomeric mixture of dithioacetal oxides **92** in 55–90% yields. This thiophilic addition was used as the key reaction in syntheses of 2-cyclopenten-1-ones **93**, as shown in Scheme 20 [59].

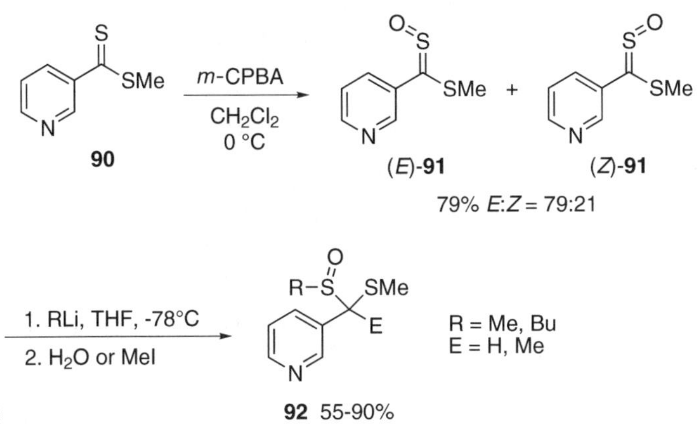

79% $E:Z$ = 79:21

R = Me, Bu
E = H, Me

Scheme 19   **92** 55-90%

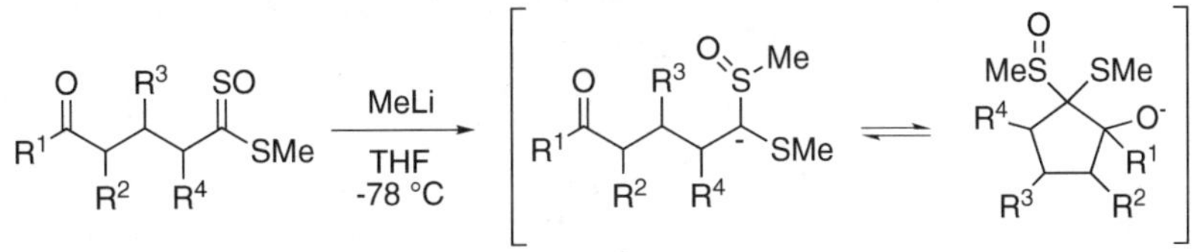

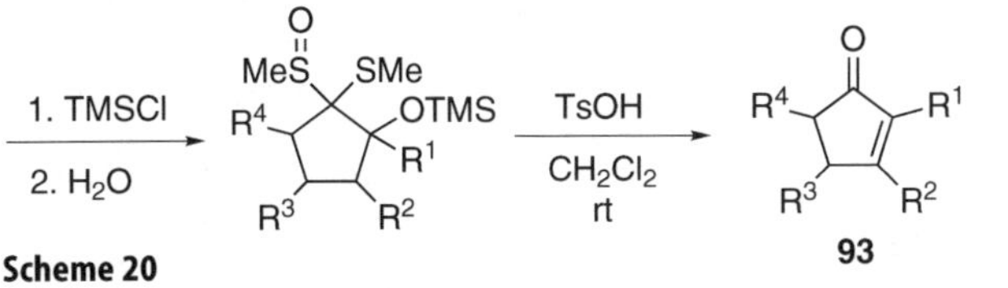

R^1	R^2	R^3	R^4
Me	Me	H	H
Me	H	Me	H
Me	Me	H	Me
-(CH$_2$)$_3$-		Me	H

Scheme 20   **93**

Oxidation of a chiral dithioate **94** with m-CPBA yielded an (E) and (Z) mixture of chiral sulfines **95**, which underwent racemization on standing for 24 h via the enesulfenic acid intermediate **96** (Scheme 21) [26].

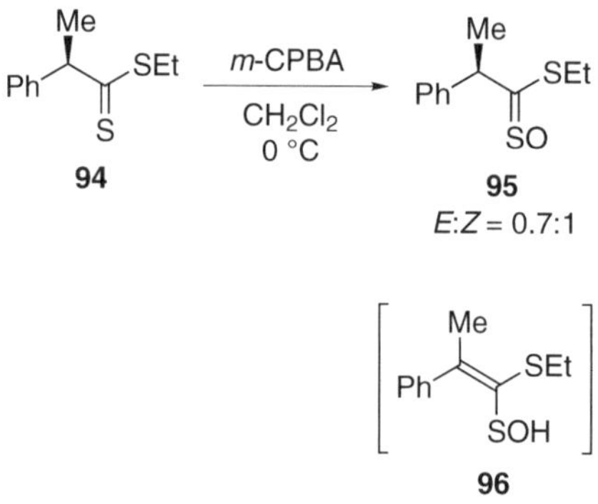

Scheme 21

The reaction of trifluorodithioacetate **77** with *t*-BuSH in the presence of a catalytic amount of *t*-BuOK in THF takes place at the thiocarbonyl sulfur atom to give the disulfide **97** with a 93% yield (Eq. 25) [41].

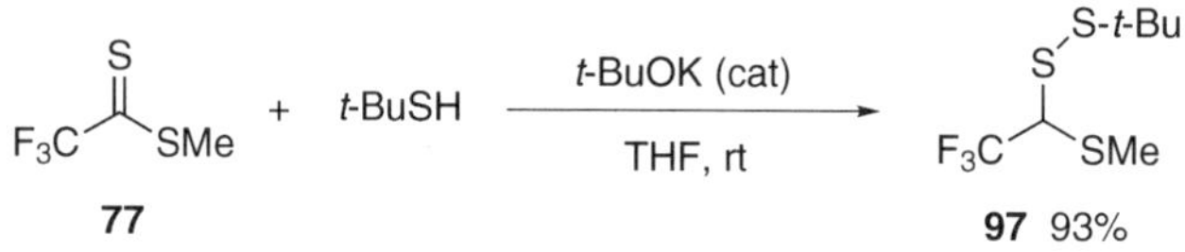

$$\text{(25)}$$

An ene-type reaction of ethyl dithioacetate and dithiopropionate with aminoborane **99** has been reported [60]. While the reaction of dithiopropionate **98** (R=Me) halted at the stage of the *S*-boryl intermediate **100** (80%), the corresponding intermediate formed from dithioacetate **98** (R=H) rearranged to *C*-boryl dithioester **101** (96%). When 2 equiv of **99** was used for **98** (R=H), **101** further reacted with **99** to give 1:2 adduct **102** (Scheme 22).

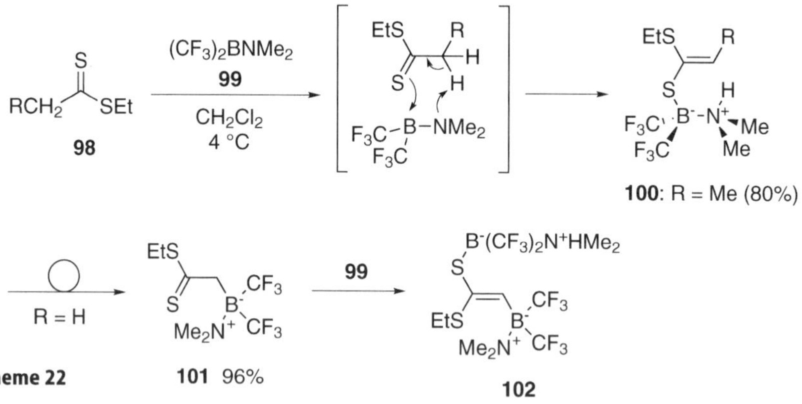

Scheme 22

3.3.2
Carbophilic Addition

3.3.2.1
Thioacylating Reagents

It is well known that reacting dithioates with alcohols and primary and secondary amines produces O-substituted thioates and thioamides, respectively, via carbophilic addition. A kinetic study of the reaction of aryl phenyldithioacetates [$PhCH_2CS_2Ar$] with benzylamine in MeCN concluded that the rate-determining step was proton transfer concurrent with the departure of the leaving group from the tetrahedral intermediate [$PhCH_2C(S^-)(SAr)N^+H_2Ph$] [61]. 2-Benzothiazoyl dithiobenzoate (**103**) is an effective thiobenzoylation agent [62]. Reacting it with primary and secondary amines and alcohols yields thioamides and O-substituted thioates, respectively, in near-quantitative yields with the recovery of 2-mercaptobenzothiazole (**104**), as shown in Eq. 26. Reactions of methyl trifluorodithioacetate (**77**) with various amines and amino acids have also been reported [41].

$$\text{(26)}$$

103   PhNH$_2$ or MeOH / CH$_2$Cl$_2$, rt   Y = PhNH, MeO   **104**

3.3.2.2
Reformatsky Reaction

The Reformatsky reagent **105** reacts with thiocarbonyl compounds – such as dithioates **106** (R^1=Ph, Et; R^2=MeS), cyclic trithiocarbonates **106** [R^1-R^2= -S(CH$_2$)$_n$S-], and thioketones **106** (R^1=Ph; R^2=Me) – via carbophilic addition to give α,β-unsaturated esters **107** (Eq. 27) [63]. In the case of dithioates, β-methylthio α,β-unsaturated esters **107** (R^2=MeS) were obtained in good yields.

$$\text{(27)}$$

106   +   BrZnCH$_2$CO$_2$Et   **105**   PhH, reflux   **107**

$$R^1 = Ph; R^2 = MeS \ (68\%)$$
$$R^1 = Et; R^2 = MeS \ (58\%)$$

$$\text{(28)}$$

108 52%   **109** 21%

The reaction of dimethyl trithiocabonate with **105** gave a mixture of methyl (ethoxycarbonyl)dithioacetate (**108**) and ketene dithioacetal **109** (Eq. 28).

3.4
Inner Salts of Dithioic Acids

Reacting nitrogen-stabilized α-carbenium dithioates **110** and **111** (inner salts) with electrophiles yields the α-carbenium dithioesters **112** and **113**, respectively, in high yields, as shown in Scheme 23 [64] and Scheme 24 [65]. Hydrolysis of **112** and **113** gives α-oxo dithioates **114** and **115**, respectively. In the case of imidazolidinium dithioate **117**, prepared from the inner salt **116**, the hydrolysis is followed by intramolecular cyclization to give a 28% yield of 3-thioxo piperazin-2-one **118** (Scheme 25) [64]. The alkylation of β-carbenium dithioate **119** proceeds similarly (Scheme 26) [66].

110 **112** **114**

R^1 = H, Me, *t*-Bu, Ph
$R^2{}_2N$ = NMe_2, pyrrolidino
R^3 = Me, Et, Bu

Scheme 23

111 **113** **115**

Scheme 24

116 **117** **118** 28%

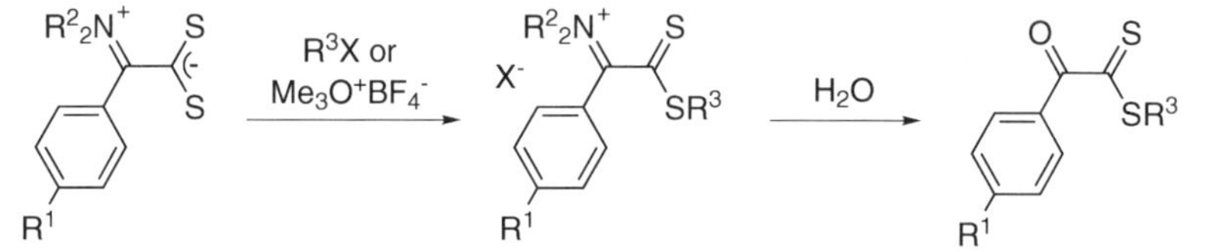

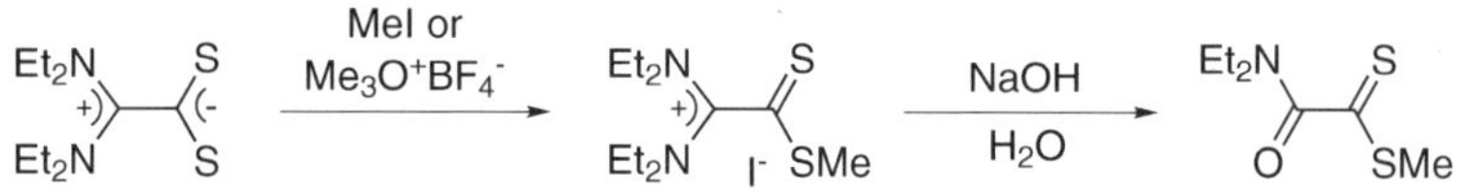

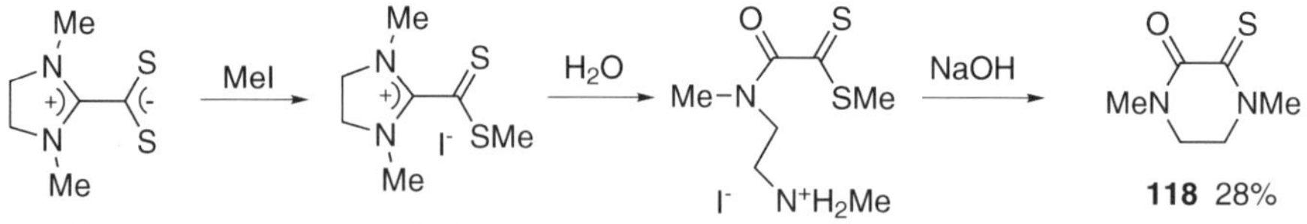

Scheme 25

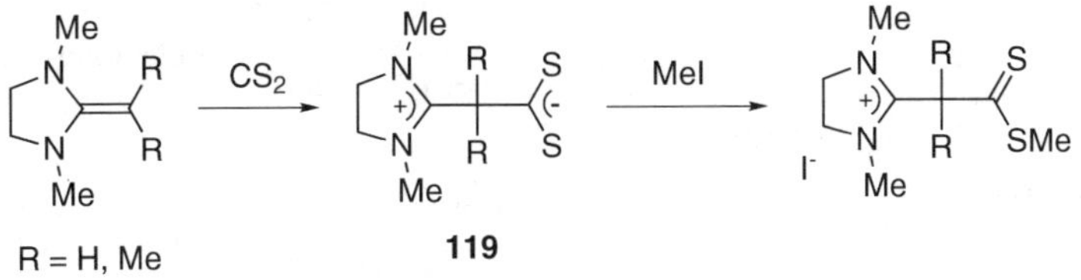

Scheme 26

The reactions of iminium dithioate **120** are summarized in Scheme 27 [64]. The reaction of **120** with aniline takes place initially at the iminium carbon, prompting an iminium exchange reaction followed by deprotonation to produce α-imino dithioate **121**. The same reaction also takes place for *p*-phenylene-diamine. The reaction with *o*-phenylenediamine occurs first at the iminium carbon and second at the dithioate carbon to give the quinoxaline **122**. When morpholine is employed, the resulting iminium salt reacts further with another molecule of morpholine (when 2 equivs of morpholine is used) or methanol (solvent) to give α,α-dimorpholino dithioate **123** or α-methoxy-α-morpholino dithioate **124**, respectively.

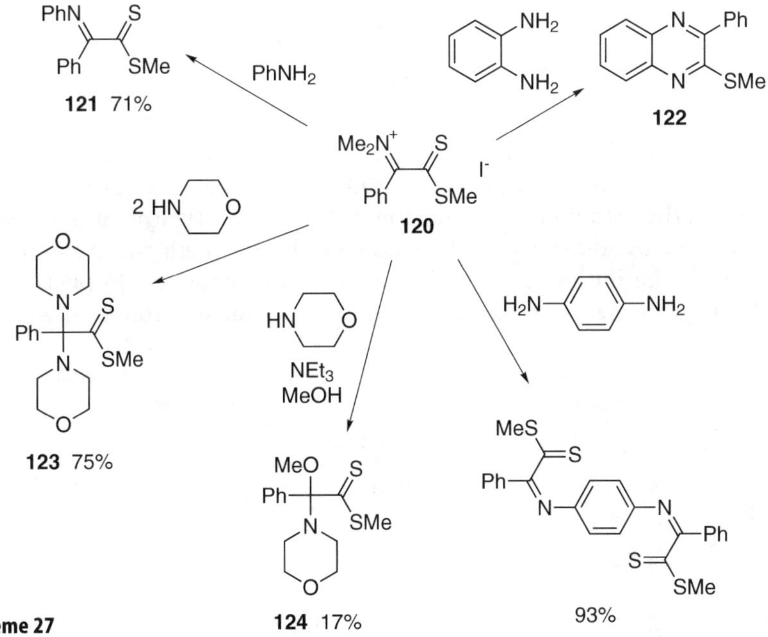

Scheme 27

The reactions of bis(diethylamino)carbenium salt **113** are summarized in Scheme 28 [65]. Nucleophilic additions of Grignard reagents, organolithium reagents, enolates, amides, thiolates, and phosphite take place regioselectively

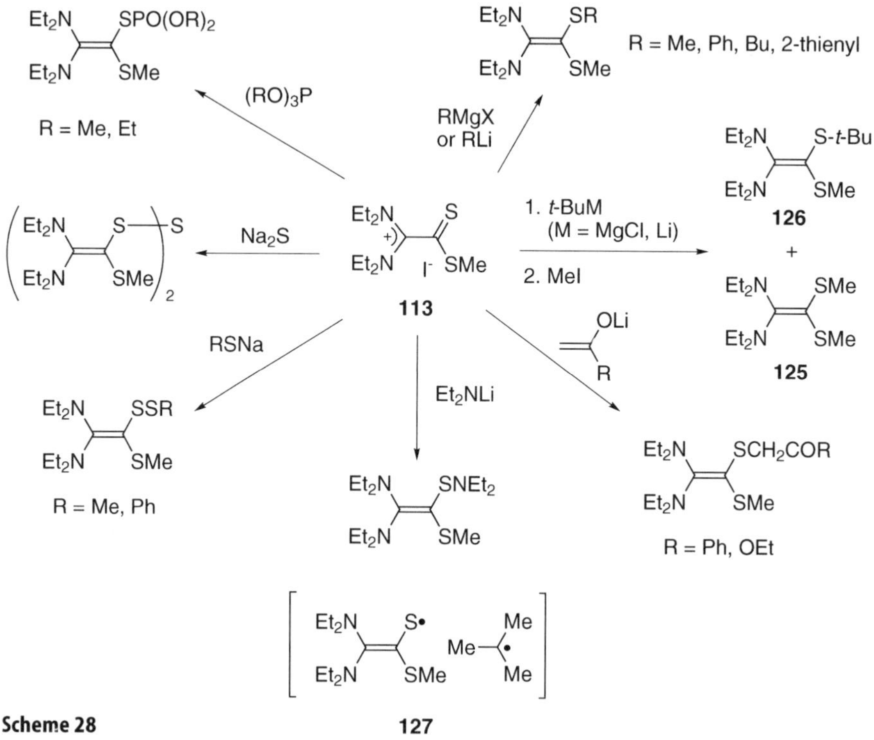

Scheme 28   **127**

at the thiocarbonyl sulfur atom to yield the corresponding alkenes in high
yields. This is in sharp contrast to reactions with oxygen nucleophiles, which
take place at the carbenium carbon atom (Scheme 24). Treatment of **113** with
t-BuLi followed by addition of methyl iodide yields bis(methylthio) compound
125 (9%) in addition to the normal *S-tert*-butyl compound **126** (85%). This
reaction suggests the occurrence of single-electron transfer from *t*-BuLi to **113**,

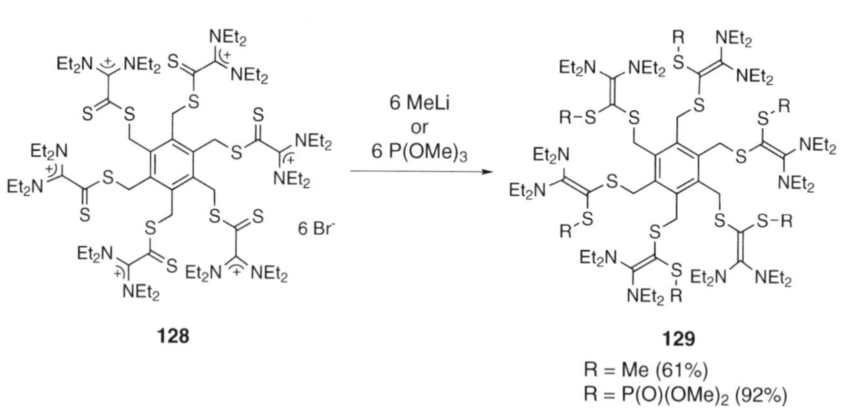

128   **129**

R = Me (61%)
R = P(O)(OMe)$_2$ (92%)

$$\tag{29}$$

giving the radical pair **127**. Combination with **127** yields **126**, while disproportionation (giving $(Et_2N)_2C=C(SMe)SH$ and 2-methylpropene) followed by reaction with methyl iodide gives **125**.

The reaction of hexa(bromomethyl)benzene with 6 equiv of **113** yields **128**, which bears six carbenium salt units (Eq. 29) [67]. The compound **128** reacts with 6 equiv of MeLi or $P(OMe)_3$ to furnish huge molecules (**129**) in high yields.

3.5
Miscellaneous

As shown in Scheme 29, aromatic and vinylic dithioester groups can be converted to trifluoromethyl groups directly by treating them with $(Bu_4N)H_2F_3$-DBH (1,3-dibromo-5,5-dimethylhydantoin) (Path a) [17]. The use of NBS or NCS (Path b) in place of DBH yields α,α-difluoro sulfides **130** [17]. This fluorination can also be performed with poly(hydrogen fluoride) pyridinium (Path c) [32].

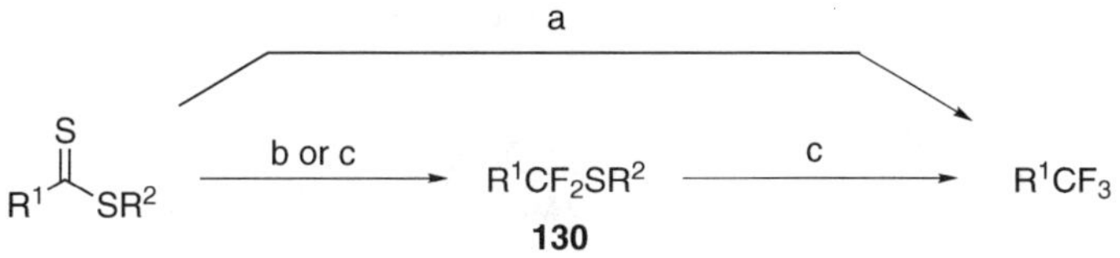

$$R^1CF_2SR^2 \qquad R^1CF_3$$
130

a: $(Bu_4N)H_2F_3$, DBH, CH_2Cl_2, 0 °C to rt [17]
b: $(Bu_4N)H_2F_3$, NBS or NIS, CH_2Cl_2, 0 °C to rt [17]
c: $(HF)_n$•py, DBH, CH_2Cl_2, 20 °C [32]

Scheme 29

It has been suggested that dithioate **131** serves as a single-electron acceptor in the reaction with LDA to yield bis(methylthio)stilbene **132** (Scheme 30) [68]. In this reaction, dianion **133**, the precursor of **132**, is believed to be formed by the dimerization of anion radical **134**, generated by a single-electron-transfer

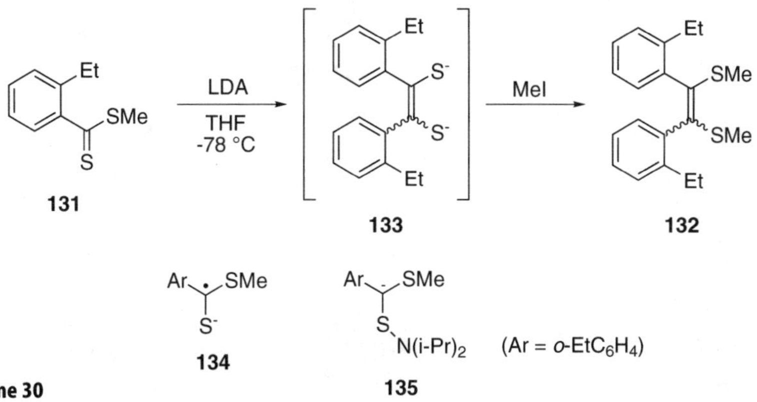

131

133

132

134

135 $(Ar = o\text{-}EtC_6H_4)$

Scheme 30

process from LDA to **131**, followed by the elimination of Me_2S_2. An alternative mechanism has been suggested, where **133** is formed by the carbophilic addition of anion **135** (generated by the thiophilic addition of LDA to **131**) to **131**, followed by elimination of Me_2S_2 and $i\text{-}Pr_2N^+$ as $(i\text{-}Pr_2N)_2$.

3.6
Metal Complexes

Dithioates with nitrogen atoms at the β- or γ-position serve as bidentate ligands for metals through their thiocarbonyl sulfur and nitrogen atoms. Thiazole-2-carbodithioate **136** reacts with $M(CO)_5(THF)$ (M=Cr, W) and $Fe(CO)_4(THF)$ to give various complexes **137** in low yields, as shown in Eq. 30 [16]. The five-membered chelete structures of these complexes have been established by X-ray crystallography. Methyl 2-(cyclohexylamino)pentene-1-carbodithioate (**138**) undergoes similar complexation with a Rh(II) complex to give the six-membered Rh(I) complex **139** (Scheme 31) [69]. The Rh(I) complex **139** can be used to obtain the

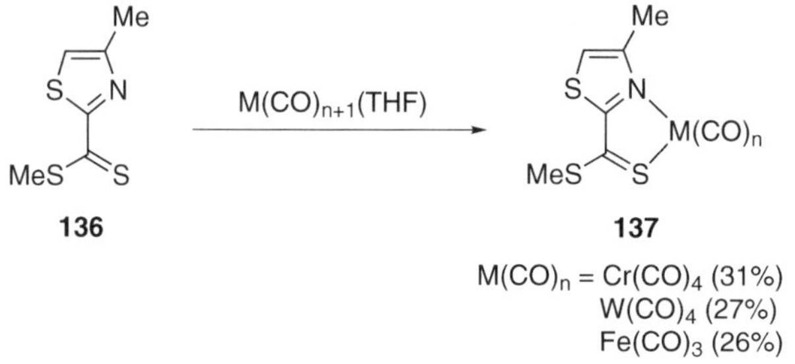

$$M(CO)_{n+1}(THF) \tag{30}$$

136   **137**

$$M(CO)_n = Cr(CO)_4\ (31\%)$$
$$W(CO)_4\ (27\%)$$
$$Fe(CO)_3\ (26\%)$$

138   **139** 40%

140 84%   **141**   **142**

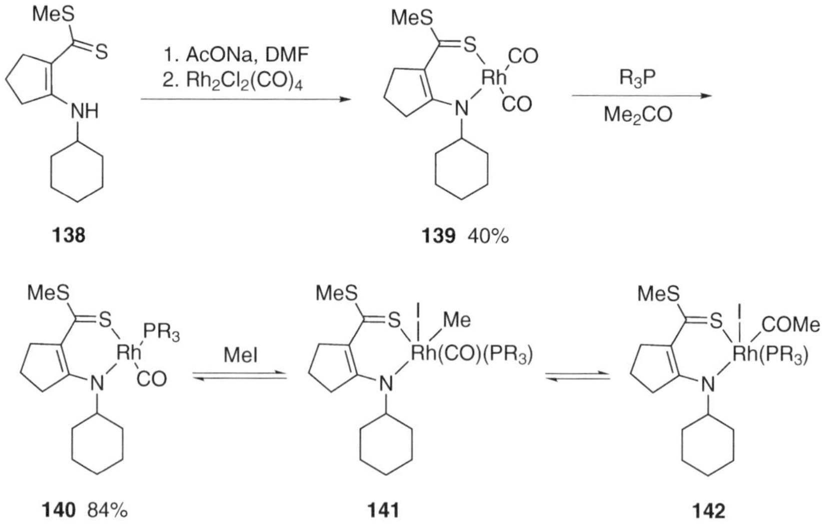

Scheme 31

phosphine complex **140**, which undergoes the oxidative addition of methyl iodide, forming the Rh(III)-alkyl complex **141** via an equilibrium step, followed by the migratory insertion of CO to furnish the Rh(III)-acyl complex **142**. A kinetic study of the oxidative-addition-migratory-insertion process has been reported.

The reaction of β-amino dithioate **143** with butyltin trichloride (0.5 equiv) under basic conditions yields the 2:1 complex **144** in 75% yield (Eq. 31) [70]. A hexa-coordinated hypervalent structure was proposed for **144**, based on spectroscopic data.

$$(31)$$

4
Cyclic Dithioates

The chemistry associated with cyclic dithioates is varied and interesting. In particular, because $3H$-1,2-dithiole-3-thiones **145** are involved in a wide varitey of biological activities, their chemistry has attracted considerable attention. A 5-(p-methoxyphenyl) derivative (anethole dithiolthione, ADT) is suggested to be a potentially efficacious chemoprevention agent for lung cancer [71], and a 5-(2-pyrazinyl)-4-methyl derivative (oltipraz) is reported to inhibit HIV-1 (AIDS) virus replication by irreversibly binding to the viral reverse transcriptase enzyme [72]. A study of metabolites of oltipraz and related compounds has been reported [73].

4.1
Five-Membered Cyclic Dithioates

The synthesis of the five-membered cyclic dithioate **65** is shown in Scheme 16 [52, 53]. The parent γ-dithiolactone (3,4,5-trihydrothiophene-2-thione) (**146**) is prepared by thionation of the corresponding thiolactone **147**, as shown in Eq. 32 [74]. Preparations for the 1,3-dithiolane-4-thione derivatives **148** have been reported [75, 76] (Scheme 32). It was proposed that the adamantanespiro compound **148a** was formed by a reaction of the thiocarbonyl ylide, generated from **149** by extrusion of N_2, with carbon disulfide [75].

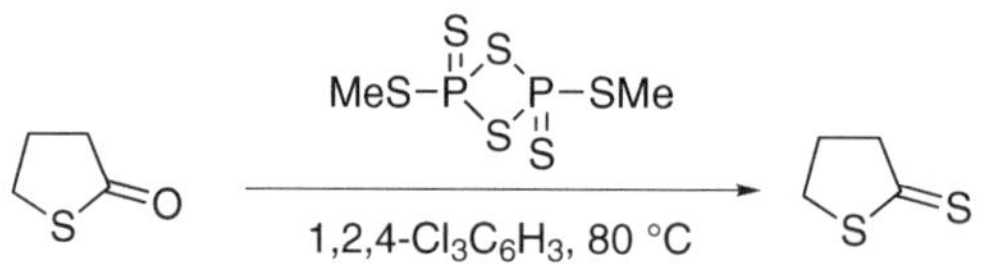

$$ (32) $$

147 → **146** 69%

MeS–P(S)(S)–P(S)(S)–SMe, 1,2,4-Cl$_3$C$_6$H$_3$, 80 °C

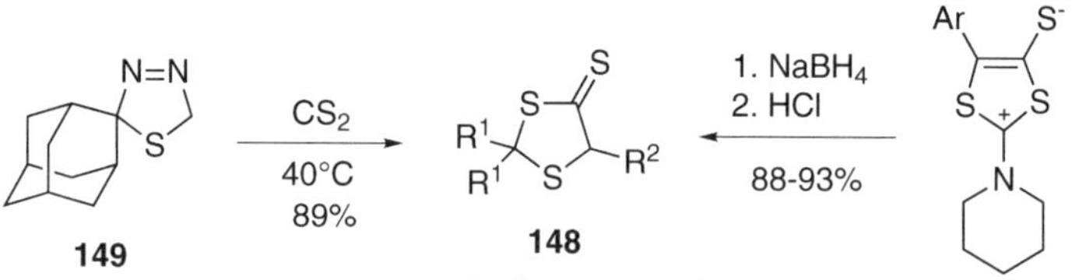

149 + CS$_2$, 40°C, 89% → **148** ← 1. NaBH$_4$ 2. HCl, 88–93%

a: R^1-R^1 = C$_9$H$_{14}$, R^2 = H [75]
b: R^1 = H; R^2 = p-XC$_6$H$_4$ (X = Me, Cl, HO$_2$C) [76]

Scheme 32

3H-1,2-Dithiole-3-thiones **145** can be prepared from enolizable ketones **150** [77–79]. Here, **150** is treated with carbon disulfide under basic conditions, and the resulting β-keto ketene dithiolate **151** is treated with hexamethyldisilathiane and then an oxidizing agent such as hexachloroethane to furnish **145** in high yield (Scheme 33) [77]. The reaction of a β-keto ketene dithiol (the protonated form of **151**) with P$_4$S$_{10}$ also gives **145** (R^1=p-PhC$_6$H$_4$; R^2=H) [80]. 5-(Alkylthio or phenylthio)-3H-1,2-dithiole-3-thiones **152** have been synthesized in low to high yields by reacting dithiomalonates **153** with P$_4$S$_{10}$-S$_8$ or LR-S$_8$ in refluxing xylene in the presence of catalytic amounts of 2-mercaptobenzothiazole (MBT) and ZnO (Eq. 33) [81].

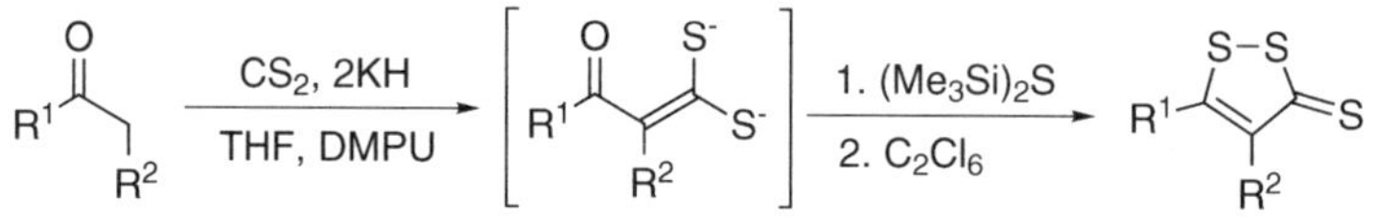

150 → [**151**] → **145**

CS$_2$, 2KH, THF, DMPU → 1. (Me$_3$Si)$_2$S 2. C$_2$Cl$_6$

R^1	R^2	Yield	R^1	R^2	Yield
Ph	H	99%	-(CH$_2$)$_3$-		73%
Ph	Me	91%	-(CH$_2$)$_4$-		77%
Ph	SMe	90%	Me	H	76%
Tol	H	88%	Me	Me	60%
t-Bu	H	91%	Et	Me	72%

Scheme 33

S-Isopropyl and N-isopropyl groups are converted to 3H-1,2-dithiole-3-thione rings by sulfuration and chlorination with disulfur dichloride [82, 83].

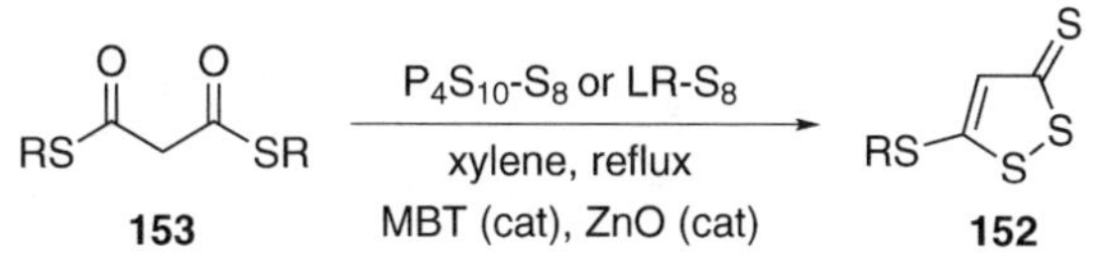

$$(33)$$

	P_4S_{10}-S_8	LR-S_8		P_4S_{10}-S_8	LR-S_8
Bu	61%	83%	Ph	44%	82%
n-C$_{12}$H$_{25}$	65%	86%	t-Bu	17%	47%
cyclopentyl	46%	75%	PhCH$_2$	41%	71%

As shown in Eq. 34, the reaction of diisopropyl sulfide with disulfur dichloride in chlorobenzene at room temperature in the presence of DABCO provides **154** and the dimeric disulfide **155** in 33% and 19% yields, respectively [82]. Similarly, treatment of *N*-isopropyl compound **156** with disulfur dichloride and DABCO and then triethylamine furnishes a 5-chloro derivative **157** in 43% yield (Eq. 35) [83].

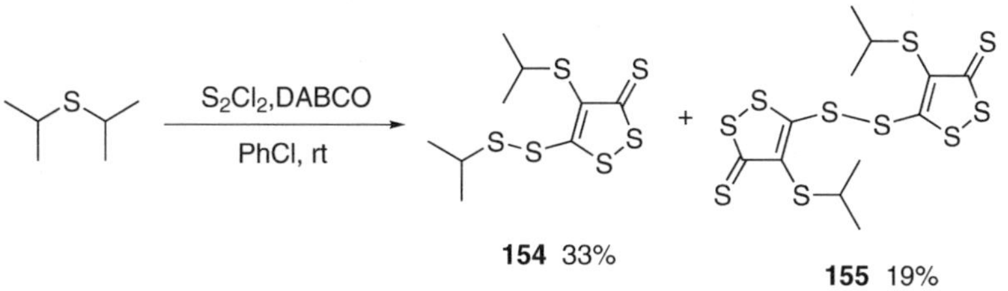

$$(34)$$

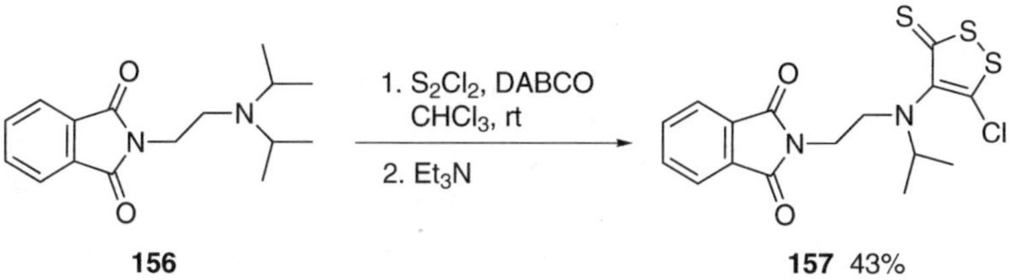

$$(35)$$

The reduction of 1,2-dithiolium salt **158** with sodium borohydride followed by oxidation yields the 4,5-bis(methylthio) derivative **159** in 95% yield (Scheme 34) [84]. 4-Fluoro-5-(fluoroalkyl) derivatives **160** are synthesized from the corresponding ketene dithioacetals **161** (Scheme 35) [85, 86]. The 4,5-bis(methoxy-

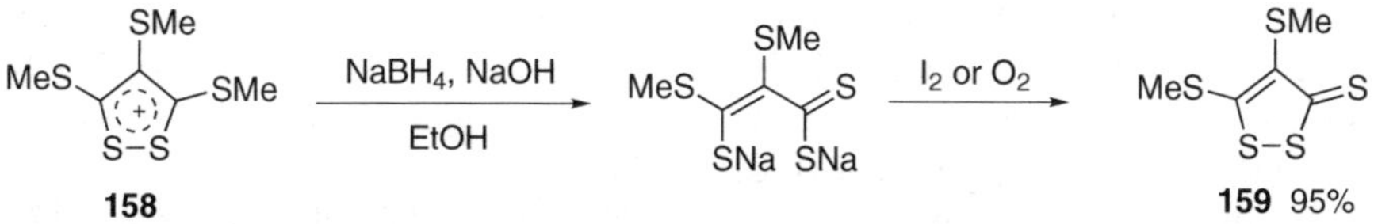

Scheme 34

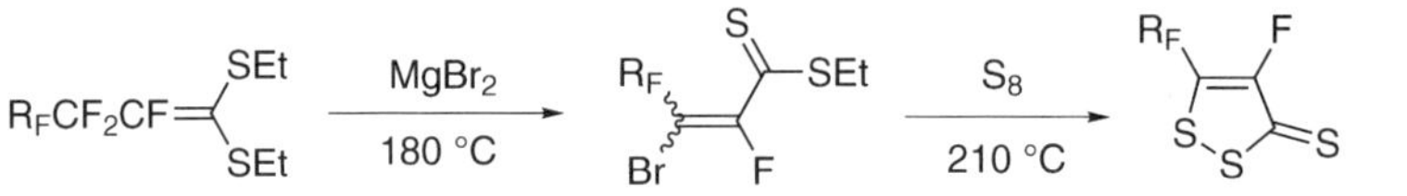

Scheme 35

RF = CF$_3$, H(CF$_2$)$_2$

161 → **160**

carbonyl) derivative **145** (R^1=R^2=CO$_2$Me) is obtained by reacting dimolybdenum μ-DMAD complexes [87].

Functionalization of the 5-alkyl substituent was achieved by treating the 4,5-dimethyl derivative **162** with NaNO$_2$ in acetic acid, followed by treatment with sodium sulfide to give the 5-(hydroxyiminoalkyl) derivative **163** (Scheme 36), which was further converted to the 5-acyl derivative [88–90]. The reaction of 4-phenyl-3H-1,2-dithiole-3-thione (**164**) with α,β-unsaturated nitriles **165** gives thiopyrano[2,3-b]pyrane **166** (Scheme 37) [91]. The synthesis and X-ray analysis of a cadmium complex with a 4,5-disulfanyl-3H-1,2-dithiole-3-thione ligand has been reported [92].

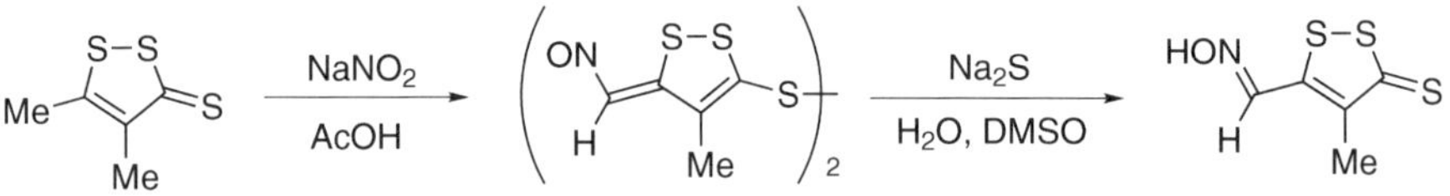

162 → **163**

Scheme 36

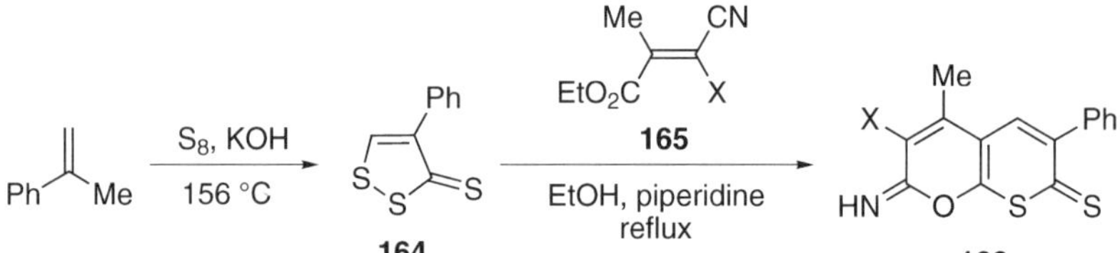

164 → **166**

X = CN (75%)
CO$_2$Et (83%)

Scheme 37

4,5-Bis(alkylthio)-3H-1,2-dithiole-3-thiones **167** can be transformed to 2,4-dimethylidene-1,3-dithietanes **168** on treatment with trialkyl phosphites [93, 94] or triphenylphosphine [82]. For example, the bis(dithiacrownether) derivatives **169** have been synthesized by this reaction [93]. The reaction of 5-(alkylthio)-3H-1,2-dithiole-3-thiones with Fischer carbene complexes gives 6-(alkylthio)-3H-1,2-dithiine-3-thione complexes with Cr and W [95].

Syntheses and reactions of bicyclic (**170** [96] and **171** [97]), tricyclic (**172** [98], **173** [99–103], **174** [104], and **175** [105, 106]), and tetracyclic (**176** [107])

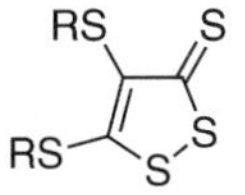

167 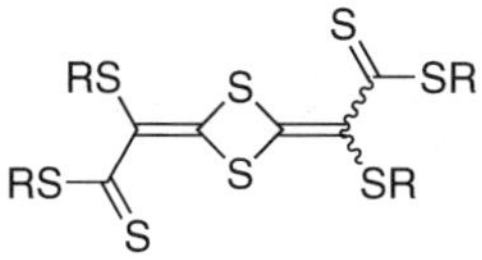   **168** 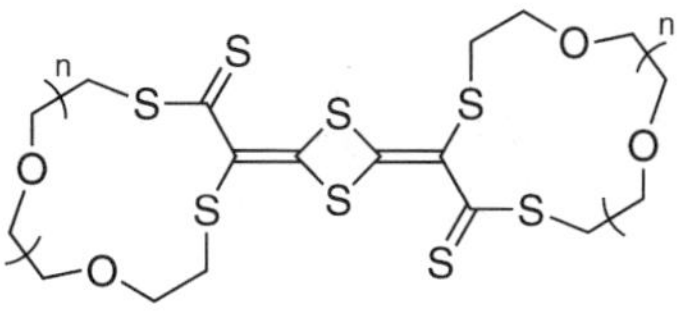   **169**

n = 1,2,3

$3H$-1,2-dithiole-3-thiones are also recent areas of interest with regard to cyclic dithioates. The chemistry of 1,3-thiazoline-5-thiones (**177** [108–110]), 5-thioxo-1,3-thiazolidin-2-ones (**178** [111]), and fused thiadiazolethiones (**179** [112]) has been reported.

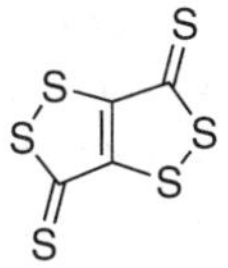

170 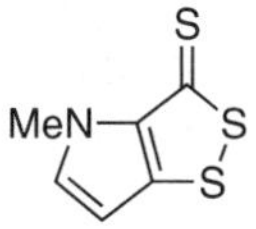   **171** 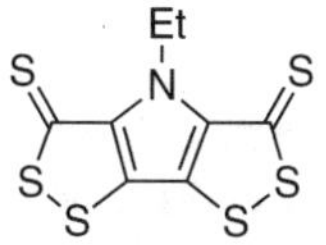   **172** 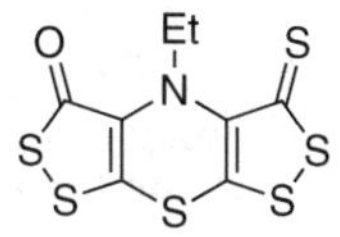   **173**

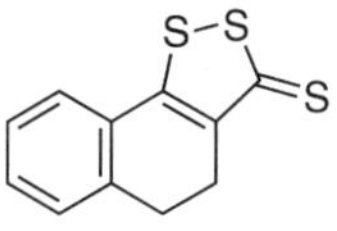

174 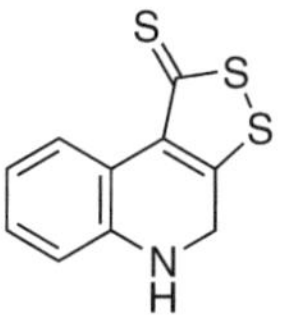   **175** 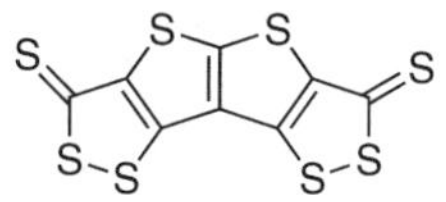   **176**

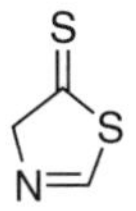

177 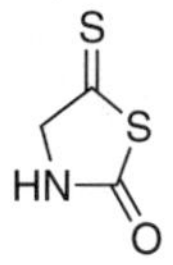   **178** 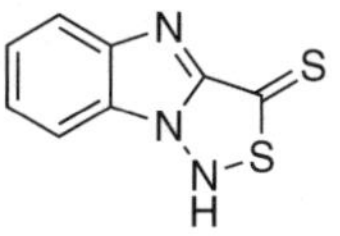   **179**

4.2
Six-Membered Cyclic Dithioates

In the thermal reaction of fluorinated fullerene ($C_{60}F_{18}$) with tetrathiafulvalene (**180**, TTF), a thiin-2-thione adduct **181** was isolated as a minor product. This is believed to form by a cycloaddition of thiin-2-thione **182** (formed by elimination of CS_2 from TTF) with the initial adduct ($C_{60}F_{16}$-TTF) of $C_{60}F_{18}$ with TTF (Scheme 38) [113]. A convenient (although low-yield) method for the preparation of 6-arylthiin-2-thione **183** has been reported (Eq. 36) [114].

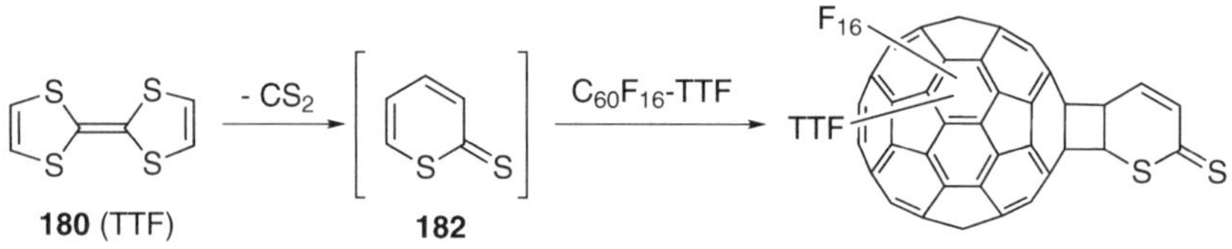

Scheme 38

180 (TTF) — **182** — **181**

(36)

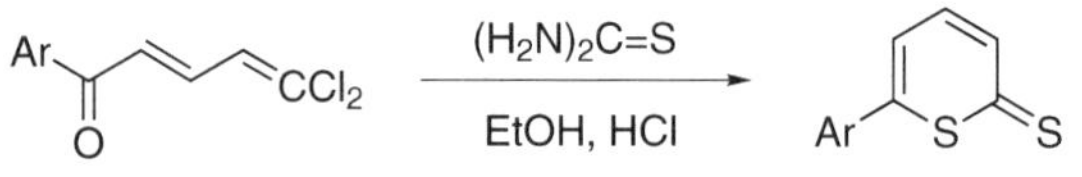

183 12-38%

The reaction of disulfide **184** with electron-deficient acetylenes **185** yields 2,3-disubstitued-4-(diethylamino)thiin-2-thiones **186** in moderate yields (Scheme 39) [115]. Syntheses of fused thiin-2-thione derivatives, **187** [91], **188** [116], **189** [117], and **190** [118] have been reported.

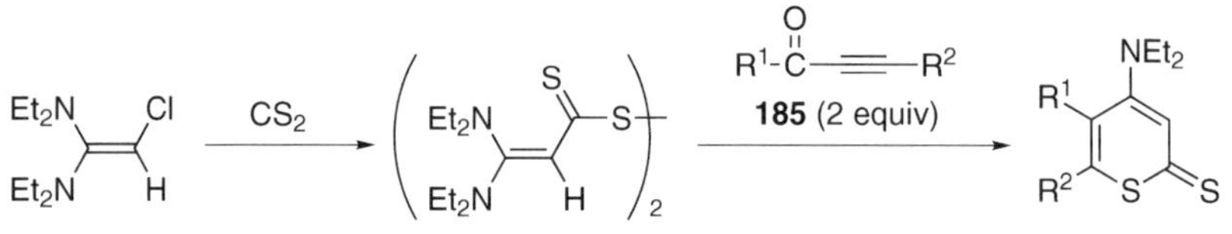

185 (2 equiv)

184

186 32-60%

R^1 = Ph, Me, MeO, EtO
R^2 = COPh, H, Ph, Me

Scheme 39

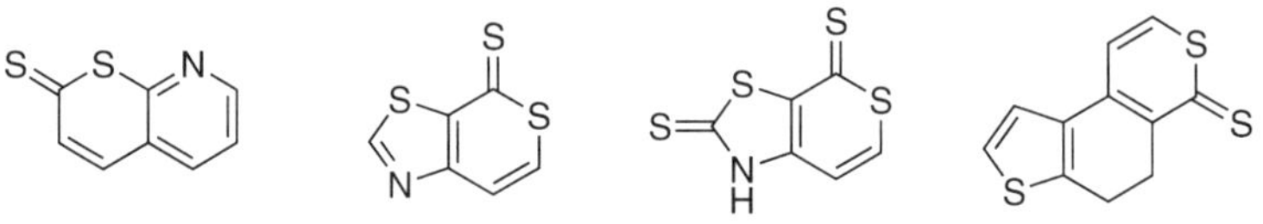

187   **188**   **189**   **190**

1,3-Thiazine-6-thiones as well as 1,3-thiazinones have been explored as possible antitumor and antibacterial agents, and antiviral agents for HIV [119]. The 1,3-thiazine-6-thiones **191** were prepared by reactions starting from 1,2,4-dithiazolium salts **192** [120] or 2,4-diamino-1-thia-3-azabutadienes **193** [119] (Scheme 40), and **194** from 1-acyl-2-bromoacetylenes **195** [121] or 2,2-dichlorovinyl trifluoromethyl ketone **196** [122] (Scheme 41).

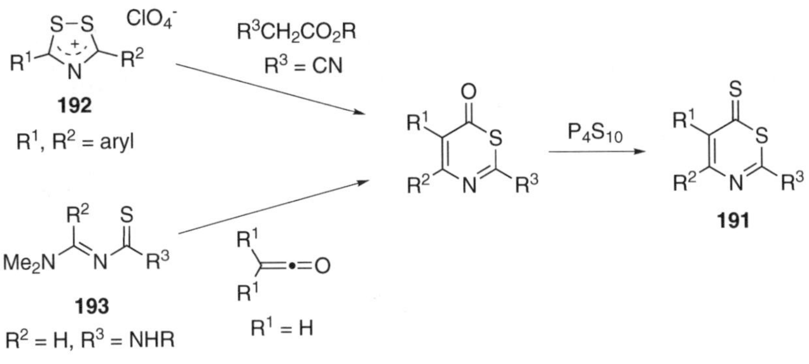

Scheme 40

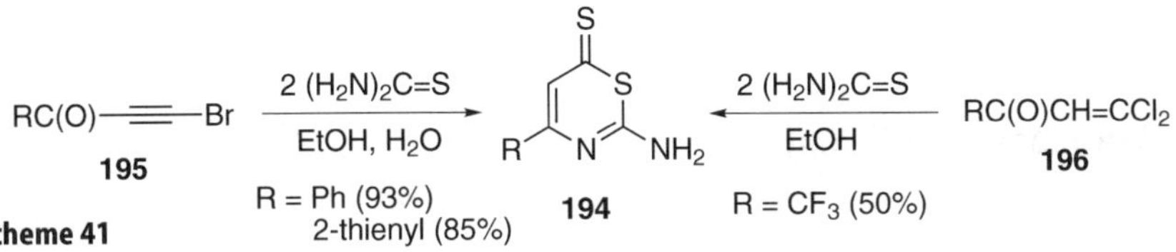

Scheme 41

R = Ph (93%)
2-thienyl (85%) **194** R = CF$_3$ (50%)

4.3
Seven-Membered Cyclic Dithioates

The reaction of inner salt **119** (R=Me) (see Scheme 26) with two molecules of DMAD yielded 3H-thiepin-2-thione **197** with a 86% yield (Eq. 37) [66].

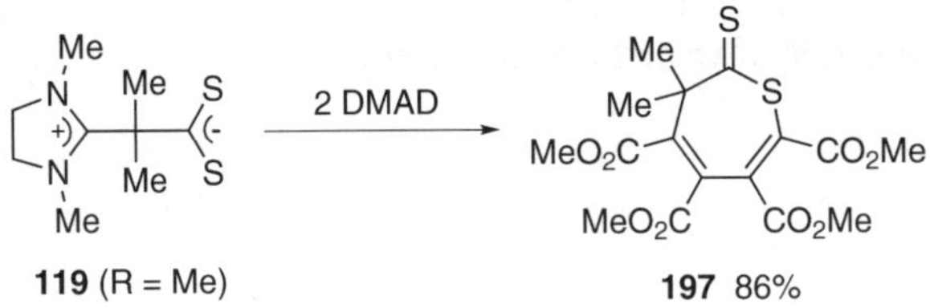

$$\text{119 (R = Me)} \quad \xrightarrow{\text{2 DMAD}} \quad \text{197 86%} \tag{37}$$

4.4
Anhydrides

Anhydrides of dithioic acids have a dithioate moiety. Thiothionophthalic anhydride **198** was prepared with a 85% yield by treating dithioacetal **199** with LDA in THF containing 0.83 equiv of HMPA (Scheme 42) [123]. The compound **198** loses a sulfur atom over 110 °C, dimerizing to **200**, like the unsubstituted thiothionophthalic anhydride [124]. An improved synthesis of and some reactions of trithio-1,8-naphthalic anhydride (**201**) has been reported (Eq. 38) [125].

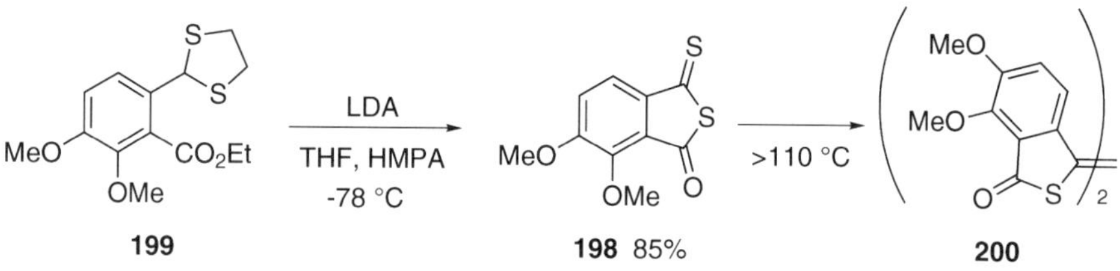

Scheme 42

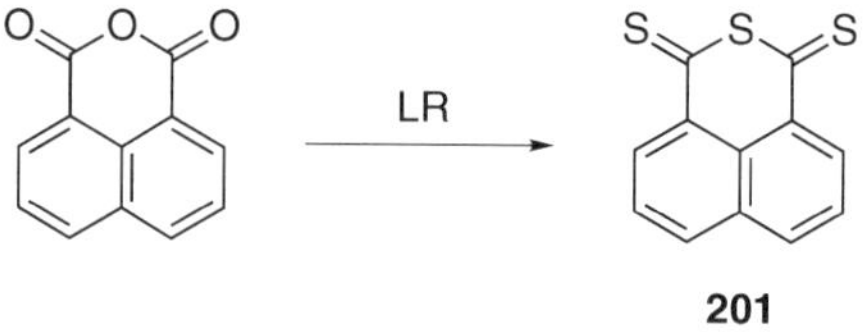

$$(38)$$

201

5
S-M Dithioates [RC(S)SM]

5.1
M=Group 14 Element (Except C)

$(MesCS_2)_2SiPh_2$ (**202**) was prepared with a 44% yield by reacting potassium 2,4,6-trimethyldithiobenzoate (**203**) with Ph_2SiCl_2, as shown in Eq. 39 [126]. The compound **202** is thermally quite stable, but highly sensitive to moisture. On X-ray analysis, **202** exhibited a slightly distorted tetrahedral structure, where the intramolecular distances between the silicon and thiocarbonyl sulfur atoms [3.315(3) and 3.341(3) Å] were 14–15% shorter than the sum of the van der Waals radii (3.90 Å), indicative of weak Si–S interaction.

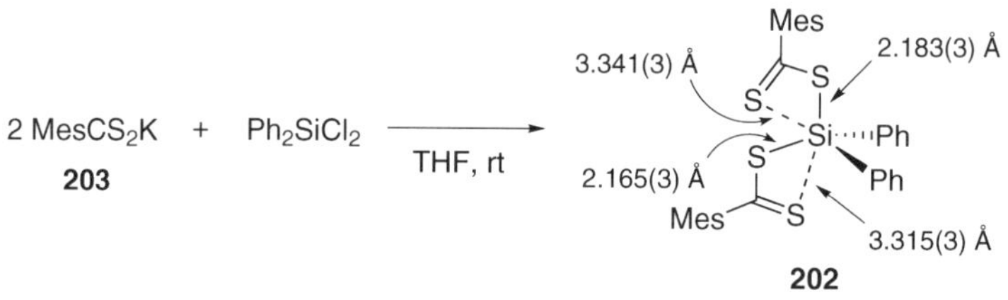

$$(39)$$

202

The *S*-germanium, *S*-tin, and *S*-lead compounds **204** were synthesized in 28–93% yields by stoichiometric reactions of piperidinium dithioates **205** with Ph_3-$GeCl$, Ph_2GeCl_2, Ph_3SnCl, Ph_2SnCl_2, $PhSnCl_3$, Ph_3PbCl, and Ph_2PbCl_2 (Eq. 40) [127].

X-Ray analyses of some of these compounds suggested weak interactions between the thiocarbonyl sulfur atoms and the Group 14 metals (M), where the dithioate groups served as anisobidentate ligands toward the M atom. In terms of example structures, $(TolCS_2)_2SnPh_2$ (**206**) took a distorted octahedral or skewed trapezoidal bipyramid and $(o\text{-}MeC_6H_4)_3SnPh$ (**207**) took a seven-co-

$$\text{205} \xrightarrow[\text{Et}_2\text{O, 20 °C}]{\text{Ph}_{4-x}\text{MCl}_x} \text{204} \tag{40}$$

205

R = Me, Ph, o-MeC$_6$H$_4$,
 p-YC$_6$H$_4$ (Y = Me, MeO, Cl)

204

M = Ge (x = 1,2)
 Sn (x = 1,2,3)
 Pb (X = 1,2)

ordinated pentagonal bipyramid in the crystalline state. Similar S–Sn interactions were observed in Ph$_2$SnCl(ACDA) (**208**), Ph$_3$Sn(ACDA) (**209**), and Bu$_2$Sn-(ACDA)$_2$ (**210**) (ACDA=2-amino-1-cyclopentene-1-carbodithioic acid), in which the orientation of ACDA resulted in NH···S intramolecular hydrogen bonding [128]. On the other hand, a normal tetra-coordinated structure for the silicon atom in N-alkyl-S-(trimethylsilyl)ACDA **211** was proposed based on IR and ^{29}Si NMR spectral data [129].

C(Ph)-Sn-C(Ph) 129(1)°
C-S 1.66(3)-1.69(3) Å

206

Sn-S 2.492(1)-2.987(1) Å
C-S 1.648(4)-1.712(4) Å
S*-Sn-C(Ph) 160.7(1)°

207

208

209

210

211

R = H, Me, Et, Bu

5.2
M=Group 15 Element

The reaction of the inner salt **111** with [N-(p-tolylsulfonyl)imino]phenyliodinane (**212**) in dichloromethane at –18 °C gives the S-nitrogen compound **213** as a red crystalline solid with a 72% yield (Eq. 41) [130]. X-ray analysis indicates that one canonical structure **213a** is a greater contributor than the other, **213b**.

S-As and S-Sb ACDA derivatives **214** were synthesized by reacting ACDA with ClM(S$_2$C$_2$H$_4$) **215** (M=As, Sb) (Eq. 42) [131]. A similar reaction employing 2-hydroxylcyclopentene-1-carbodithioic acid instead of ACDA in Eq. 42 gave **216** [132].

$$\text{(41)}$$

111   **212**   **213a**   **213b**

$$\text{(42)}$$

ACDA (R = NH$_2$)   **215**   **214**

M = As, Sb

M = As; R = H, Me
M = Sb; R = H, Me

X-ray analysis of **214** showed the dithioate group serving as a monodentate ligand with a lack of nitrogen participation in the coordination [131], while it was claimed that the dithioate group in a related S-As compound **217** acted as an anisobidentate ligand (As–S^1 2.272(2) Å; As–S^2 3.125(3) Å; C–S^1 1.759(8) Å; C–S^2 1.685(8) Å) with intramolecular NH···S hydrogen bonding [133]. Incidentally, asymmetric dithioate ligand binding was observed at the Bi atom in Bi(N-ethyl ACDA)$_3$ (short Bi–S 2.617(2)–2.647(1) Å; long Bi–S 2.963(5)–3.108(2) Å) [134].

216   **217**

M = As, Sb

(Thioaroylthio)phosphines [135a] and (thioacylthio and thioaroylthio)arsines [135b] of the type (RCS$_2$)$_n$MPh$_{3-n}$ **218** (M=P, As) were synthesized in low to high yields and characterized by IR and NMR spectroscopies and X-ray crystallography (Eq. 43).

$$\text{(43)}$$

218

X = Na, ⟨ ⟩NH$_2$, Y = Cl, Br, M = P, As, n = 1,2,3

S-P compounds **218** (M=P) were unstable thermally and moisture-sensitive, so **218** (M=P) hydrolyzed gradually to the corresponding dithioic acid upon exposure to air [135a], while *S*-As compounds **218** (M=As) were stable thermally and toward oxygen and water [135b]. X-ray analysis of $(TolCS_2)_2PPh$ (**219**) showed that the two dithioate ligands exist in the same plane and that the distances between the thiocarbonyl sulfur atom and the phosphorus atom are shorter than the sum of their van der Waals radii (3.66 Å) [134]. Similar weak interactions between the thiocarbonyl sulfur atom and the As atom were observed in $(ArCS_2)_nAsPh_{3-n}$. The interaction was found to be an n(S)-σ*(As-C or As-S) orbital interaction by ab initio calculations [135b].

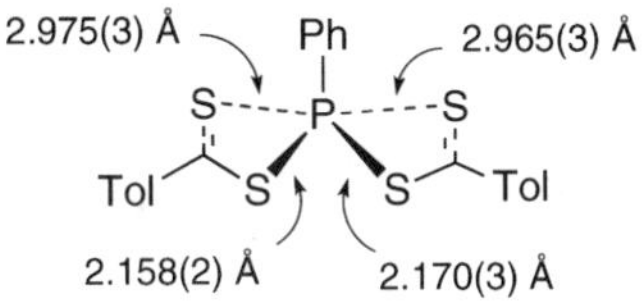

219

5.3
M=Group 16 Element

A variety of (dithioperoxo)thioic acid esters **220** were synthesized in moderate to good yields by reacting dithioate magnesium halides **221** with *S*-alkyl *p*-toluenethiosulfonates **222** or *S*-methyl methanethiosulfonates (**223**) (Scheme 43) [136, 137]. Oxidation of **220** with *m*-CPBA gave the sulfines **224** as an *E*-*Z* geometrical mixture [136]. These sulfines isomerized to acyl trisulfides ($R^1C(O)$-S_3R^2) after 10 days at room temperature.

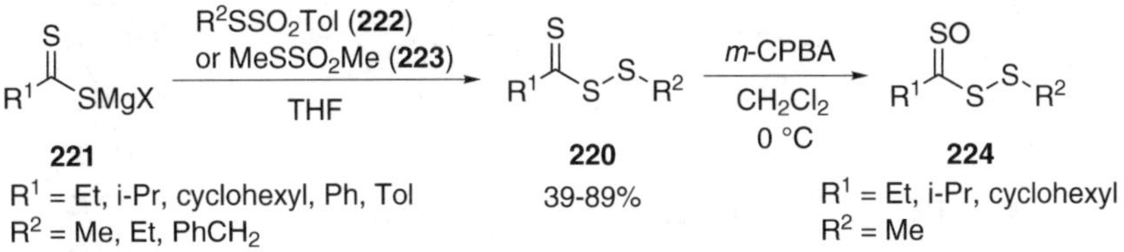

R^2SSO_2Tol (**222**)
or $MeSSO_2Me$ (**223**)

THF

m-CPBA

CH_2Cl_2
0 °C

221
R^1 = Et, i-Pr, cyclohexyl, Ph, Tol
R^2 = Me, Et, $PhCH_2$

220
39-89%

224
R^1 = Et, i-Pr, cyclohexyl
R^2 = Me

Scheme 43

Oxidation of dithioic acids or their alkaline metal salts with XeF_2 in acetonitrile or with iodine in methanol provides bis(thioacyl) disulfides **225** in moderate to good yields (Eq. 44) [138]. X-Ray analyses of $(o\text{-}MeC_6H_4CS_2)_2$ and $(TolCS_2)_2$ show that the distances between the thiocarbonyl sulfur atom and the γ-sulfur atom (3.254(7)-3.321(1) Å) are within the sum of their van der Waals radii (3.6 Å). Oxidation of inner salt **111** with $NOPF_6$ or Br_2 initially results in $[(Et_2N)_2CC(S)]_2S_2 \cdot X_2$ (X=PF_6^- or Br_3^-) which loses a sulfur atom under these conditions to furnish $[(Et_2N)_2CC(S)]_2S \cdots X_2$ as the final product in high yields

$$\text{(44)}$$

225

M = H, Na, K, Rb, Cs

For XeF$_2$: R = Ph, o-XC$_6$H$_4$ (X = Me, MeO), p-XC$_6$H$_4$ (X = Me, MeO, Cl)

For I$_2$: R = Me, Et, i-Pr, Pr, Bu, n-C$_5$H$_{11}$

[139]. Preparation of (PhO)$_2$P(S)SSC(S)R **226** (R=Tol and Et) has been reported, as shown in Eq. 45 [140].

$$\text{(45)}$$

226

R = Tol (86%), Et (80%)

5.4
M=Group 4–12 Element

Sulfur-coordinated transition metal complexes have attracted considerable interest due to their electronic and structural properties, which are useful in both industrial catalysis and biological systems [141, 142]. Dithioate ions act as bidentate ligands in Group 4–12 transition metal complexes, such those with Ti (**227** and **228**) [141], V (**229**) [141], Mo and W (**230, 231**) [143, 144], Mn (**232**)

227

Ar = 2,6-MesC$_6$H$_3$

228

229

Ar = 2,6-(2,4,6-i-Pr$_3$C$_6$H$_2$)C$_6$H$_3$

230

M = Mo, W

Ar = Ans

231

M = Mo, W

pz = pyrazol-1-yl

Ar = Ans

232

233: M = Tc
234: M = Re

235

R =

R' = Re(CO)$_4$, NHEt

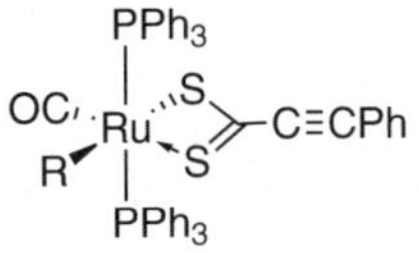

236

R = H, PhCH=CPh,
MesC≡C

$R_F = 2,4,6-(CF_3)_3C_6H_2$

$(R_FCS_2)M(THF)$

237: M = Co
238: M = Ni

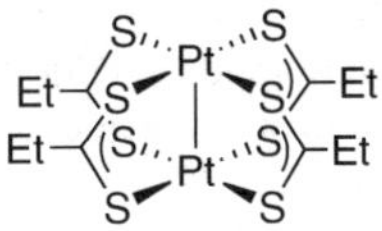

239

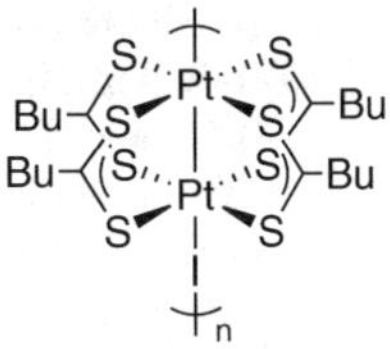

240

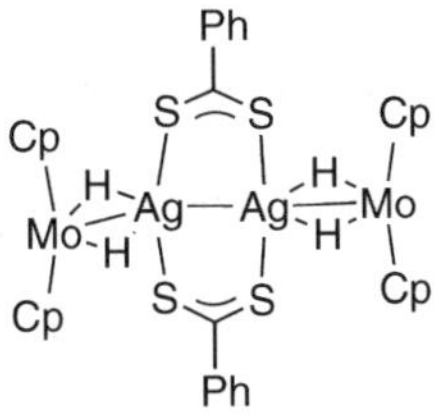

241

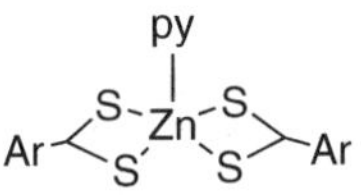

242

Ar = 3,5-t-Bu$_2$C$_6$H$_3$

[145], Tc (**233**), Re (**234**) [146], (**235**) [147, 148], Ru (**236**) [149], Co (**237**) [150], Ni (**238**) [150], Pt (**239** and **240**) [151,152], Ag (**241**) [153], and Zn (**242**) [154].

6
Conclusions

While most possible combinations of the dithioate substituents R^1 and R^2 have already been synthesized, the development of more convenient synthetic methods for them continues. The practical use of dithioates as building blocks in organic synthesis has been widely studied. Dithioates, in particular cyclic dithioates, have attracted considerable attention due to their biological activity, and dithioic acids can be used as bidentate ligands in a wide variety of transition metal and heavy main group element complexes. Therefore, dithioic acids and their esters are becoming increasingly important, not only in organic chemistry, but also in the industrial and medical sciences.

References

1. (a) Murai T, Kato S (1995) Functions with two chalcogens other than oxygen. In: Katritzky AR, Meth-Cohn O, Rees CW (eds) Comprehensive organic functional group transformations. Pergamon, Oxford, vol 5, p 545; (b) Metzner P (1992) Synthesis 1185; (c) Metzner P (1999) Top Curr Chem 204:127
2. Castro A, Hernandez L, Dorronsoro I, Saenz P, Perez C, Kalko S, Orozco M, Luque FJ; Martinez A (2002) Med Chem Res 11:219
3. Marzouk MI, El-Hashash M, Eissa AMF (2002) Int J Chem 12:157

4. Elgemeie GH, Ali HA, Elghandour AH, Hussein AM (2003) Synthetic Commun 33:555
5. Abdelhamid AO, Rateb NM, Dawood KM (2000) Phosphorus Sulfur 167:251
6. Abdelhamid AO, Mohamed GS (1999) Phosphorus Sulfur 152:115
7. Kukaniev MA, Shukurov SS, Khodzhibaev Y, Kurbonova N, Bandaev SG (1998) Russ Chem Bull (Translation of Izv Akad Nauk, Ser Khim) 47:2300
8. Barton DHR, Swift KAD, Tachdjian C (1995) Tetrahedron 51:1887
9. Alayrac C, Nowaczyk S, Lemarie M, Metzner P (1999) Synthesis 669
10. Zhang Q, Zhao Y-L, Shi Y, Wang L-X, Liu Q (2002) Synthetic Commun 32:2369
11. Barillier D, Benhida R, Vazeux M (1993) Phosphorus Sulfur 78:83
12. Lacova M, Ticha E, Varkonda S, Konecny V (1995) SK Patent 90–1605 (Chem Abstr 126: 305580)
13. Trofimov BA, Sobenina LN, Mikhaleva AI, Demenev AP, Tarasova OA, Ushakov IA, Zinchenko SV (2000) Tetrahedron 56:7325
14. Trofimov BA, Sobenina LN, Mikhaleva AI, Sergeeva MP, Polovnikova RI (1993) Sulfur Lett 15:219
15. Sobenina LN, Demenev AP, Mikhaleva AI, Elokhina VN, Mal'kina AG, Tarasova OA (2001) Synthesis 293
16. Raubenheimer HG, Marais EK, Cronje S, Esterhuysen C, Kruger GJ (2000) J Chem Soc Dalton 3016
17. Furuta S, Kuroboshi M, Hiyama T (1999) B Chem Soc Jpn 72:805
18. Hagiwara T, Ishizuka M, Fuchikami T (1998) Nippon Kagaku Kaishi 750
19. Ennis DS, Armitage MA (1995) Tetrahedron Lett 36:771
20. Hartke K, Rettberg N, Dutta D, Gerber HD (1993) Liebigs Ann Chem (1993) 1081
21. Alayrac C, Fromont C, Metzner P, Anh NT (1997) Angew Chem Int Edit 36:371
22. Alayrac C, Nowaczyk S, Lemarie M, Metzner P (1999) Phosphorus Sulfur 153/154:317
23. Lee PH, Cho M, Han I-S, Kim S (1999) Tetrahedron Lett 40:6975
24. Rettberg N, Wagner U, Hartke, K (1993) Arch Pharm (Weinheim Ger) 326:977
25. Makomo H, Masson S, Saquet M (1993) Phosphorus Sulfur 80:31
26. van der Linden JB, Timmermans JL, Zwanenburg B (1995) Recl Trav Chim Pays-B 114:91
27. Josse O, Labar D, Marchand-Brynaert J (1999) Synthesis 404
28. Keun OH, Kyung KS, Whang LH, Lee I (2001) New J Chem 25:313
29. Kanagasabapathy S, Sudalai A, Benicewicz BC (2001) Tetrahedron Lett 42:3791
30. Brouwer WG (1999) US Patent 98–39769 (Chem Abstr 131:243178)
31. Sudalai A, Kanagasabapathy S, Benicewicz BC (2000) Org Lett 2:3213
32. Josse O, Labar D, Georges B, Gregoire V, Marchand-Brynaert J (2001) Bioorg Med Chem 9:665
33. Fromont C, Metzner P (1994) Phosphorus Sulfur 95/96:421
34. Beslin P, Perrio S (1994) Phosphorus Sulfur 95/96:383
35. Beslin P, Perrio S (1993) Tetrahedron 49:3131
36. Sreekumar R, Padmakumar R (1997) Tetrahedron Lett 38:2413
37. Jung ME, Parker MH (1997) J Org Chem 62:7094
38. Nair SK, Asokan CV (1999) Synthetic Commun 29:791
39. Nair SK, Samuel R, Asokan CV (2001) Synthesis 573
40. Minami T, Okauchi T, Matsuki H, Nakamura M, Ichikawa J, Ishida M (1996) J Org Chem 61:8132
41. Laduron F, Nyns C, Janousek Z, Viehe HG (1997) J Prakt Chem/Chem-Zeit 339:697
42. Neugebauer W, Pinet E, Kim M, Carey PR (1996) Can J Chem 74:341
43. Villemin D, Thibault-Starzyk F (1993) Synlett 148
44. Nakayama J, Choi KS, Akiyama I, Hoshino M (1993) Tetrahedron Lett 34:115

45. Portella C, Shermolovich YG, Tschenn O (1997) B Soc Chim Fr 134:697
46. Charbonnel-Jobic G, Guemas J-P, Adelaere B, Parrain J-L, Quintard J-P (1995) B Soc Chim Fr 132:624
47. Mandal SS, Chakraborty J, De A (1999) J Chem Soc Perkin T 1 2639
48. El-Sayed I, Ali OM, Fischer A, Senning Al (2003) Heteroatom Chem 14:170
49. Bouhadir G, Legrand N, Quiclet-Sire B, Zard SZ (1999) Tetrahedron Lett 40:277
50. Thang SH, Chong YK, Mayadunne RTA, Moad G, Rizzardo E (1999) Tetrahedron Lett 40:2435
51. Beslin P, Vallee Y (1985) Tetrahedron 41:2691
52. Eames J, Warren S (1999) J Chem Soc Perkin T 1: 2783
53. Eames J, Jones RVH, Warren S (1996) Tetrahedron Lett 37:707
54. Al-Badri H, Collignon N, Maddaluno J, Masson S (2000) Tetrahedron 56:3909
55. Middleton WJ (1965) J Org Chem 30:1390
56. Ishida M, Sakamoto M, Hattori H, Shimizu M, Inagaki S (2001) Tetrahedron Lett 42:3471
57. Lempereur C, Ple N, Turck A, Queguiner G, Corbin F, Alayrac C, Metzner P (1998) Heterocycles 48:2019
58. Portella C, Shermolovich Y (1997) Tetrahedron Lett 38:4063
59. Corbin F, Alayrac C, Metzner P (1999) Tetrahedron Lett 40:2319
60. Bürger H, Pawelke G, Rothe J (1994) J Organomet Chem 474:43
61. Oh HK, Kim SK, Cho IH, Lee HW, Lee I (2000) J Chem Soc Perkin T 2: 2306
62. Yeo SK, Choi BG, Kim JD, Lee JH (2002) B Korean Chem Soc 23:1029
63. Chandrasekharam M, Bhat L, Ila H, Junjappa H (1993) Tetrahedron Lett 34:6439
64. Hansen J, Heinemann FW (1996) Phosphorus Sulfur 118:155
65. Nakayama J, Otani T, Sugihara Y, Ishii A (1997) Tetrahedron Lett 38:5013
66. Nakayama J, Akimoto K, Sugihara Y (1998) Tetrahedron Lett 39:5587
67. Nakayama J, Otani, T, Tadokoro T, Sugihara Y, Ishii A (2001) Chem Lett 768
68. Hartke K, Bien N, Weller, F (1996) Z Naturforsch B 51:1173
69. Steyn GJJ, Roodt A (2001) S Afr J Chem (2001) 54:242
70. Sharma RK, Singh Y, Rai AK (2000) Phosphorus Sulfur 166:221
71. Lam S, MacAulay C, le Riche JC, Dyachkova, Y, Coldman A, Guillaud M, Hawk E, Christen M-O, Gazdar AF (2002) J Natl Cancer Inst 94:1001 (Chem Abstr 137:134689)
72. (a) Prochaska HJ, Yeh Y, Baron P, Polsky B (1993) P Natl Acad Sci USA 90:3953 (Chem Abstr 119:357); (b) Prochaska HJ, Rubinson L, Yeh Y, Baron P, Polsky B (1994) Mol Pharmacol 45:916 (PubMed 8190108); (c) Prochaska HJ, Fernandes CL, Pantoja RM, Chavan SJ (1996) Biochem Biophys Res Commun 221:548 (Chem Abstr 124:309255)
73. Navamal M, McGrath C, Stewart J, Blans P, Villamena F, Zweier J, Fishbein JC (2002) J Org Chem 67:9406
74. Kurihara H, Kobayashi A, Kumamoto T (1996) Nippon Kagaku Kaishi 875
75. Mloston G, Huisgen R, Polborn, K (1999) Tetrahedron 55:11475
76. Lorcy D, Le Paillard MP, Robert A (1993) Tetrahedron Lett 34:5289
77. Curphey TJ, Libby AH (2000) Tetrahedron Lett 41:6977
78. Curphey TJ, Joyner H (1993) Tetrahedron Lett 34:7231
79. Curphey TJ, Joyner HH (1993) Tetrahedron Lett 34:3703
80. Zayed SE, Hussin IA (1993) Phosphorus Sulfur 84:191
81. Aimar ML, Kreiker J, de Rossi RH (2002) Tetrahedron Lett 43:1947
82. Rees CW, Rakitin OA, Marcos CF, Torroba T (1999) J Org Chem 64:4376
83. Garcia-Valverde M, Pascual R, Torroba T (2003) Org Lett 5:929
84. Singh JD, Singh HB, Das K, Sinha UC (1995) J Chem Res-S 312
85. Timoshenko VM, Bouillon J-P, Shermolovich YG, Portella C (2002) Tetrahedron Lett 43:5809

86. Timoshenko VM, Bouillon J-P, Chernega AN, Shermolovich YG, Portella C (2003) Eur J Org Chem 2471
87. Abbott A, Bancroft MN, Morris MJ, Hogarth G, Redmond SP (1998) Chem Commun 389
88. Bertrand HO, Abazid M, Christen MO, Burgot JL (1994) Sulfur Lett 17:231
89. Abazid, M, Bertrand HO, Christen MO, Burgot JL (1994) Phosphorus, Sulfur, Silicon 88:195
90. Abazid M, Bertrand HO, Christen MO, Burgot JL (1994) Chem Commun 131
91. Barsy MA, Abd El Latif FM, Ahmed SM, El-Maghraby MA (2000) Phosphorus Sulfur 165:1
92. Wang D-Q, Dou J-M, Niu M-J, Li D-C, Liu Y (2003) Acta Chim Sinica 61:551 (Chem Abstr 139:190047)
93. Rudershausen S, Drexler H-J, Holdt H-J (1998) J Prakt Chem/Chem-Zeit 340:450
94. Amzil J, Catel J-M, Le Coustomer G, Mollier Y, Sauve J-P (1988) B Soc Chim Fr 101
95. Granados AM, Kreiker J, de Rossi RH (2002) Tetrahedron Lett 43:8037
96. Steudel R, Kustos M, Munchow V, Westphal U (1997) Chem Ber/Recl 130:757
97. Schachtner JE, Nienaber J, Stachel H-D, Waisser K (1999) Pharmazie 54:335
98. Barriga S, Fuertes P, Marcos CF, Miguel D, Rakitin OA, Rees CW, Torroba T (2001) J Org Chem 66:5766
99. Barriga S, Garcia N, Marcos CF, Neo AG, Torroba T (2002) ARKIVOC 212 (Chem Abstr 138:368832)
100. Barriga S, Marcos CF, Riant O, Torroba T (2002) Tetrahedron 58:9785
101. Amelichev SA, Barriga S, Konstantinova LS, Markova TB, Rakitin OA, Rees CW, Torroba T (2001) J Chem Soc Perkin T 1 2409
102. Konstantinova LS, Obruchnikova NV, Rakitin OA, Rees CW, Torroba T (2000) J Chem Soc Perkin T 1 3421
103. Marcos CF, Polo Cecilia, Rakitin OA, Rees CW, Torroba T (1997) Chem Commun 879
104. Ando T, Mita S, Kanematsu A, Sakai K (1997) JP Patent 95–251175 (Chem Abstr 127:17684)
105. Shikhaliev KS, Medvedeva SM, Pigarev VV, Solov'ev AS, Shatalov GV (2000) Russ J Gen Chem (Translation of Zh Obshch Khim) 70:450
106. Shikhaliev KS, Pigarev VV, Shmyreva ZV, Medvedeva SM, Ermolova GI (1998) Izv Vyssh Uchebn Zaved Khim Khim Tekhnol 41:48 (Chem Abstr 130:223195)
107. Gompper R, Knieler R, Polborn K (1993) Z Naturforsch B 48:1621
108. Gebert A, Heimgartner H (2002) Helv Chim Acta 85:2073
109. Shi J, Tromm Peter, Heimgartner H (1994) Helv Chim Acta 77:2133
110. Shi J, Linden A, Heimgartner H (1994) Helv Chim Acta 77:1903
111. Mohamed NR (2000) Phosphorus Sulfur 161:123
112. Tumkevicius S, Labanauskas L, Bucinskaite V, Brukstus A, Urbelis G (2003) Tetrahedron Lett 44:6635
113. Darwish AD, Avent AG, Boltalina OV, Gol'dt I, Kuvytchko I, Da Ros T, Street JM, Taylor R (2003) Chem Eur J 9:2008
114. Kozhinov DV, Woznesensky SA, Dudinov AA, Krayushkin MM (2000) Mendeleev Commun 77
115. Akimoto K, Masaki K, Nakayama J (1996) B Chem Soc Jpn 69:2091
116. Flaig R, Hartmann H (1997) Monatsh Chem 128:1051
117. Kandeel MM (2000) Phosphorus Sulfur 156:225
118. Geies AA (1999) Phosphorus Sulfur 148:201
119. Landreau C, Deniaud D, Reliquet F, Reliquet A, Meslin JC (2000) Heterocycles 53:2667 (and references cited therein)
120. Holzer Max, Dobner B, Briel D (1994) Liebigs Ann Chem 895

121. Glotova TE, Komarova TN, Nakhmanovich AS, Lopyrev VA (2000) Russ Chem Bull (Translation of Izv Akad Nauk Ser Khim) 49:1917
122. Levkovskaya GG, Bozhenkov GV, Larina LI, Evstaf'eva IT, Mirskova AN (2001) Russ J Org Chem (Translation of Zh Org Khim) 37:644
123. Morrison CF, Burnell DJ (2000) Org Lett 2:3891
124. Raasch MS, Huang N-Z, Lakshmikantham MV, Cava MP (1988) J Org Chem 53:891
125. Huang TB, Qian X, Tao Z-F, Wang K, Song G-H, Liu L-Fe (1999) Heteroatom Chem 10:141
126. Kano N, Nakagawa N, Kawashima T (2001) Angew Chem Int Edit 40:3450
127. Kato S, Tani K, Kitaoka N, Yamada K, Mifune H (2000) J Organomet Chem 611:190
128. Tarassoli A, Sedaghat T, Neumuller B, Ghassemzadeh M (2001) Inorg Chim Acta 318:15
129. Sharma RK, Singh Y, Rai AK (1998) Phosphorus Sulfur 142:249
130. Nakayama J, Zhao L, Otani T, Sugihara Y, Ishii A (2000) Chem Lett 1248
131. Cea-Olivares R, Toscano RA, Luna A, Cervantes F (1993) Main Group Met Chem 16:121
132. Cea-Olivares R, Toscano RA, Lopez M, Garcia P (1993) Heteroatom Chem 4:313
133. Cea-Olivares R, Toscano R-A, Estrada M, Silvestru C, Garcia PG, Lopez-Cardoso M, Blass-Amador G (1995) Appl Organomet Chem 9:133
134. Bharadwaj PK, Musker WK (1987) Inorg Chem 26:1453
135. (a) Tani K, Matsuyama K, Kato S, Yamada K, Mifune H (2000) B Chem Soc Jpn 73:1243; (b) Tani K, Hanabusa S, Kato S, Mutoh S, Suzuki S, Ishida M (2001) J Chem Soc Dalton 518
136. Leriverend C, Metzner P (1994) Tetrahedron Lett 35:5229
137. Sukhai RS, Brandsma L (1979) Synthesis 971
138. Niyomura O, Kitoh Y, Nagayama K, Kato S (1999) Sulfur Lett 22:195
139. Otani T, Sugihara Y, Ishii A, Nakayama J (1998) Heteroatom Chem 9:703
140. Shi M, Kato S (2002) Helv Chim Acta 85:2559
141. Ives C, Fillis EL, Hagadorn, JR (2003) J Chem Soc Dalton 527
142. Mevellec F, Tisato F, Refosco F, Roucoux A, Noiret N, Patin H, Bandoli G (2002) Inorg Chem 41:598
143. Cook DJ, Hill AF (2003) Organometallics 22:3502
144. Cook DJ. Hill AF (1997) Chem Commun 955
145. Shikhaliev KS, Leshcheva EV, Medvedeva SM (2002) Chem Heterocycl Compd 38:755 (Chem Abstr 138:238044)
146. Mévellec F, Roucoux A, Noiret N, Patin H, Tisato F, Bandoli G (1999) Inorg Chem Commun 2:230
147. Adams RD, Chen L, Huang M (1995) J Chin Chem Soc (Taipei Taiwan) 42:11 (Chem Abstr 123:9616)
148. Adams RD, Chen L, Wu W (1994) Angew Chem Int Edit 33:568
149. Adams H, McHugh PE, Morris MJ, Spey SE, Wright PJ (2001) J Organomet Chem 619:209
150. Belay M, Edelmann FT (1994) J Organomet Chem 479:C21
151. Mitsumi M, Yoshinari T, Ozawa Y, Toriumi K (2000) Mol Cryst Liq Cryst 342:127
152. Mitsumi M, Kitamura K, Morinaga A, Ozawa Y, Kobayashi M, Toriumi K, Iso Y, Kitagawa H, Mitani T (2002) Angew Chem Int Edit 41:2767
153. Brunner H, Hollman A, Zabel M (2001) J Organomet Chem 630:169
154. Darensbourg DJ, Adams MJ, Yarbrough JC (2002) Inorg Chem Commun 5:38

Top Curr Chem (2005) 251: 227–246
DOI 10.1007/b101010
© Springer-Verlag Berlin Heidelberg 2005

Carboselenothioic and Carbodiselenoic Acid Derivatives and Related Compounds

Akihiko Ishii (✉) · Juzo Nakayama

Department of Chemistry, Faculty of Science, Saitama University, Sakura-ku, Saitama 338–8570, Japan
ishiiaki@chem.saitama-u.ac.jp

Abstract In this chapter we consider derivatives of carboselenothioic and carbodiselenoic acids, and compounds related to them. Ammonium salts and inner salts of selenothioic and diselenoic acids have been prepared, while their free acids are yet to be isolated. Selenothioic acid *Se*-esters are prepared by reacting bis(thioacyl) sulfides with thiols and thioic acid *O*-esters with Me$_2$AlSeMe. Selenothioic acid *S*-esters are synthesized by three reactions: (a) reaction of *Se*-alkynyl selenoates with thiols; (b) reaction of lithium alkyneselenolate with thiols; (c) reaction of *O*-methyl selenoates with Me$_2$AlSR. Diselenoic acid esters are synthesized by reacting *O*-methyl selenoates with Me$_2$AlSeMe. Mo and W complexes of diselenoic and selenothioic acids have also been synthesized. Selenothioic acid *Se*-esters act as thioacylating agents for amines and alcohols. Reactions of selenothioic acid *S*-esters with electrophiles in basic conditions are reviewed in detail; these reactions take place at the selenium atom or at the α-carbon atom, depending on the esters and electrophiles chosen. Other reactions such as cycloaddition, oxidation, and reduction are also reported.

Keywords Selenothioic acid · Diselenoic acid · Se-substituted selenothioate · Se-substituted selenothioate · Diselenoate

Abbreviations

Ac	acetyl
Ar	aryl
Bu	butyl
s-Bu	*sec*-butyl
t-Bu	*tert*-butyl
cat	catalyst
DMAD	dimethyl acetylenedicarboxylate
Et	ethyl
h	hour(s)
Me	methyl
Mes	2,4,6-trimethylphenyl
min	minute(s)
m-CPBA	*m*-chloroperoxybenzoic acid
Ph	phenyl
Ans	4-methoxylphenyl (anisyl)
PNB	4-nitrobenzyl
Pr	propyl
i-Pr	isopropyl
rt	room temperature
TBAF	tetrabutylammonium fluoride
Tf	trifluoromethanesulfonyl (triflyl)
TFA	trifluoroacetic acid
THF	tetrahydrofuran
TMS	trimethylsilyl
Tol	4-methylphenyl

1
Introduction

Selenothioic and diselenoic acids and their derivatives were once a rare class of compounds; indeed, a review of them [1] published in 1995 summarized their chemistry in just a few pages. Since then, however, their chemistry has developed rapidly, and several types of derivatives have been prepared by appropriate methods. Syntheses of these compounds were exhaustively described in the review [2] published in 1998. In this article, selenothioic and diselenoic acids and their derivatives are reviewed in terms of theoretical techniques, syntheses, and reactions.

selenothioic *Se*-acid   selenothioic *S*-acid   diselenoic acid

selenothioic acid *Se*-ester   selenothioic acid *S*-ester   diselenoic acid ester

2
Selenothioic and Diselenoic Acids and Their Salts

2.1
Theoretical Techniques

Theoretical analyses of HC(X)YH (X, Y=O, S, Se, Te) have been performed using MP2 and B3LYP methods with the 6–311G(2D) basis set for H, C, O, and S, and Huzinaga's basis set for Se (433111/43111/411) and Te (4333111/433111/4311) [3]. The optimized structures of selenothioformic *Se*-acid (**1**), selenothioformic *S*-acid (**2**), and diselenoformic acid (**3**) are summarized in Fig. 1.

Figure 2 shows the tautomerism (intramolecular 1,3-H shift) between *Se*-acid **1** and *S*-acid **2** and between **3** and **3'**, with the respective transition states (TS). Calculated relative energies for the ground states and the TSs are also shown. At the MP2 level, the *S*-acid form (C=Se double bond form) **2** was more stable than the *Se*-acid form (C=S double bond form) **1** by 1.3 kcal mol^{-1}. The energy difference diminished to 0.1 kcal mol^{-1} at the B3LYP level. The energy barriers for the two tautomerizations were calculated to be ~20 kcal mol^{-1}, and these values were much smaller (by ~10 kcal mol^{-1}) than those for the tautomerism between HC(=O)XH and HC(=X)OH (X=O, S, Se, Te), illustrating that the barrier for tautomerism reduces as the electronegativity of the chalcogen decreases.

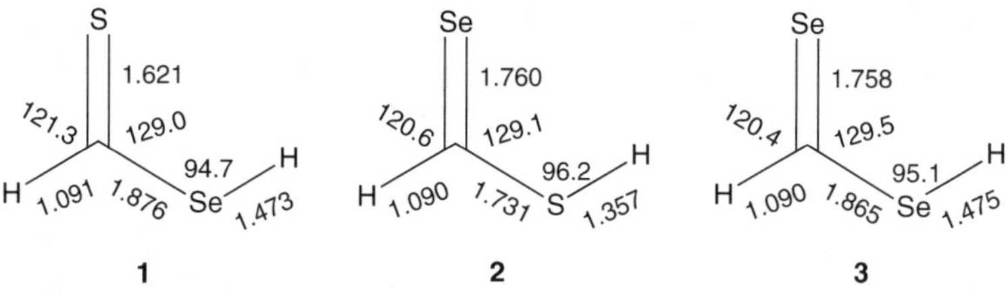

Fig. 1 Optimized geometries of selenothioformic *Se*-acid (**1**), selenothioformic *S*-acid (**2**), and diselenoformic acid (**3**) at the MP2 level; bond lengths (Å) and bond angles (deg)

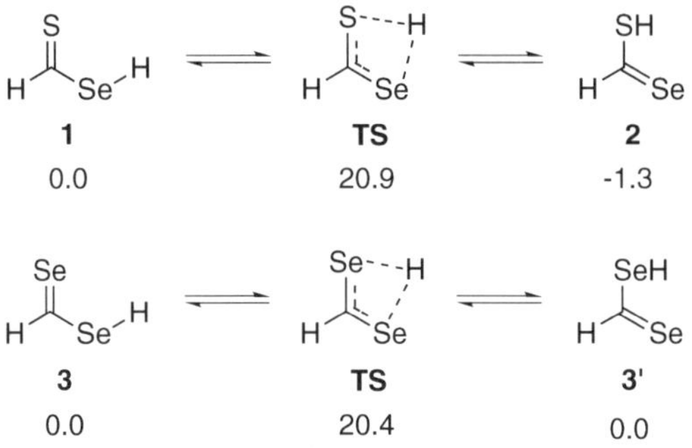

Fig. 2 Tautomerism of selenothioformic acids (**1** and **2**) and diselenoformic acid (**3**), and relative energies (kcal mol^{-1}) of the ground states and the transition states (TS) at the MP2 level

2.2
Syntheses and Reactions of Selenothioic Acids and Salts

Treatment of *S*-2-(trimethylsilyl)ethyl selenothioates **4** with tetraalkylammonium fluoride in THF yielded the ammonium salts **5** in high yields (Eq. 1) [4]. The salts **5a–5d** (R′=Bu) were obtained as green to purple oils (λ_{max}=583–597 nm in THF) and **5b–5d** (R′=Me) as green to purple solids. The structure of **5b** (R′=Me) was established by X-ray crystallography. These salts were thermally stable but sensitive toward water and oxygen.

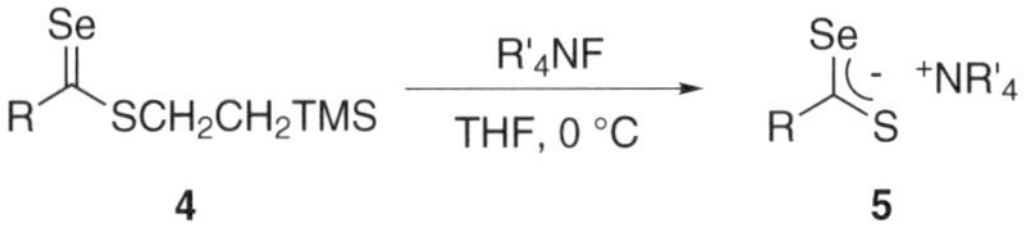

$$R = Ph\ (\mathbf{a}),\ o\text{-}MeC_6H_4\ (\mathbf{b}),\ Tol\ (\mathbf{c}),\ C(CH_2CH=CH_2)_3\ (\mathbf{d})$$
$$R' = Bu,\ Me$$

(1)

Acidification of **5d** (R′=Bu) with hydrochloric acid in ether at 0 °C gave selenothiolactone **6** with 37% yield, probably through the selenothioic acid **7/7′** (Scheme 1) [4]. The intervention of **7/7′** was verified during spectroscopic observations of the acidification of **5d** (R′=Me) at –78 °C; ^{1}H NMR δ=8.36 (S*H*); ^{13}C NMR δ=251.4 (*C*=S) at –20 °C; λ_{max}=603 nm (n→π*). In ^{77}Se NMR, the signal could not be observed probably due to the peak-broadening by rapid

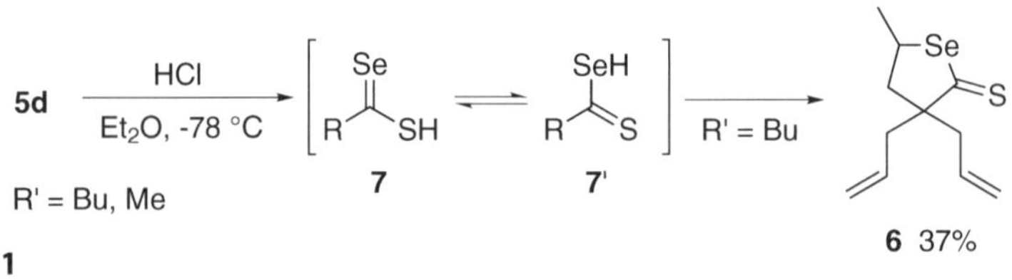

Scheme 1

tautomerization between **7** and **7′**. The acylation and thiocarbamoylation of **5d** (R′=Bu) gave **8** and **9**, respectively (Scheme 2).

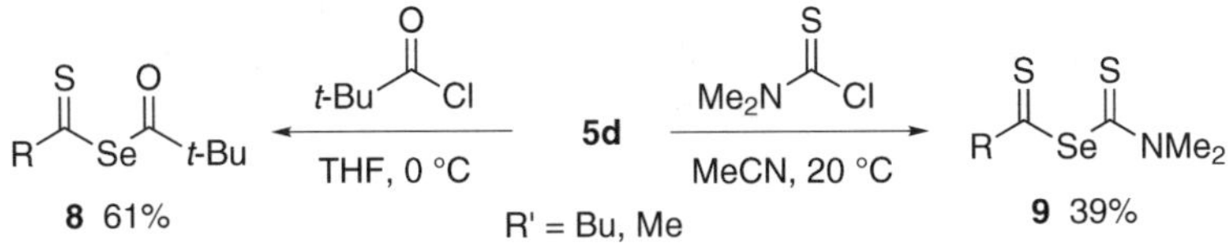

8 61%   **5d**   **9** 39%

Scheme 2

2.3
Syntheses and Reactions of Diselenoic Acids and Salts

Similar to selenothioic acid and the ammonium salts (Sect. 2.2), the reaction of the 2-(trimethylsilyl)ethyl diselenoates **10** with Me_4NF provided the ammonium salts **11** as green solids, which were characterized by NMR, UV spectroscopy and X-ray crystallography (for **11a**) (Eq. 2) [5]. The salts **11** were thermally stable, and, in particular, **11b** was stable in air not only in the solid state but also in a solution of acetonitrile at –20 °C. The double bond character of the C–Se bond in **11** was suggested by data from an X-ray analysis of **11a** [C–Se 1.828(4) and 1.831(4) Å], ^{13}C and ^{77}Se NMR ($\delta_{C\text{-}Se}$=256.1–263.6 and δ_{Se}=1363–1493, respectively), ^{13}C–^{77}Se coupling constants (J=208.7–214.9 Hz), and UV-Vis spectroscopy [λ_{max}=417–453 nm (π-π*) and 634–690 nm (n-π*) in THF].

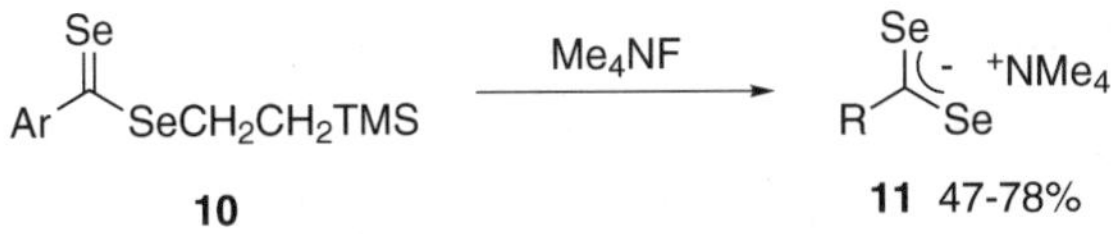

10   **11** 47-78%   (2)

Ar = Ph (**a**), o-MeC$_6$H$_4$ (**b**), p-BrC$_6$H$_4$ (**c**), Ans (**d**)

On acidification with TfOH or HCl/Et$_2$O in THF-d_8 at –70 °C, a light green suspension of **11b** turned to a green solution, which, however, did not give any NMR signals. However, addition of methyl vinyl ketone to the solution gave the diselenoate **12** with 21% yield, implying the generation of diselenoic acid **13** (see Eq. 3, [5].

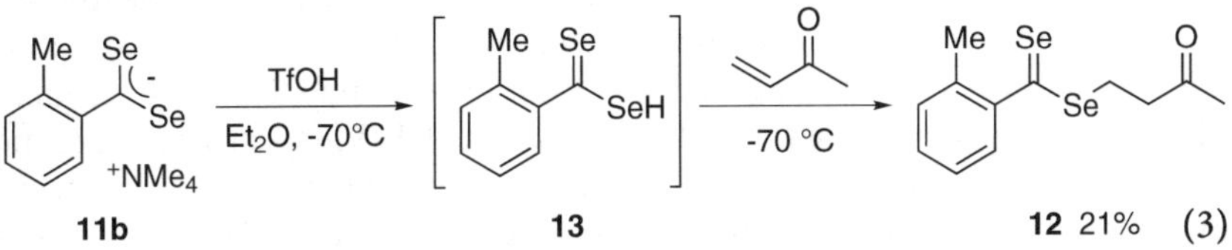

11b   **13**   **12** 21%   (3)

The reaction of cyclic enediamine **14** with Se$_2$Cl$_2$ in the presence of triethylamine yielded a diselenoate inner salt **15** as a dark green solid (λ_{max}=508 and

635 nm in CH_2Cl_2) (Eq. 4) [6]. The structure was established by X-ray crystallography. In the solid state, the diselenoate part was nearly perpendicular (86.3°) to the plane of the carbenium ion part, which suggested an attractive interaction between the negatively charged selenium atoms and the carbenium carbon atom. Acidification of **15** with acids brought about immediate decomposition with the formation of red selenium.

$$\text{(4)}$$

The selenium atoms in **15** were replaced stepwise with sulfur (Eq. 5) [6]. Stirring **15** with elemental sulfur in $CDCl_3$ at room temperature gave a mixture of diselenoate (**15**), selenothioate (**16**), and dithioate (**17**) inner salts in the ratio of 2:2:1.

$$\text{(5)}$$

The reactions of **15** with MeI (or $Me_3O^+BF_4^-$), DMAD (2 equiv), and TsN=IPh produced **18**, **19**, and **20**, respectively.

The reaction of acyclic enediamine **21a** with Se_2Cl_2 did not give the inner salt like **15**, but hexaselenacyclooctane **22a** (Eq. 6) [6]. The same type of compound **22b** was obtained by the reaction of chloroenediamine **21b** with elemental selenium in the presence of triethylamine [7]. The UV-Vis spectrum of **22b** measured in a polar solvent (CH_2Cl_2, $CHCl_3$, and DMF) exhibited characteristic absorption maximum at ~440 nm, which, together with its NMR data, suggested the dissociation of **22b** into the inner salts **23** or **24** (Eq. 7) [7]. Compound **22b** gave the corresponding α-carbenium methyl diselenoate on treatment with methyl iodide in 84% yield.

$$\textbf{21a}: R_2N = \text{morpholino}, R' = H$$

$$\textbf{21b}: R_2N = Et_2N, R' = Cl$$

$$\textbf{22a}: R_2N = \text{morpholino}$$

$$\textbf{22b}: R_2N = Et_2N$$

(6)

(7)

3
Syntheses of Selenothioic and Diselenoic Acid Esters

3.1
Selenothioic Acid *Se*-esters

A limited number of selenothioic acid *Se*-esters **25** [8], **26** [9], **27** [10], and **28** [11] were known before 1988.

25

26

27

28

In 1988, Kato reported that the bis(thioacyl) sulfides **29** reacted with sodium areneselenolate **30** to yield the *Se*-aryl selenothioates **31** (Eq. 8) [12]. Thioacyl

(8)

R	Ar	Yield/%	R	Ar	Yield/%
Me	Ph	26	Ph	Ph	44
Et	Ph	51	Tol	Ph	73
Pr	Ph	68	Tol	Tol	81
i-Pr	Ph	77	Tol	p-ClC$_6$H$_4$	70
Bu	Ph	60	Mes	Ph	0
n-C$_5$H$_{11}$	Ph	69	Ans	Ph	68
$cyclo$-C$_6$H$_{11}$	Ph	83	p-ClC$_6$H$_4$	Ph	78

chlorides [RC(S)Cl] could be used instead of **29**, and the reaction of RC(S)Cl with ArSeTMS also yielded **31** in some cases. A limitation of this method is that sterically hindered selenothioate **31** (R=Mes; Ar=Ph) could not be prepared. Alkaneselenothioates **31** (R=alkyl; Ar=Ph) prepared in this way showed absorptions due to the C=S stretching vibration ($v_{C=S}$) around 1170–1210 and 831–878 cm^{-1} in IR spectroscopy and absorption maxima (λ_{max}) due to the n-π* transition of the C=S group around 486–493 nm (hexane) in UV-Vis spectroscopy. The corresponding values for arenecarboselenothioates **31** (R=aryl) were $v_{C=S}$=1238–1250 cm^{-1} and λ_{max}=528–542 nm (hexane), respectively. In ^{13}C NMR spectroscopy, thiocarbonyl carbons appeared at δ=240.1–250.2 for **31** with R=alkyl, and δ=229.5–232.2 for **31** with R=aryl.

After that, *Se*-methyl selenothioates **32** were prepared by the reaction of *O*-methyl (or ethyl) thioates **33** with Me$_2$AlSeMe (**34**) in high yields (Eq. 9) [13]. In ^{13}C NMR spectroscopy, **32** (R^1=alkyl) displayed a signal from the thiocarbonyl carbon at δ=241.3–245.1 and **32** (R^1=Ph, Ans) at δ=230.3–233.2, in a similar region to **31**. The thiocarbonyl carbon in 2-thiophenecarboselenothioate **32** (R^1=2-thienyl) resonated at a somewhat higher field (δ 216.1).

$$\underset{\textbf{33}}{R^1\text{-}C(=S)\text{-}OR^2} \; + \; \underset{\textbf{34}}{Me_2AlSeMe} \quad \xrightarrow{Et_2O,\ rt} \quad \underset{\textbf{32}}{R^1\text{-}C(=S)\text{-}SeMe} \qquad (9)$$

R^1	R^2	Yield/%	R^1	R^2	Yield/%
n-C$_5$H$_{11}$	Et	96	Ph	Me	91
n-C$_7$H$_{15}$	Me	98	Ans	Me	87
n-C$_{15}$H$_{31}$	Me	90	2-thienyl	Me	89
PhCH$_2$	Et	94			

Both *Se*-aryl (**31**) and *Se*-methyl (**32**) selenothioates are orange or red compounds and are fairly stable in air at room temperature.

3.2
Selenothioic Acid *S*-esters

The reaction of *Se*-alkynyl selenoacetates **35** with alkanethiols (2 equiv) in the presence of a catalytic amount of TFA gave *S*-alkyl alkaneselenothioates **36** as blue-violet liquids in 30–70% yields (Eq. 10) [14]. The selenothioates **36** were stable to heat and moisture, but were extremely sensitive toward oxygen and decomposed quickly on exposure to air, with the liberation of red selenium. The relevant spectroscopic data for the *S*-alkyl alkaneselenothioates **36** are as follows: ^{13}C NMR δ(*C*=Se)=237.5–240.0; ^{77}Se NMR δ(C=*Se*)=1538.8–1569.7 [δ=2294.8 for **36** (R^1=Ph, R^2=*t*-Bu)]; UV-Vis λ_{max}=571–591 nm (cyclohexane).

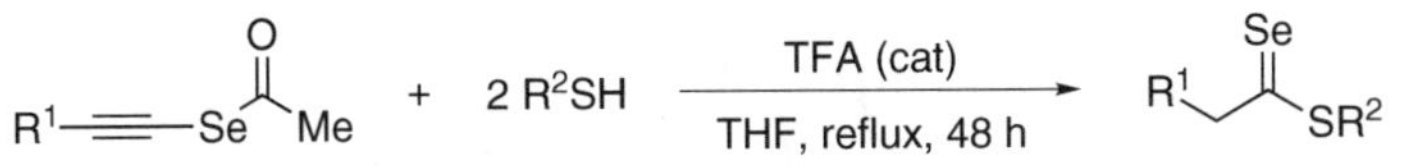

$$R^1-\!\!\equiv\!\!-\underset{35}{Se}\overset{O}{\underset{}{C}}Me \quad + \quad 2\ R^2SH \quad \xrightarrow[\text{THF, reflux, 48 h}]{\text{TFA (cat)}} \quad R^1\underset{36}{\overset{Se}{C}}SR^2 \qquad (10)$$

R^1 = Ph; R^2 = Et (48%)
R^1 = Ph; R^2 = Pr (70%)
R^1 = Ph; R^2 = i-Pr (54%)
R^1 = Ph; R^2 = t-Bu (33%)
R^1 = H; R^2 = Pr (30%)

In the synthesis shown in Eq. 10, selenoketene ($R^1CH=C=Se$) was proposed as the intermediate [14]. This consideration led to a more straightforward and convenient synthesis of **36** [15]. Therefore, lithium alkyneselenolate **37** was treated with a thiol (R^2SH) to give the selenoketene intermediate **38**, which further reacted with lithium thiolate (R^2SLi) to yield S-substituted selenothioate **36** after protonation, as shown in Scheme 3. Quenching this reaction with allyl bromide yielded α-allyl selenothioates [15c]. When (trimethylsilyl)acetylene was used, the trimethylsilyl group was detached from **36** during purification with silica-gel column chromatography to give the S-alkyl selenothioacetates **39**. X-Ray crystallographic analyses of **36** (R^1=Ph$_3$Si, R^2=i-Pr) and **36** (R^1=Ph$_3$Si, R^2=2,6-Me$_2$C$_6$H$_3$) were performed [15a]. S-Aryl selenothioates were less stable than S-alkyl selenothioates; **36** (R^1=Ph$_3$Si, R^2=Ph) decomposed gradually even below –10 °C under argon to give 1,3-diselenetane **40**, the dimer of selenothioate **39** (Scheme 4) [15a].

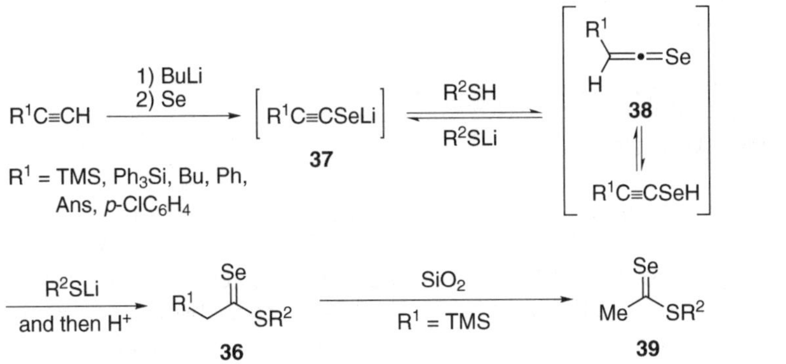

$$R^1C\equiv CH \xrightarrow[\text{2) Se}]{\text{1) BuLi}} \left[R^1C\equiv CSeLi\right] \underset{37}{\overset{R^2SH}{\underset{R^2SLi}{\rightleftarrows}}} \left[\begin{array}{c} R^1 \\ ^{H}\!\!\diagup\!\!=\!\bullet\!=Se \\ \mathbf{38} \\ \updownarrow \\ R^1C\equiv CSeH \end{array}\right]$$

R^1 = TMS, Ph$_3$Si, Bu, Ph, Ans, p-ClC$_6$H$_4$

$$\xrightarrow[\text{and then H}^+]{R^2SLi} \quad R^1\underset{36}{\overset{Se}{C}}SR^2 \quad \xrightarrow[R^1 = \text{TMS}]{\text{SiO}_2} \quad Me\underset{39}{\overset{Se}{C}}SR^2$$

36: R^1 = Ph$_3$Si; R^2 = i-Pr (24%), Ph (23%)
R^1 = Bu; R^2 = Bu (58%)
R^1 = Ph; R^2 = Bu (29%)
R^1 = Ans; R^2 = Bu (19%)
R^1 = p-ClC$_6$H$_4$; R^2 = Bu (13%)

39: R^2 = Et (51%), Bu (98%),
s-Bu (54%), t-Bu (42%),
PhCH$_2$ (83%),
CH$_2$CO$_2$Et (20%)

Scheme 3

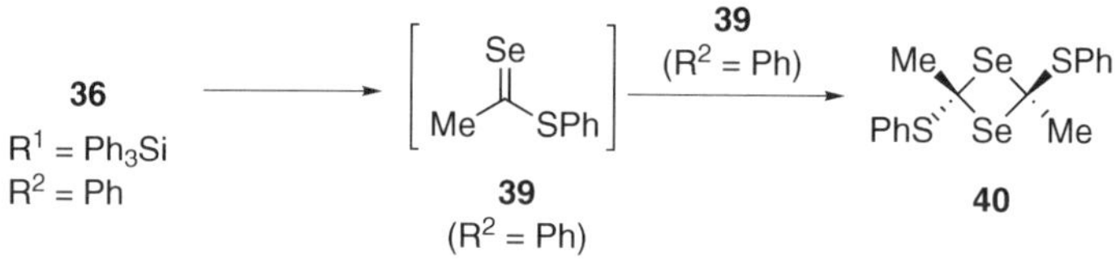

Scheme 4

S-Alkyl arenecarboselenothioates **41–44** were prepared by transesterification of *O*-methyl selenoates **45** with Me$_2$AlSR **46** in 40–92% yields (Eq. 11) [16]. No *S*-phenyl ester **41** (R=Ph) was obtained by this method because of its decomposition during aqueous work-up. While **41** (R=Me) and **42** were highly labile and quite susceptible to oxidation in air, the *p*-methoxy and *o*-methyl substituents on the phenyl group stabilized the esters **43** and **44**, respectively, to air. Relevant spectroscopic data for **41–44** are as follows: ^{13}C NMR $\delta(C{=}Se){=}$ 231.5–237.9; ^{77}Se NMR $\delta(C{=}Se){=}$1623.8 for **41** (R=Me); UV-Vis λ_{max}=597–629 nm (cyclohexane) [499 nm for **44** (R=Me)].

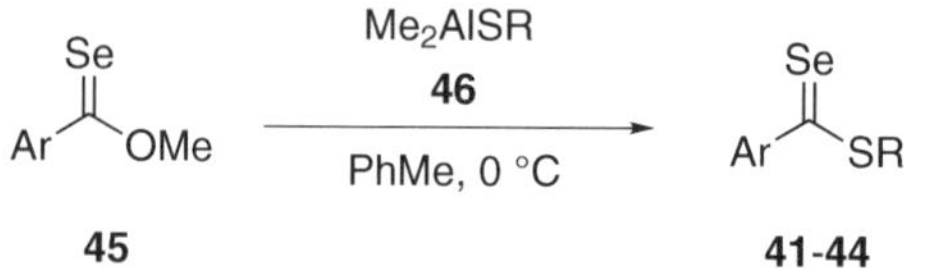

$$(11)$$

41: Ar = Ph; R = Me (82%), Bu (75%), *s*-Bu (42%), Et (49%)
42: Ar = Tol; R = *t*-Bu (81%)
43: Ar = Ans; R = Me (54%)
44: Ar = *o*-MeC$_6$H$_4$; R = Me (92%), Bu (80%)

3.3
Diselenoic Acid Esters

Transesterification of *O*-methyl arenecarboselenoates **47** with Me$_2$AlSeMe **48** yielded the methyl diselenoates **49** as green compounds in 37–91% yields (Eq. 12) [17]. In the ^{13}C NMR spectra of **49**, the selenocarbonyl carbons appeared at δ=233.3–240.5, and in the UV-Vis spectra, absorption maxima due to π-π* and n-π* transitions were observed in the regions of 356–383 and 615–619 nm, respectively.

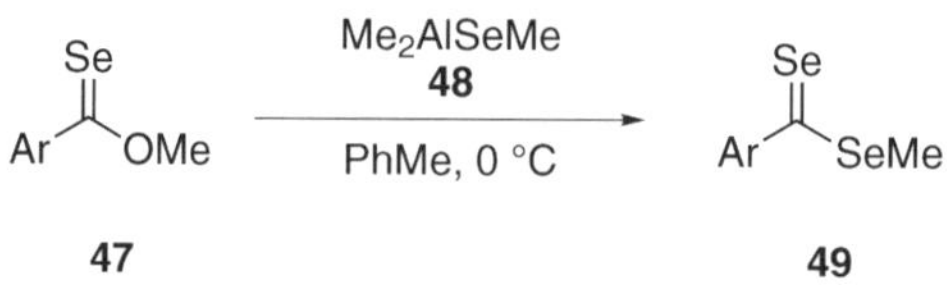

$$(12)$$

49: Ar = Ph (91%), Tol (64%), *o*-MeC$_6$H$_4$ (37%), Ans (73%),
        p-BrC$_6$H$_4$ (51%), *p*-CF$_3$C$_6$H$_4$ (64%)

When *O*-methyl alkaneselenoates **50** [R=Bu, Cl(CH$_2$)$_3$, Ph, Tol, Ans, *p*-ClC$_6$H$_4$, 1-naphthyl] were treated with **48**, only **51** was successfully obtained in the pure form as a deep purple oil with 61% yield [17].

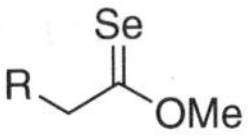

50

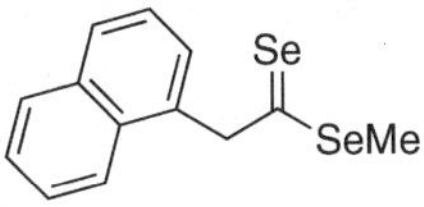

51

3.4
Metal Complexes

A few metal complexes of selenothioic and diselenoic acids have been reported. In a study on the reactivity of metal-carbon triple bond compounds toward chalcogens, it was found that molybdenum (**52**) and tungsten (**53**) carbyne complexes reacted with elemental selenium to yield the corresponding diselenoate complexes **54** and **55**, respectively (Eqs. 13 and 14) [18].

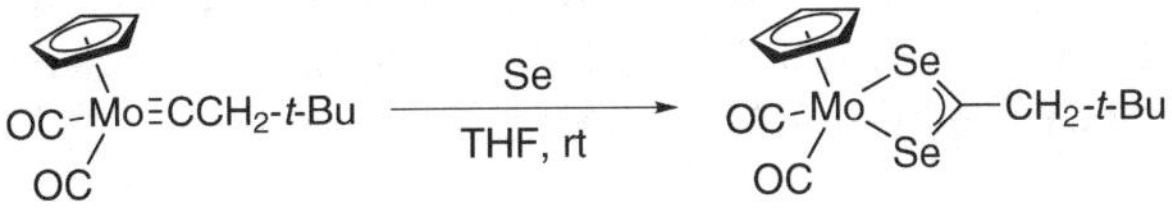

$$ \text{(13)} $$

52 **54** 78%

$$ \text{(14)} $$

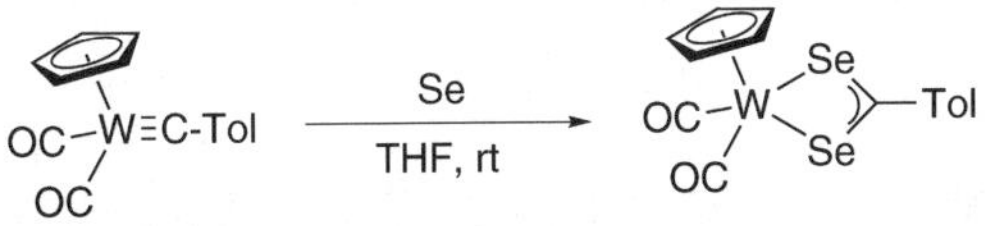

53 **55** 94%

The stoichiomeric reaction of another molybdenum alkylidyne complex **56** with 2-methylthiirane as the sulfur source yielded the thioacyl complex **57**. A further reaction of **57** with Se$_8$-LiBHEt$_3$ furnished the selenothioate complex **58** in high yield (Scheme 5) [19].

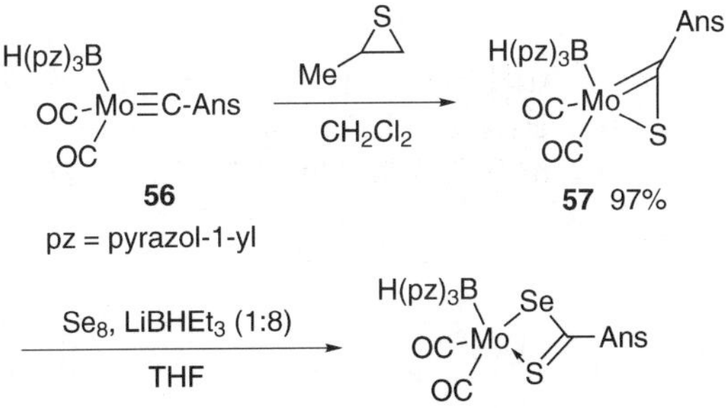

56

pz = pyrazol-1-yl

57 97%

58 87%

Scheme 5

4
Reactions of Selenothioic and Diselenoic Acid Esters

Selenothioic acid *Se*-esters serve as thioacylating agents toward amines and alcohols. Similarly, selenothioic acid *S*-esters and diselenoic acid esters can be used as selenoacylation agents. In addition, their thiocarbonyl and selenocarbonyl groups act as dienophiles. The reactions of *S*-substituted selenothioates with electrophiles under basic conditions were investigated in detail.

4.1
Selenothioic Acid *Se*-esters

4.1.1
Reactions with Amines

The reaction of *Se*-phenyl *p*-toluenecarboselenothioate (**31**: R=Tol, Ar=Ph) with aniline gave thioamide **59** and PhSeSePh, while that with aliphatic secondary amines yielded the ammonium dithioates **60** (Scheme 6) [12]. The formation mechanism of **60** was not clarified [12]. On the other hand, the reaction of *Se*-methyl selenothioates **32** with dimethylamine in dichloromethane provided the corresponding thioamides **61** and MeSeSeMe in quantitative yields (Eq. 15) [13].

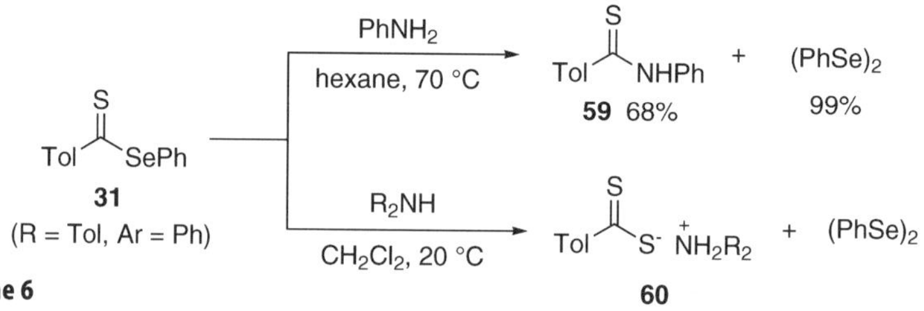

Scheme 6

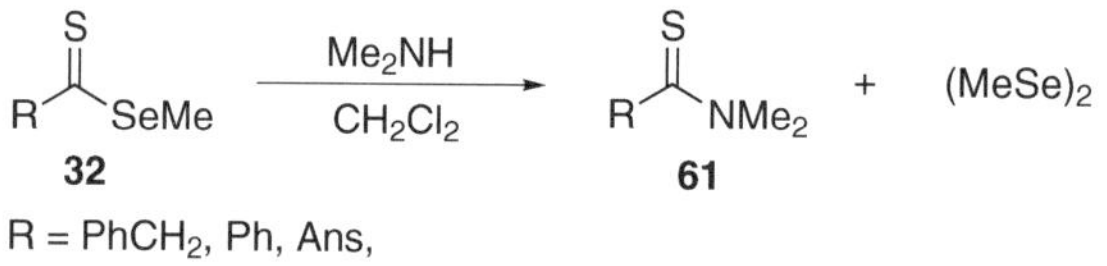

Reactions of **31** (R=Tol, Ar=Ph) with NH_2NH_2 in ethanol at 20 °C, NH_2NH_2 in refluxing ethanol, and $H_2NC(S)NHNH_2$ provided the thiohydrazide **62** (78%), the 1,3,4-thiadiazole **63** (84%), and the 3,4-dihydro-2*H*-1,3,4-triazole-2-thione **64** (89%), respectively [12].

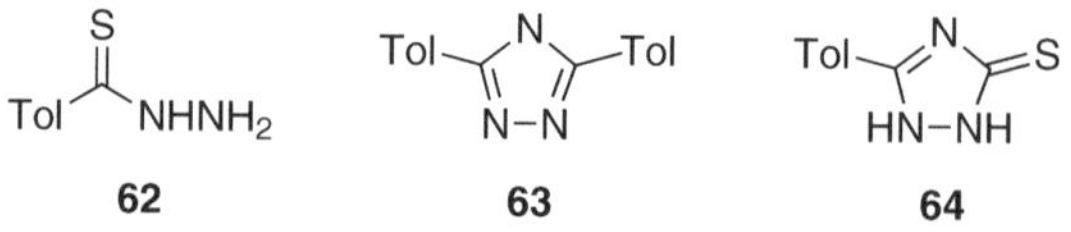

4.1.2
Reactions with Alcohols

The reaction of *Se*-phenyl selenothioates **31** (R=Tol, Ar=Ph) with RONa in ROH (R=Me, Et) at room temperature furnished the *O*-alkyl thioates **65** (R=Me, Et) and PhSeSePh in high yields (Eq. 16) [12]. *O*-Phenyl thioates **65** (R=Ph) also could be prepared by reacting **31** (R=Tol, Ar=Ph) with PhONa in ether at room temperature.

$$\text{(16)}$$

31
(R = Tol, Ar = Ph)

65
R = Me (86%), Et (91%),
Ph (79%)

4.1.3
Oxidation

Oxidation of *Se*-phenyl selenothioates **31** (R=i-Pr, Bu, Tol) with *m*-CPBA in ether gave a mixture of *E*- and *Z*-*S*-oxides (sulfines) **66**, phenylseleno acyl sulfides **67**, diphenyl diselenide, and diacyl disulfides **68** (only in the case of R=Tol) (Eq. 17) [12]. It was proposed that compounds **67** were formed by rearrangements of oxathiirane intermediates **69** [12].

31
R = i-Pr, Bu, Tol
Ar = Ph

E-**66**   *Z*-**66**   **67**

68   **69**

$$\text{(17)}$$

4.2
Selenothioic Acid *S*-esters

4.2.1
Reactions with Electrophiles under Basic Conditions

The reactions of *S*-substituted selenothioates **36** with electrophiles under basic conditions were investigated in detail [14, 20–22]. An α-proton of **36** was readily abstracted by triethylamine [14, 20], LDA [14, 21], and TBAF [22], and the resulting eneselenolates **70** reacted with various types of electrophiles. The reaction took place at the selenium atom in the case of alkyl halides [Scheme 7,

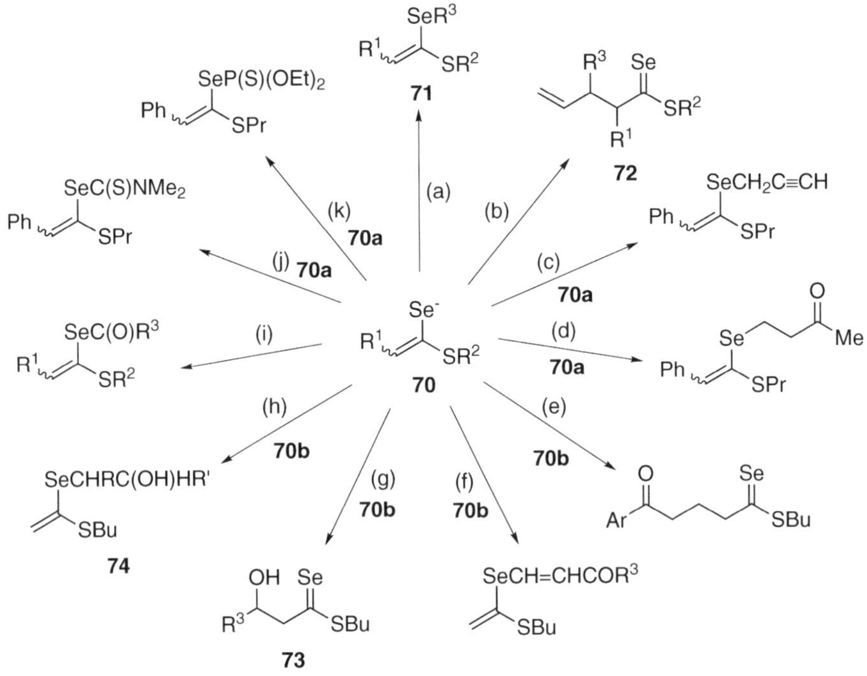

Central: **70** R^1C(Se$^-$)=SR^2

- (a) → **71**: R^1C(SeR^3)=SR^2
- (b) → **72**: R^3CH(–CH=CH$_2$)–CH(R^1)–C(Se)=SR^2 **70a**
- (c) → Ph–C(SeCH$_2$C≡CH)=SPr **70a**
- (d) → Ph–C(Se–CH$_2$CH$_2$C(O)Me)=SPr **70a**
- (e) → Ar–C(O)–CH$_2$CH$_2$–C(Se)=SBu **70b**
- (f) → H$_2$C=C(SeCH=CHCOR^3)–SBu **70b**
- (g) → **73**: R^3CH(OH)–CH$_2$–C(Se)=SBu **70b**
- (h) → **74**: H$_2$C=C(SeCHRC(OH)HR′)–SBu **70b**
- (i) → R^1C(SeC(O)R^3)=SR^2
- (j) → Ph–C(SeC(S)NMe$_2$)=SPr **70a**
- (k) → Ph–C(SeP(S)(OEt)$_2$)=SPr **70a**

70a: R^1 = Ph, R^2 = Pr; **70b**: R^1 = H, R^2 = Bu

Reagents: (a) R^3I; (b) R^3CH=CHCH$_2$Br; (c) HC≡CCH$_2$Br; (d) H$_2$C=CHCOMe;
(e) H$_2$C=CHCOAr; (f) HC≡CCOR^3 (g) RCHO and then H$_2$O; (h) epoxides;
(i) R^3COCl or [R^3C(O)]$_2$O; (j) Me$_2$NC(S)Cl; (k) (EtO)$_2$P(S)Cl

Scheme 7

(a) and (c)], methyl vinyl ketone (d), alkynyl ketone (f), epoxides (h), acyl chloride or acid anhydride (i), Me$_2$NC(S)Cl (j), and (EtO)$_2$P(S)Cl (k) to furnish the corresponding ketene selenothioacetals, as exemplified by the formation of **71**. In the case of the allyl bromides (b), the initial *Se*-adducts underwent [3,3]-sigmatropic rearrangement (seleno-Claisen rearrangement) to yield the selenothioates **72** as the final products. While **70a** reacted with methyl vinyl ketone at the Se atom (d) [14], the reaction of **70b** with aryl vinyl ketone took place at the α-carbon atom (e) [22a]. Interestingly, as shown in Scheme 8, the initial adduct (**73**) with aldehydes was in equilibrium with another eneselenolate **70′**, and quenching of the reaction with methyl iodide or allyl bromide gave **75** or **76**, respectively, the latter of which was formed via seleno-Claisen rearrangement with high diastereoselectivity [21a]. The ketene *Se*-β-hydroxy selenothioacetals **74** [Scheme 7, (h)] were converted to the 1,3-oxaselenolanes **77** on treatment with TFA (Eq. 18) [21b].

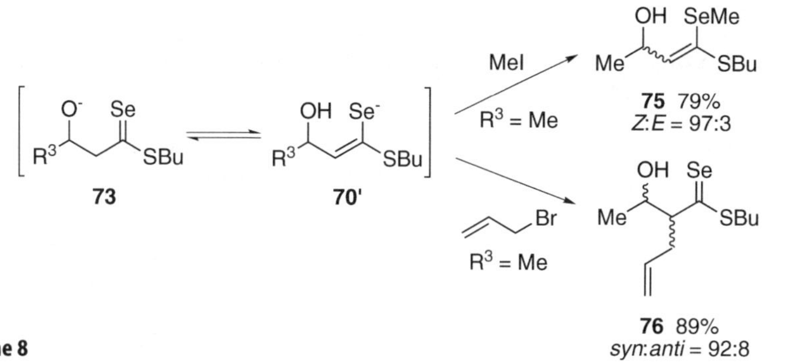

Scheme 8

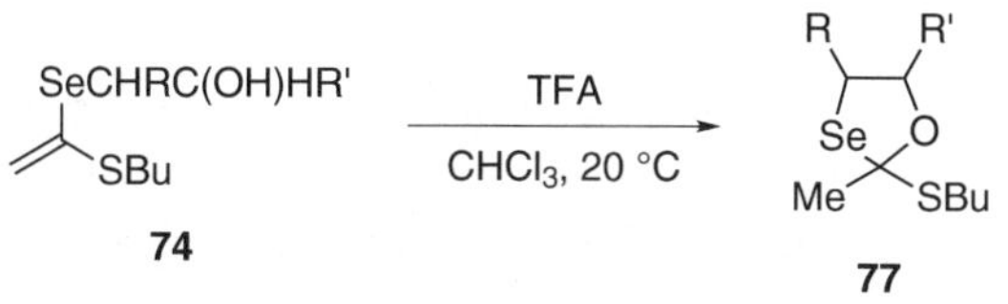

$$(18)$$

4.2.2
Cycloaddition

S-Butyl selenothioacetate serves as a 1,3-dipolarophile [23]. Therefore, the azomethine ylide **78**, generated by the thermolysis of oxazolidinone **79**, was trapped with the selenothioacetate to give the cycloadduct **80** as a 1:1 mixture of stereoisomers with a 45% yield (Scheme 9). Similar reactions took place when selenoketones and *O*-methyl selenobenzoate were used as 1,3-dipolarophiles. However, *Se*-methyl diselenobenzoate, selenoamides, and selenourea failed to give the corresponding cycloadducts.

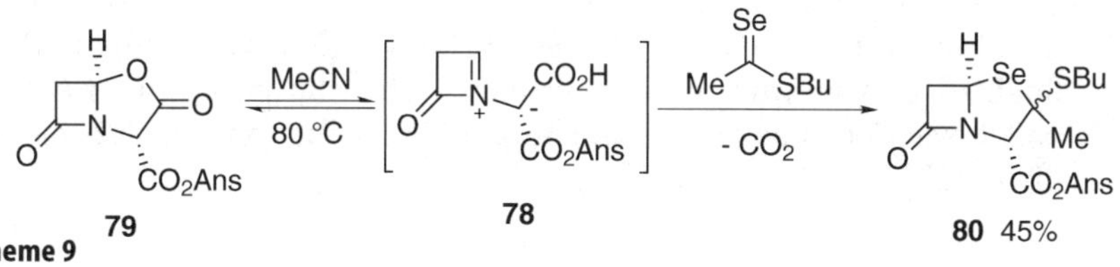

Scheme 9

The reaction of *S-sec*-butyl selenothioacetate with DMAD in refluxing toluene afforded 1,3-thiaselenole **81**, where a stepwise mechanism was proposed (Scheme 10) [20c].

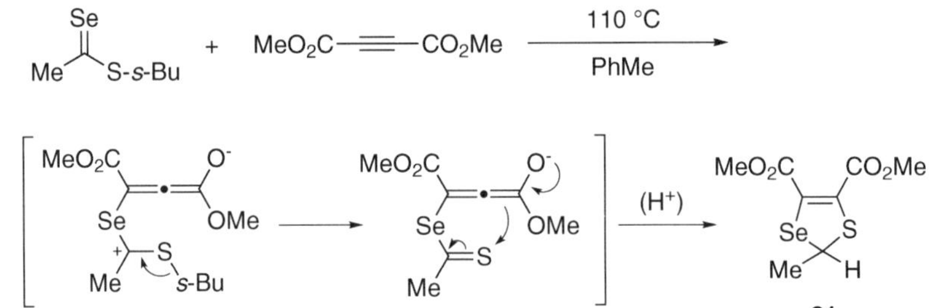

Scheme 10

4.2.3
Oxidation and Reduction

Oxidation of S-propyl phenylselenothioacetate with m-CPBA was reported [14]. The oxidation at $-25\,°C$ gave the diselenide **82**, while that at $100\,°C$ yielded the bis(selenothioate) **83**, which was formed via [3,3]-sigmatropic rearrangement of **82** (Scheme 11).

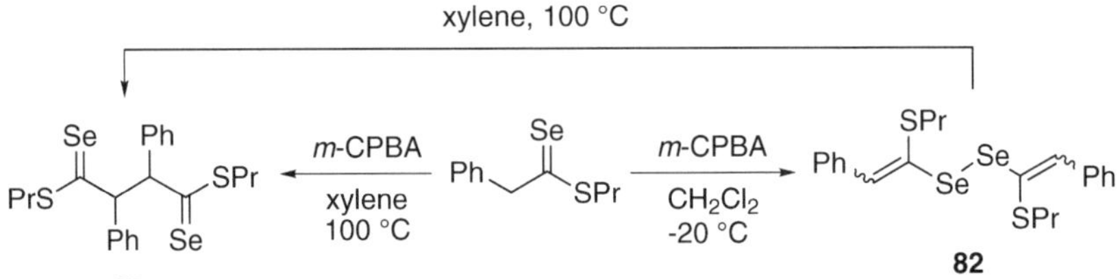

Scheme 11

Reduction of γ,δ-unsaturated S-butyl selenothioates **84** was employed for the synthesis of the tetrahydroselenophenes **85** (Eq. 19) [24]. The compounds **85** were obtained as a mixture of stereoisomers. The SBu group of **84** (R^1=H, R^2=R^3=allyl) was detached during the course of the reduction to give 2-methyl-4,4-diallyltetrahydroselenophene.

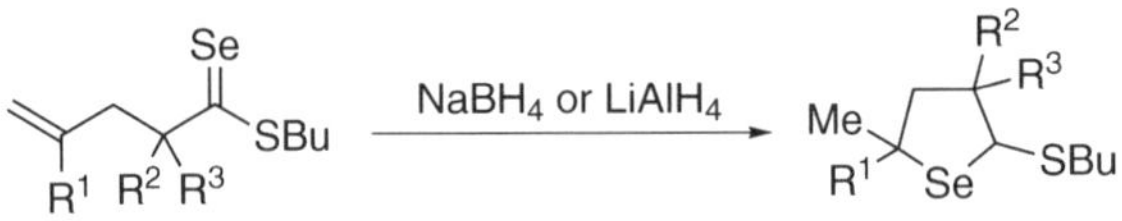

$$\text{(19)}$$

4.2.4
Miscellaneous

As in the case of *Se*-substituted selenothioates (4.1.1), the reaction of *S*-substituted selenothioates with primary and secondary amines in THF provided the corresponding selenoamides in 40–84% yields [14].

The reactions of *S*-butyl selenothioates with trivalent phosphorus reagents were examined [25]. As a result, the reaction yielded α-phosphoryl sulfide **86** as the main product and, in some cases, selenothioacetal **87** as the minor product (Eq. 20).

$$R^1 = H, CH_2=CHCH_2, CH_2=CBrCH_2$$
$$R^2R^3P(OMe): P(OMe)_3, PhP(OMe)_2, Ph_2P(OMe)$$

The formation of **86** was explained in terms of the selenophilic attack of $R^2R^3P(OMe)$ giving **88**, followed by elimination of the selenium atom to yield **89**, protonation of which provided **86**. On the other hand, the formation of **87** was rationalized by the carbophilic attack of $R^2R^3P(OMe)$ to produce **90** followed by Me-migration to afford **91**, and the elimination of $R^2R^3P(OH)$ which furnished **87** [25a].

Symmetrically substituted selenophenes **92** were obtained when *S*-butyl selenothioates were treated with Et_3N and $Cd(OAc)_2$ (for R=Me, allyl) or $NiCl_2$ (for R=H), as shown in Eq. 21 [20d].

$$R = H, Me, CH_2=CHCH_2$$

Use of ZnI_2 instead of $NiCl_2$ for *S*-butyl selenothioacetate (R=H) resulted in the formation of **93**. The formation mechanism for **92** was not clear; the initial for-

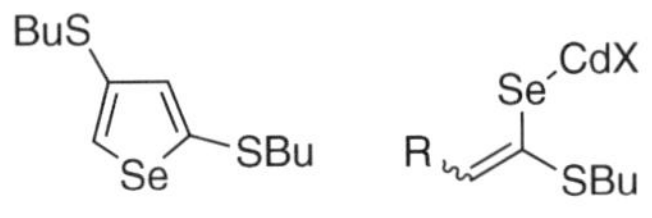

93

94

mation of the cadmium salt **94** was proposed based on experimental observations that the reaction in the presence of alkyl halide gave selenothioacetals like **71**.

4.3
Diselenoic Acid Esters

Reactions of methyl diselenobenzoate have been reported (Scheme 12) [17]. Cycloaddition of the diselenoate with cyclopentadiene gave the cycloadduct **95** as a stereoisomeric mixture in 67% yield. The reductive coupling of the same diselenoate mediated with copper yielded the alkene **96** with 39% yield.

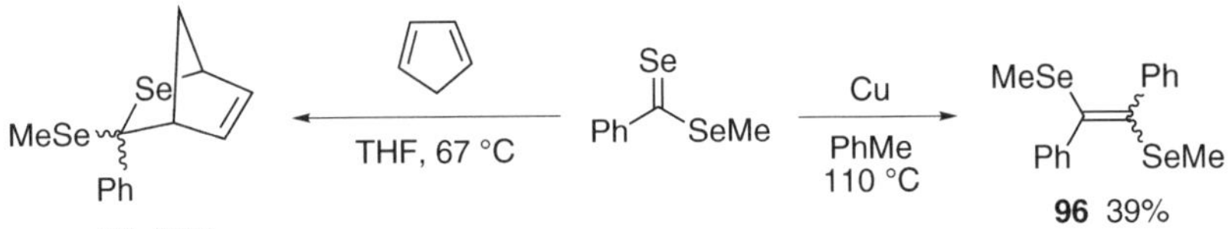

95 67%

THF, 67 °C

Ph—SeMe

Cu

PhMe 110 °C

96 39%

Scheme 12

Reactions of diselenoates with trivalent phosphorus reagents took place in a manner similar to those of the *S*-butyl selenothioates (Eq. 20) to give α-phosphoryl selenides **97** (Scheme 13) [25].

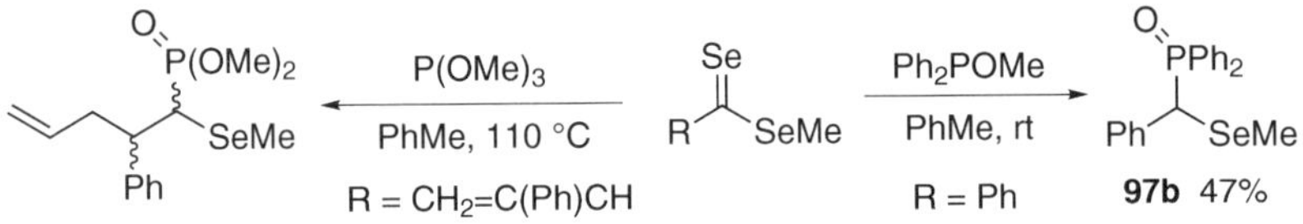

97a
84% (60:40)

P(OMe)$_3$

PhMe, 110 °C

R = CH$_2$=C(Ph)CH

R—SeMe

Ph$_2$POMe

PhMe, rt

R = Ph

97b 47%

Scheme 13

5
Conclusions

Although ammonium salts and inner salts of selenothioic and diselenoic acids have been synthesized, isolation of their free acids has yet to be achieved. Eight selenothioates I–VIII and four diselenoates IX–XII are possible if we take all of the combinations of substituents R (alkyl) and Ar (aryl) into account. Out of these twelve, three esters, VIII, X, and XII, have not been synthesized yet. The esters that have been synthesized are generally highly colored and present characteristic spectroscopic properties. Whereas most of the esters are sensitive to air, the esters are stable enough to heat to be handled at ambient temperatures under inert atmospheres. Therefore, considering their high reactivities, they could be employed as substrates in several organic reactions.

$$
\begin{array}{cccc}
\overset{S}{\underset{}{\|}} & \overset{S}{\underset{}{\|}} & \overset{S}{\underset{}{\|}} & \overset{S}{\underset{}{\|}} \\
R{-}C{-}SeR & R{-}C{-}SeAr & Ar{-}C{-}SeR & Ar{-}C{-}SeAr \\
\text{I} & \text{II} & \text{III} & \text{IV} \\[2ex]
\overset{Se}{\underset{}{\|}} & \overset{Se}{\underset{}{\|}} & \overset{Se}{\underset{}{\|}} & \overset{Se}{\underset{}{\|}} \\
R{-}C{-}SR & R{-}C{-}SAr & Ar{-}C{-}SR & Ar{-}C{-}SAr \\
\text{V} & \text{VI} & \text{VII} & \text{(VIII)} \\[2ex]
\overset{Se}{\underset{}{\|}} & \overset{Se}{\underset{}{\|}} & \overset{Se}{\underset{}{\|}} & \overset{Se}{\underset{}{\|}} \\
R{-}C{-}SeR & R{-}C{-}SeAr & Ar{-}C{-}SeR & Ar{-}C{-}SeAr \\
\text{IX} & \text{(X)} & \text{XI} & \text{(XII)}
\end{array}
$$

R = alkyl, Ar = aryl

References

1. Murai T, Kato S (1995) Functions with two chalcogens other than oxygen. In: Katritzky AR, Meth-Cohn O, Rees CW (eds) Comprehensive organic functional group transformations. Pergamon, Oxford, UK, vol 5, p 545
2. Murai T, Kato S (1998) Sulfur Rep 20:397
3. Jemmis ED, Giju KT, Leszczynski J (1997) J Phys Chem A 101:7389
4. Murai T, Kamoto T, Kato S (2000) J Am Chem Soc 122:9850
5. Tani K, Murai T, Kato S (2002) J Am Chem Soc 124:5960
6. Nakayama J, Kitahara T, Sugihara Y, Sakamoto A, Ishii A (2000) J Am Chem Soc 122:9120
7. Nakayama J, Akiyama I, Sugihara Y, Nishio T (1998) J Am Chem Soc 120:10027
8. Renson M, Collinene R (1964) B Soc Chim Belg 73:491
9. (a) Rosenberg P, Mautner HG (1967) Science 155:1569; (b) Chu SH, Mautner HG (1968) J Med Chem 11:446
10. Raash MA (1972) J Org Chem 37:1347
11. Oae S, Sakaki K, Fukumura M, Tamagaki S, Matsuura Y, Kakudo M (1982) Heterocycles 19:657
12. Kato S, Yasui E, Terashima K, Ishihara H, Murai T (1988) B Chem Soc Jpn 61:3931
13. Khalid M, Ripoll JL, Vallée Y (1991) Chem Commun 964
14. Kato S, Komuro T, Kanda T, Ishihara H, Murai T (1993) J Am Chem Soc 115:3000
15. (a) Murai T, Kakami K, Hayashi A, Komuro, T, Takada H, Fujii M, Kanda T, Kato S (1997) J Am Chem Soc 119:8592; (b) Murai T, Hayashi A, Kanda, T, Kato S (1993) Chem Lett 1469; (c) Murai T, Kakami K, Itoh N, Kanda T, Kato S (1996) Tetrahedron 52:2839
16. Murai T, Ogino Y, Mizutani T, Kanda T, Kato S (1995) J Org Chem 60:2942

17. Murai T, Mizutani T, Kanda T, Kato S (1993) J Am Chem Soc 115:5823
18. Gill DS, Green M, Marsden K, Moore I, Orpen AG, Stone FGA, Williams ID, Woodward P (1984) J Chem Soc Dalton 1343
19. Cook DJ, Hill AF (1997) Chem Commun 955
20. (a) Murai T, Takada H, Kakami K, Fujii M, Maeda M, Kato S (1997) Tetrahedron 53:12237; (b) Murai T, Takada H, Kanda T, Kato S (1995) Chem Lett 1057; (c) Murai T, Takada H, Kanda T, Kato S (1994) Tetrahedron Lett 35:8817; (d) Murai T, Fujii M, Kanda T, Kato S (1996) Chem Lett 877
21. (a) Murai T, Endo H, Ozaki M, Kato S (1999) J Org Chem 64:2130; (b) Murai T, Fujii M, Kato S (1997) Chem Lett 545
22. (a) Murai T, Kondo T, Kato S (2004) Heteroatom Chem 15:187; (b) Murai T, Hayakawa S, Kato S (2001) J Org Chem 66:8101; (c) Murai T, Hayakawa S, Kato S (2000) Chem Lett 368
23. Brown GA, Anderson KM, Murray M, Gallagher T, Hales NJ (2000) Tetrahedron 56:5579
24. Murai T, Maeda M, Matsuoka F, Kanda T, Kato S (1996) Chem Commun 1461
25. (a) Murai T, Izumi C, Itoh T, Kato S (2000) J Chem Soc Perkin T 1 917; (b) Murai T, Izumi C, Kato S (1999) Chem Lett 105

Top Curr Chem (2005) 251: 247–272
DOI 10.1007/b101011
© Springer-Verlag Berlin Heidelberg 2005

Thio-, Seleno-, Telluro-Amides

Toshiaki Murai (✉)

Department of Chemistry, Faculty of Engineering, Gifu University, Yanagido,
Gifu 501-1193, Japan
mtoshi@cc.gifu-u.ac.jp

Abstract Recent developments regarding synthetic methods of producing chalogen isologues of amides are reviewed. Among exhaustive studies on the syntheses, properties, and broad utilities of thiamides, we describe methods of synthesizing them via three-component coupling reactions of aldehydes, amines, and sulfur, via thionation of amides, and via thiolysis of iminium salts. The use of thioacylating agents, and functional group manipulations of thioamides leading to other thioamides are also presented. The discussion focuses on synthetic methods for selenoamides starting from amides, and functional group manipulations of selenoamides leading to functionalized selenoamides. The review ends with some examples of how to synthesize telluroamides.

Keywords Thioamides · Willgerodt-Kindler Reaction · Lawesson's Reagents · Selenoamides · Telluroamides

1
Introduction

Amides are less reactive than most carboxylic acid derivatives, but are often found in natural products, and they play important roles in biological and material sciences and in industrial applications. There are three types of chalcogen amides (as with chalcogen acids and esters). Exhaustive papers have reported on the sulfur isologues of amides (thioamides) **1**. Increasing attention is also being paid to selenium isologues of amides, **2**, and new ways of synthesizing them have been developed. This is partly because of the biological interest of organoselenium compounds. In contrast, tellurium isologues of amides (telluroamides) **3** are still rare, mainly because of their instability in air.

In this chapter we review synthetic methods for these calcogen amides.

2
Thioamides

Thioamides have become a ubiquitous but important class of compounds – much like ordinary organic compounds such as aldehydes, ketones, and carboxylic acid derivatives – and extensive studies into their syntheses and reactions have been carried out in recent years. When an oxygen atom in an amide is replaced with a sulfur atom, the compounds become less polar and more soluble in various organic solvents. They are reactive but still stable enough to be handled in air. Therefore, a tremendous number of studies have been made into their fundamental properties, biological aspects, and broad applicabilities. Theoretical studies into the rotational barriers that exist in the thioformamides **4** and the thioacetamides **5** have also been made (for example [1]).

Compared to the corresponding formamides and acetamides, the non-bonding interaction between the sulfur atom and the hydrogen atom of the methyl group attached to the nitrogen atom is important. Two rotamers of the α,β-unsaturated thioamides **6** and **7** are observed in ^{1}H NMR spectra, and their rotational barriers can be elucidated. The isomers **6**, where the thiocarbonyl and benzyl groups are located in a *cis* position with respect to the C–N single bond, are more stable than the other isomers 7 [2a,b]. Nevertheless, these two isomers are in equilibrium. The free energy of activation for the rotation of a C–N bond is estimated to be 18.4~19.5 kcal/mol. X-Ray molecular structure analyses of 6 have shown that alkenyl groups are almost perpendicular to the thioamide group. Thioamide linkages have been introduced to various peptides, and conformational studies have been carried out on them [3]. Increasing the ionic radius of a sulfur atom and the length of the bond between sulfur and carbon influences the conformational changes of peptides, and peptides containing introduced thioamides 8 are expected to be useful in drug design and a potential probe of local conformations of peptides because of their special spectroscopic properties. The thioamide NH is a stronger hydrogen donor and the sulfur atom is a weaker hydrogen bond acceptor compared to those of the corresponding amides, and this influences conformations [4]. As examples of biological interest, the binding affinities of thiobenzamides 9 toward the estrogen receptor [5] and the anti-influenza activities of thiobenzamides 10 have been elucidated [6]. Additionally, the thioamides 11 exhibit higher in vitro antimycobacterial and antifungal activity than the corresponding amides [7]. Various thioamides have been used for the synthesis of metal complexes (for example [8]) and sulfur- and nitrogen-containing heterocycles [9]. These results have proved that thioamides are widely applicable in organic syntheses. Photochemical reactions of thioamides, particularly α,β-unsaturated thioamides, are well-documented [2]. Some achiral α,β-unsaturated thioamides provide chiral crystals, and their solid-state photoreactions give optically active thiolactams. Synthetic methods for sugar thioamides, their properties [10], and various

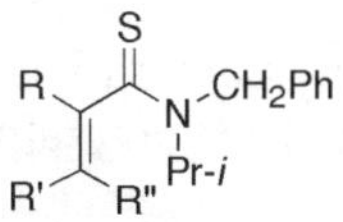

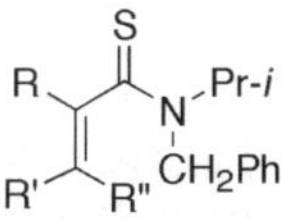

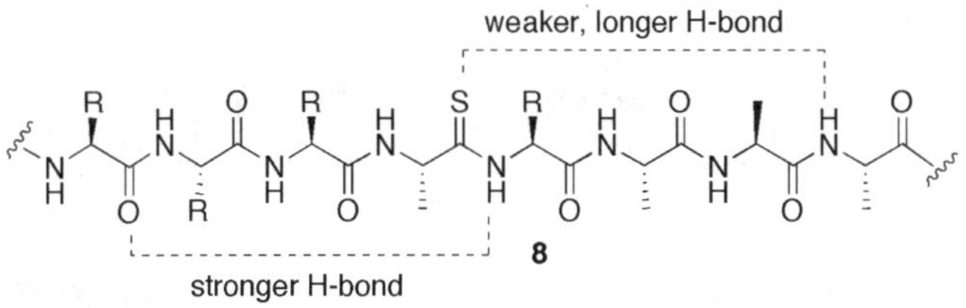

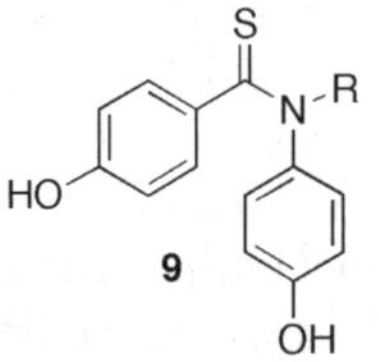

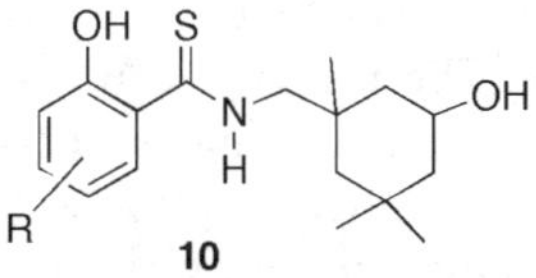

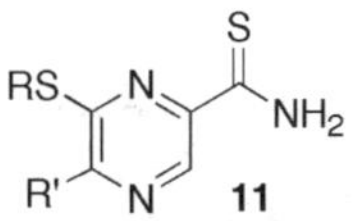

reactions [11] have also been reviewed. In this chapter, we give examples of thioamide syntheses reported over the last few years, as well as discussing reactions of thioamides that lead to functionalized thioamides.

2.1
Syntheses

2.1.1
Willgerodt-Kindler Reaction

Three component coupling reactions of aldehydes or ketones, amines, and elemental sulfur – called Willgerodt-Kinder reactions – are used to synthesize various thioamides, although the protocol of this reaction was reported over 80 years ago. However, the reaction procedure has been developed. For example, the reaction of dialdehydes, diamines and sulfur was investigated for the preparation of polythioamides **12** (Eq. 1) [12].

$$(1)$$

The reaction is carried out with 2.5 equiv elemental sulfur and DMF or DMA as a solvent. Performing the reaction at a temperature higher than the melting point of sulfur (112 °C) leads to polymers with higher molecular weights. The polythioamides **12** exhibit good solubility in organic solvents compared to analogous polyamides. The mechanism of the reaction has also been proposed. Schiff base polymers **13** are initially formed. Then polysulfide anions **14** (formed by the cleavage of elemental sulfur) and amines attack **13** to form thioamides **12** and degraded polysulfide anions **15** (Eq. 2). This process can take place repeatedly to give the polythioamides **12**.

$$(2)$$

A microwave-assisted Willgerodt-Kindler reaction was developed to synthesize α-aryl thioacetamides **16** from aromatic ketones, morpholine, and elemental sulfur (Eq. 3) [13]. The reaction is carried out without solvents and completes within 4 min. The reaction can also be used for aldehydes. The reaction temperature and times are dependent on the substituents in the aldehydes and amines [13, 14]. The use of secondary cyclic amines gives the corresponding thioamides in good to high yields. In the primary amine reaction, higher reaction temperatures are necessary. Primary thioamides are formed by reacting ammonia and aromatic aldehydes at temperatures higher than 160 °C for 20 min, albeit in moderate yields. The reaction of aniline at 200~250 °C gives the product in yields of 30% at the most, along with 2-phenylbenzothiazole. For aliphatic aldehydes and ortho-substituted aromatic aldehydes, the corresponding products are obtained in lower yields.

$$\text{(3)}$$

Ar = Ph, 4-MeC$_6$H$_4$, 4-ClC$_6$H$_4$

Instead of aldehydes and ketones, benzyl -thiol, -amine, [15] and nitriles [16] can be used under solvent-free conditions. The condensation reaction of benzyl -thiol or amine with morpholine and elemental sulfur for 4 h gives thiobenz-amide in high yields. Microwave-irradiation facilitates the reaction. The reaction of aromatic nitriles, morpholine and elemental sulfur is also assisted by micro-wave-irradiation to give the aromatic thioamides **17** in good yields (Eq. 4), except in the reaction of 4-chlorobenzonitrile [16]. In the reaction of benzylnitrile, the sulfur atom is introduced to the benzyl carbon atom and the cyano group is replaced with morphorine. On the other hand, the use of 4-chlorobenzylni-trile gives a mixture of α-4-chlorophenyl thioacetoamide and 4-chloro thio-benzamide.

$$\text{(4)}$$

Ar = Ph, 4-MeC$_6$H$_4$, 4-EtOC$_6$H$_4$

2.1.2
Thionation of Amides

Direct conversion of ordinary amides to thioamides has been developed. One of the most relevant methods for this conversion is to use commercially available 2,4-bis(4-methoxyphenyl)-1,3-dithia-2,4-adiphosphetane 2,4-disulfide (Lawes-son's reagent) **18**. The β-thiopeptides **19** and **20** are synthesized by treating the corresponding amides with Lawesson's reagent **18** in THF for 2.5~4 h [17]. Con-densation of two β-thiopeptides **19** and **20** gives the precursor β-thiopeptide **21**, which is again treated with **18**. In these thionation reactions, the alkoxycarbonyl group is inert and the oxygen atom of the aminocarbonyl group is selectively replaced with the sulfur atom. Similarly, thiopeptides **22** [18], thioamides **23** [19], **24** [20], and sugar thioamides **25** [21] are prepared with **18** in toluene or benzene at about 70~80 °C. The reaction was tested using peptides bearing various alkoxycarbonyl groups. The results have shown that the thionation of the aminocarbonyl group with **18** proceeds with high site-selectivity, and the alkoxycarbonyl group is not competitively thionated. When the products **22** and **23** are reduced with Raney nickel, the thiocarbonyl groups are converted to methylene groups. The synthesis of thioamide **26** is achieved with **18** in THF at room temperature [22]. 1 equiv of pyridine is added during the synthesis of **27** [23] to neutralize the phosphoric acids that are the by-products derived from **18**.

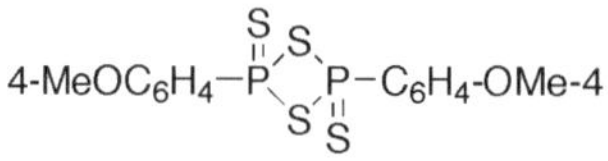

18

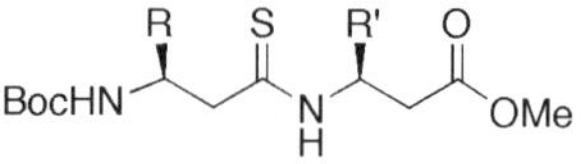

19 R = Me_2CH, R' = Me
20 R = Me_2CHCH_2, R' = Me_2CH

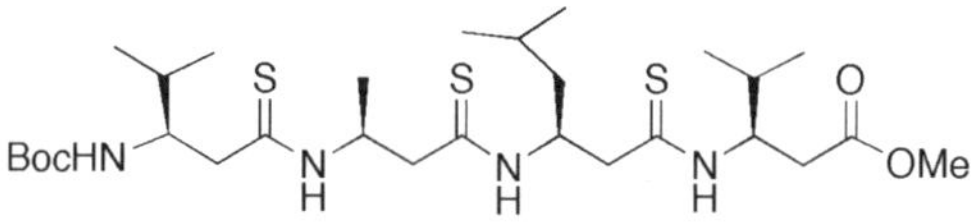

21

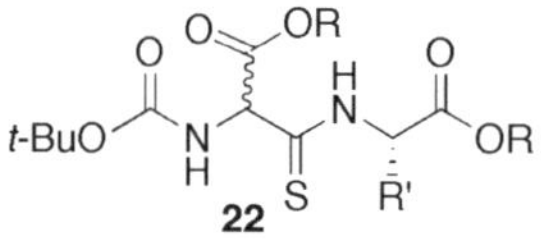

22

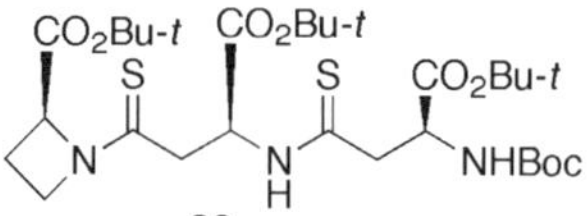

23

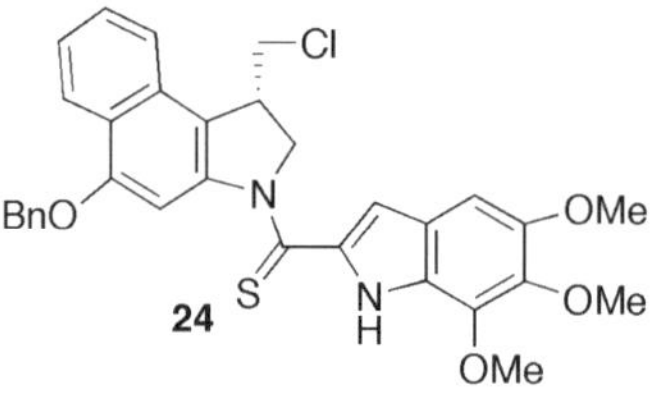

24

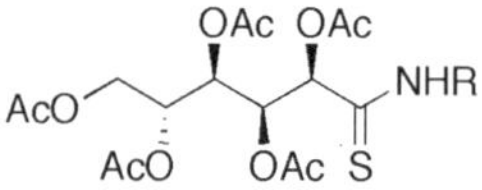

25

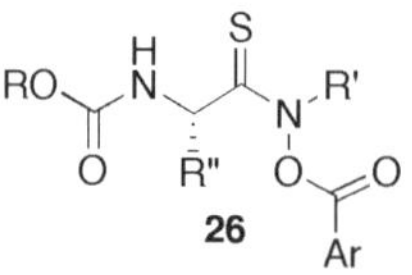

26

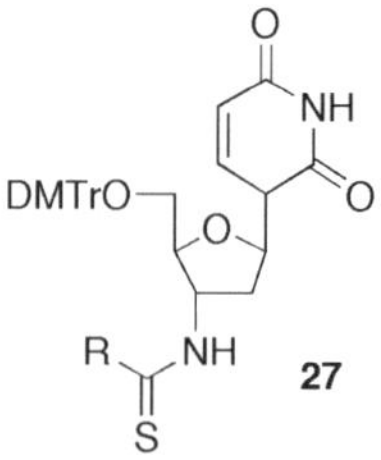

27

Thionations of the pyridazinecarboxamides **28** with **18** leading to **29** have been tested (Eq. 5) [24]. The reaction of **28** (X=Cl, MeO) is carried out with 1.1 equiv of **18**, and the chlorine and methoxy group in **28** are replaced with an SH group during the thionation to furnish **30**. The reaction of **28** (X=SMe) with

$$\text{(5)}$$

X = Cl, H, OMe, SMe (compound **28**) + **18** $\xrightarrow{\text{toluene}\ 110\,°C}$ (compound **29**, X = Cl, H, OMe, SMe) + (compound **30**)

1.1 equiv of **18** requires a longer reaction time, but the corresponding thioamides are obtained in high yields. No replacement of the MeS group is observed. α,β-Unsaturated thioamides are synthesized by reacting the corresponding amides with **18** [2, 10a,b, 25]. It is worth noting that the geometric isomerization from *E*-**31** to *Z*-**32** proceeds during the thionation (Eq. 6) [2c]. As for the thionation of ω-hydroxy amides, the reaction is accompanied by the cyclization and elimination of H_2O, depending on the starting amides [25]. Dichloromethane and HMPA have been suggested as better solvents than THF for the thionation of amides and lactams bearing simple carbon skeletons [26]. For example, nicotinamide is converted to the corresponding thioamide with **18** in HMPA at 80 °C for 7 h. The thionation of amides with **18** under microwave-irradiated conditions was examined with *N,N*-dimethyl formamide, primary and secondary acetamides, aromatic amides and lactams [27]. The solvent-free reaction is complete within 8 min, and the elongation of the reaction time causes the thioamides formed to decompose. The thionation of resin-bound amides was investigated [28]. To confirm the consumption of the starting amides, the amide carbonyl stretch at about 1670 cm^{-1} was used [28a]. A wide variety of solvents were examined in parallel solid-phase synthesis at elevated temperatures [28b]. Benzyl benzoate was suggested as a superior high-boiling solvent. The resin-bound thioamide was cleaved with 30% trifluoroacetic acid in DMA. The thioamide moiety remained intact during the cleavage.

$$\text{(6)}$$

(compound **31**) + **18** $\longrightarrow$ (compound **32**)

A polymer-supported thionating reagent was devised [29]. The reagent **35**, which can easily be handled and stored for several months below 0 °C under an inert atmosphere, is synthesized by reacting commercially available diamine resin **33** and dichlorothiophosphate **34** (Eq. 7) in pyridine at 0 °C. The thionation of secondary and tertiary amides and lactams with more than 3 equiv of **35** is carried out to give the corresponding thioamides in quite high yields, although it requires more than 30 h. In contrast, the reaction of primary amides with **35** gives nitriles, except in the case of benzamide. To shorten the reaction

$$\text{(7)}$$

(compound **33**) + (compound **34**) $\xrightarrow{\text{pyridine}}$ (compound **35**) time, the thionation with **35** is irradiatied with microwaves. As a result, the reaction of tertiary amides is complete within 15 min in toluene at 200 °C in a sealed tube. The reagent **35** is thermally stable, and no decomposition is observed during the reaction. Small amounts of an ionic liquid (1-ethyl-3-methyl-1H-imidazolium) are added to distribute heat efficiently. As for the reaction of secondary amides and lactams, more than 4.4 equiv of **35** are used.

Alternatively, we can use P_4S_{10} as an amide thionating agent. The amides **36** can be thionated with P_4S_{10} and anhydrous Na_2CO_3 in THF at 0 °C for 2 h [30]. For the β-phenyl and β-methyl α,β-unsaturated amides **37**, the thionation takes place accompanied by intramolecular cyclization to afford 1,5-benzodiazepines **38** (Eq. 8). In the thionation of an acrylamide, a polymerization product is formed. Hexamethyldisiloxane ($Me_3SiOSiMe_3$) is used as an additive in the thionation of amides with P_4S_{10} [26]. The reaction is performed in CH_2Cl_2, $CHCl_3$, HMPA, and benzene with high efficiency, and much less P_4S_{10} is necessary compared with the reaction without hexamethyldisiloxane.

$$\textbf{(8)}$$

2.1.3
Thiolysis of Iminium Salts and Nitriles

In thionation with Lawesson's reagent **18** and P_4S_{10}, large amounts of phosphorus-containing co-products are formed, and separating them from the desired thioamides is troublesome in some cases. Alternatively, we can convert activated amides such as iminium salts, which are reactive toward sulfur-containing nucleophiles, into thioamides. For example, to activate the amides, they are treated with triflic anhydride in the presence of pyridine to form iminium salts **39** (Eq. 9) [31]. Then the thiolysis of **39** is performed with H_2S, NaSH or ammonium sulfide (($NH_4)_2S$). Among them, the slow addition of aqueous ($NH_4)_2S$ to the reaction mixture containing the iminium salts **39** has given the corre-

$$\textbf{(9)}$$

sponding thioamides in the best yields. Thiolysis of iminium salts derived from amides and oxalyl chloride or phosphoric trichlorides with ammonium tetra-thiomolybdate **40** occurs below room temperature within 2 h (Eq. 10) [32]. It is advantageous that the by-products are easily separated, although H_2S is needed in order to prepare **40**.

$$(10)$$

The addition reaction of hydrogen sulfide to nitrile has been well documented as a synthetic procedure for primary thioamides. It generally requires elevated temperatures and in some cases high pressures. The reaction with hydrogen sulfide in the presence of anion-exchange resin (Dowex 1X8 SH^- form) is carried out at room temperature and ambient pressure in MeOH or $EtOH/H_2O$ [33]. This reaction is suitable for compounds containing an aromatic ring and compounds that do not dissolve in organic solvents. The efficiency of the conversion of aliphatic nitriles is moderate to good.

2.1.4
Friedel-Crafts Reaction with Metal Thiocyanates

Alkylation of metal thiocyanates has been known to produce isothiocyanates, and this can be followed by treatment with carbon nucleophiles to give thioamides. Alternatively, metal thiocyanates can be used in the Friedel-Crafts reactions of aromatic compounds. For example, the reactions of aromatic compounds bearing alkoxy groups and heteroaromatic compounds with potassium thiocyanate under acidic conditions are carried out at 30 °C (Eq. 11) [34]. In the reaction a large excess of methanesulfonic acid is used.

$$(11)$$

2.1.5
Thioacylation of Amines

Thioacylation of amines have been developed as a synthetic method for thioamides. Various thioacylating agents can be used. The dithioic acid salts **41**, which are generated in situ from organolithium or Grignard reagents with CS_2,

R-M + CS$_2$ ⟶ (**41**) + **42** ⟶ (12)

M = Li, MgBr

R = Ph, C$_6$H$_4$-Cl-4, Pr-*i*, CH=CH$_2$

43 98% ee + HNMe$_2$ ⟶ (THF, rt) **44** 96% ee (13)

major **45** + minor **45** + HNMe$_2$ ⟶ minor **46** + major **46** (14)

47 R = Ph, CH$_2$Ph + **48** ⟶ (NaOH, H$_2$O) **49** (15)

are reacted with 2,4-dinitrobenzenesulfonamides **42** to give the corresponding thioamides (Eq. 12) [35]. When THF is removed from the mixture involving **41**, and DMF is added to the resulting suspension prior to the addition of **42**, the reaction proceeds at 80 °C. Thioic acid *O*-esters [36] and dithioic acid esters [37–39] are also used as thioacylating agents. Although the former requires higher temperatures, acylation with dithioic acid esters is highly efficient even at room temperature. For example, enantiopure dithioic ester **43** is reacted with dimethylamine in THF at room temperature for 20 min to quantitatively give the thioamide **44** without any significant loss of enantiopurity (Eq. 13) [37a]. On the other hand, dynamic kinetic resolution is observed in the reaction of dithioic acid esters **45** with HNMe$_2$. The minor isomers of the starting materials **45** react with HNMe$_2$ more quickly than the major isomers of **45** to give mainly the thioamides **46** (Eq. 14) [37b]. Dithioic acid carboxymethyl esters **47** are used for the thioacylation of Fmoc-protected 4-aminophenyl-substituted alanine **48** in the presence of NaOH in H$_2$O to give thioamides **49** (Eq. 15) [38]. The applicability of **49** to solid-phase peptide synthesis has also been shown. Thioacylating aliphatic amines using dithiobenzoic acid 2-benzothiazolyl ester **50** under reflux in benzene completes within 10 min, whereas thioacylating aniline and *t*-butylamine requires a longer reaction time [39]. The thioacyl dithiophosphates **52** generated via several steps from acid chlorides and dithiophosphoric acid **51** are used as efficient thioacylating agents for amines (Eq. 16) [40]. The starting acid **51** is prepared in 80% yields from 2,2-dimethyl-1,3-propanediol and P_4S_{10} in benzene at 50 °C. Notably, aromatic, aliphatic acid chlorides and even pyvaloyl chloride react with acid **51** to form the corresponding dithiophosphates **52**. The dithiophosphates **52** are purified by column chromatography on silica gel or by crystallization, but they are used without further purification as thioacylating agents. Thioacylation with **52** is applicable to a wide range of amines. Ammonia, aromatic, aliphatic primary and secondary amines are used, and the products are obtained in yields higher than 85%. The reaction of *N*-hydroxy alkylamine hydrochloric acid salt also proceeds smoothly to afford the thiohydroxamic acids **53** with about 70% yield. Triethylgermanium sodium **55** mediated thiocarbamoylation of 2-fluoro-2-phenylthio-2-phenylacetonitrile **54** with *N,N*-dialkylthiocarbamoyl chloride is reported to afford α-fluoro, α-cyano thioamides **56** with 90% yield (Eq. 17) [41]. The ring-opening of the compounds **57** with aliphatic amines proceeds in absolute ethanol at room temperature to afford *N*-alkyl-1,2-dihydro-2-thioxo-3-pyridinecarbothioamides **58** (Eq. 18) [42]. Various aminoalkylamines are used, and the thioamides obtained are converted to the corresponding hydrochloric acid salts as water-soluble thioamides.

2.1.6
Via Functionalization of Thioamides

In this section, functional group manipulations of thioamides to form products that still contain the thioamide moieties are described.

The Michael addition of amines to α,β-unsaturated thioamides has been carried out [43]. The reaction of tertiary α,β-unsaturated thioamides **59** with cyclic amines such as pyrrolidine, morpholine, and piperidine proceeds slowly (Eq. 19). After five days, the adducts **60** are obtained as stable compounds in moderate yields. On the other hand, the products derived from the addition of pyrrolidine to secondary α,β-unsaturated thioamides **61** are unstable. The stereochemistry of the Michael addition to cyclic α,β-unsaturated thioamides **62** is dependent on the substituents on the nitrogen atom.

$$(19)$$

Selective generation of various carbanions from thioamides is achieved probably because the thiocarbonyl group finely attenuates the acidity in the compounds. One of the most well known groups of carbanions are the enethiolates, which are formed by deprotonation at the position α to the thiocarbonyl group in thioamides. Potassium t-butoxide is used to deprotonate thioacetamide **63**, and then arylation and alkylation with iodobenzene and iodoadamantane are carried out in the presence of a catalytic amount of $FeBr_2$ (Eq. 20) [44]. The reaction is believed to involve the iron enethiolate.

$$(20)$$

The optically active α-cyclohexylsulfinyl thioamide **64** is prepared by reacting the enethiolate, generated from the thioacetamide and NaHMDS, with the optically active sulfinates **65** (Eq. 21) [45]. Two optical isomers **64** are selectively

$$\text{(21)}$$

obtained by using different bases in the reaction of the racemic sulfinyl chloride with diacetone-D-glucose (DAG). The use of pyridine gives R-**64**, whereas performing the reaction in the presence of N,N-diisopropylethylamine furnishes S-**65**. α-Cyclohexylsulfinyl thioamide **64** is further deprotonated with t-BuLi, and allylation of the resulting enethiolate gives the keteneaminothioacetals **66** which undergo thio-Claisen rearrangement [46] to form α-sulfinyl γ,δ-unsaturated thioamides **67** (Eq. 22). The sulfinyl group in **64** is expected to enhance the acidity of the proton α to the thiocarbonyl group, but deprotonation from **64** with other bases fails. Nevertheless, deprotonation with t-BuLi generates the Z-enethiolate, and only Z-isomers of **66** are formed. High stereoselectivity is also observed for thio-Claisen rearrangement of **66**. This can be understood by assuming that the model **68**, where the cyclohexyl group is oriented at the outside position, is more favorable than the model **68′**. Highly stereoselective constructions of three successive carbon centers are achieved by using the keteneaminothioacetals **69** derived from crotyl and cinnamyl bromides. Notably, the major isomers in the thio-Claisen rearrangement of **69** are reversed, with the substituents at the terminal position of the allylic group (Eq. 23).

$$\text{(22)}$$

$$\text{(23)}$$

Thioamide-bearing (+)-*trans*-2,5-diphenylpyrrolidine **70** shows high diastereoselectivity during the thio-Claisen rearrangement (Eq. 24) [47]. Two stereogenic centers are constructed with the allylic bromides **71**. Thio-Claisen rearrangement of the keteneaminothioacetals **72** derived from the Z-allylic bromides **71** (R=H) requires higher temperatures than if they are derived from the E-allylic bromides (**71**) (R′=H). The use of **71** (R, R′=Et, Me) instead gives thioamides bearing asymmetric quarternary carbon centers at the position β to the thiocarbonyl group. Although a longer reaction time is necessary, the rearrangement proceeds with high *syn:anti* selectivity and high asymmetric induction.

$$(24)$$

Remote chiral induction is performed with bicyclic thiolactams [48]. For example, the thiolactam **73** is deprotonated with LDA, and this is followed by allylation to form thiolactam **74** (Eq. 25) [48c]. The reaction with **71** (R=R′=H) shows poor selectivity, whereas that with cinnamyl and prenyl bromides gives the corresponding products with high diastereoselectivity.

$$(25)$$

The enethiolates generated from the axially chiral thioamides **75** are allylated (Eq. 26) [49]. The allylation is sluggish at room temperature, and is carried out under reflux in THF. As a result, the keteneaminothioacetals **76**, which should be formed as intermediates but are not detected, undergo thio-Claisen rearrangement to form two rotamers of the γ,δ-unsaturated thioamides **77** in a ratio of about 80:20. Higher selectivity is observed for **76** derived from metallyl iodide

$$(26)$$

and prenyl bromide. The introduction of the *tert*-butyl group to the meta-position does not affect the stereoselectivity. The ratio of the two rotamers is consistent and independent of time and temperature. This is assumed to be due to kinetic control, and the barrier to rotation being high.

The deprotonation of secondary aromatic thioamides derived from Girgnard reagents and isothiocyanates has been studied [50]. The treatment of thioamides **78** with *s*-BuLi in THF at −78 °C for 30 min generates the yellow monoanion **79**, which is then acylated with acid chlorides to give the *N*-acyl thioamides **80** in good to high yields (Eq. 27). The acylation takes place selectively at the nitrogen atom of **79**, whereas methylation of **79** proceeds at the sulfur atom. Further deprotonation of **79** is carried out with 2 equiv of *s*-BuLi. The benzylic proton in **79** is selectively abstracted to form the red dianion **81** (Eq. 28). Methylation of **81** with 1 equiv of MeI occurs selectively at the benzylic carbon atom to form thioamides **82**. A similar treatment of **81** with various carbon electrophiles and heteroatom-containing electrophiles effectively gives the thioamides; the electrophiles are introduced to the benzylic carbon atom. Aldehydes and α,β-unsaturated ketones are also used as electrophiles. In the latter case, 1,2-addition products are obtained. Deprotonation from *O*-ethyl thiobenzamide and 2-methyl thio-1-naphthamide proceeds in a similar fashion to form the corresponding dianions.

$$(27)$$

$$(28)$$

When deprotonation with excess base is applied to the secondary *N*-benzyl thioamides **83**, the deprotonation proceeds in a different fashion. The treatment of **83** with 2 equiv of BuLi gives a purple solution, which involves the dianions **84** (Eq. 29) [51]. The dianions **84** can be regarded as synthetically useful carbanions adjacent to a nitrogen atom. The selective formation of **84** is in marked contrast to a similar deprotonation of the corresponding amides, where ortho-lithiation at the benzene ring also takes place. Alkylation of **84** takes place

$$(29)$$

selectively at the carbon atom adjacent to a nitrogen atom to give the thioamides **85**. The reaction of dianions with oxiranes and aldehydes also takes place at the carbanionic center of **84**. The dianions **84** are stable under inert atmosphere and are observed using NMR spectroscopy.

2.1.7
Miscellaneous

The Ugi four- or five-component condensation using CS_2 and COS leading to thioamides has been reported [52]. Isocyanides and CS_2 are added to a reaction mixture of primary amines and aldehydes, and the stirring is maintained for 12–18 h. Thioamides **86** are formed in 22~92% yields along with thioureas as by-products (Eq. 30). The reaction involves the nucleophilic attack, by isocyanides, of imines formed in situ. This yields the intermediates **87**, and then the dithiocarbamic acid salts **88** are added. The reaction of primary amines, aldehydes, isocyanides, COS, and methanol also gives thioamides, but two or three types of thioamides are formed in nearly equal amounts.

$$(30)$$

Reduction of *N*-arylimino-1,2,3-dithiazoles **89**, prepared from dithiazolium chloride and primary amines, using lithium aluminum hydride furnishes the *N*-aryldithioxamides **90** (Eq. 31) [53]. The reaction is carried out for 15 min at room temperature, and the yields of **90** are moderate. It should be noted that the thioamide moiety, which is easily reduced with lithium aluminum hydride, remains in the products.

$$(31)$$

Ring opening of 1,3-thiazole-5(4*H*)-one **91** with α-aminoalkyl esters **92** in the presence of 1-hydroxybenzotriazole and *N*-ethyl diisopropylamine gives the endothiotripeptide **93** (Eq. 32) [54]. The starting compound **91** is prepared by the ($\pm$)-camphor-10-sulfonic acid-catalyzed cyclization of endodipeptide. The reaction with **92**, where a bulky alkyl group is introduced to the carbon atom

CSA: (±)-camphor-10-sulfonic acid

$$\text{(32)}$$

α to the carbonyl group, requires over seven days. The molecular structure of the endothiotripeptide has been elucidated.

3
Selenoamides

Compared to thioamides, less attention has been paid to selenoamides. Nevertheless, studies of selenoamides have gradually increased in number. Various heterocycles have been synthesized [55]. In particular, primary selenoamides are good precursors of selenazoles and selenazines. Their biological activities have been partly elucidated [55h]. Primary selenoamides are used as selenium transfer reagents in reactions leading to selenourea [56], diacyl selenides [57], and diselenides [58]. Alkylation of selenoamides with carbon nuleophiles [59] and carbon electrophiles [60] has also been investigated. In the former case, the selenoamides are converted to ketones, and the latter reaction gives selenoiminium salts. Synthetic procedures established before 1998 have been reviewed previously [61]. In the following sections, we describe recent advances in selenoamide synthesis.

3.1
Syntheses

3.1.1
Selenonation of Amides

Selenonation of aromatic amides is carried out with 1,2,3,5,7-pentaselena-4,6,8-triphosphocane **95**, which is prepared from phosphine **94** and elemental selenium, but the corresponding selenoamides are obtained in only low yields (Eq. 33) [62]. The reaction of $(PhP)_5$ with 10 equiv of elemental selenium gives 2,4-bis(phenyl)-1,3-diselenadiphosphetane-2,4-diselenide **96**, and this is isostructural with Lawesson's reagent **18** [63]. **96** is then used for the direct con-

$$\text{ArPH}_2 \xrightarrow{\text{Se}} \mathbf{95} \xrightarrow{} \text{selenoamide} \tag{33}$$

94 **95**

version of aromatic amides to the corresponding selenoamides. Reacting with tertiary and secondary amides gives the products in moderate to good yields, whereas the conversion of benzamide is less efficient. **96** is applicable to the

96

selenonation of formamides, tertiary aliphatic amides, and lactones, when the reaction is carried out with benzene as the solvent at room temperature for 20 h [64]. It is characteristic that all four selenium atoms in **96** are used during the selenonation of amides. This is in sharp contrast to Lawesson's reagent **18** where only two of four sulfur atoms are used. Alternatively, selenonation with reagents bearing an aluminum-selenium bond has been developed. The reagent **97** is prepared by heating a mixture of diisobutylaluminum hydride and elemental selenium for 1 h under an argon atmosphere (Eq. 34) [65]. The amides are then reacted with **97** to form selenoamides. The reaction has been applied to syntheses of various selenoformamides, which are obtained in yields >50%. In contrast, the selenonations of acetamide and benzamide with **97** give the corresponding selenoamides in lower yields.

$$i\text{-Bu}_2\text{AlH} + \text{Se} \xrightarrow[120 \sim 130\,^\circ\text{C}]{\text{toluene}} \left[\begin{array}{c} (i\text{-Bu}_2\text{AlSe})_n \\ (i\text{-Bu}_2\text{AlSe})_2 \\ \mathbf{97} \end{array} \right] \longrightarrow \text{selenoamide} \tag{34}$$

3.1.2
Selenolysis of Iminium Salts

Amides are converted to iminium salts **98** with oxalyl chloride. Selenolysis of **98** with the reagent **99** generated from lithium aluminum hydride and elemental selenium will give selenoamides (Eq. 35) [66]. The reaction is effective

$$\text{LiAlH}_4 + \text{Se} \xrightarrow{\text{THF}} \left[\text{LiAlHSeH}\right] \xrightarrow{\mathbf{98}} \text{selenoamide} \tag{35}$$

99

for synthesizing tertiary aromatic selenoamides, except for ortho-substituted selenoamides and selenoamides with bulky groups at their nitrogen atoms. Alternatively, tetraethylammonium tetraselenotungstate **100** can be used as a reagent transferring the selenium atom to iminium salts [67]. The reagent **100** is applicable to the syntheses of various selenoamides. Tertiary aliphatic and aromatic amides are efficiently converted to the corresponding selenoamides. *N*-ethyl selenoacetamide is obtained with 76% yield, whereas the reaction with primary amides fails. The reaction of lactams with **100** successfully gives the selenolactams **101** (Eq. 36).

$$(36)$$

The $POCl_3$-mediated reaction of acetamides with DMF followed by treatment with $HClO_4$ forms the iminium salts **102** (Eq. 37) [68]. The reaction of **102** with sodium selenide gives the β-amino-α,β-unsaturated selenoamides **103** as thermally stable compounds, although the yields of **103** are lower than those of thioamides derived by a similar reaction.

$$(37)$$

Addition reactions of a selenium source to nitriles is a well-known method for synthesizing primary selenoamides. The selenoic acid potassium salt **104** is employed as the selenium source, and reactions with nitriles lead to the corresponding selenoamides in moderate to good yields [69]. The reaction is useful for synthesizing less stable aliphatic primary selenoamides. The salt **104** mainly undergoes Michael addition to α,β-unsaturated nitriles accompanied by addition of **104** to the cyano group. The selectivity of the reaction is not high, and two types of selenoamides are formed. Notably, dicyano compound reactions have also been examined [70]. Both the products where two cyano groups are converted to selenoamide moieties and where one cyano goup is converted to

$$(38)$$

selenoamide moiety are formed, but stepwise conversion of two cyano groups successfully gives the diselenoamide **105** (Eq. 38).

3.1.3
Via Functionalization of Selenoamides

N-Alkylation of selenoamides is achieved by reacting with the *N*-hydroxyalkyl benzotriazoles **106** generated from benzotriazole and aliphatic aldehydes (Eq. 39) [71]. The reaction of primary aromatic selenoamides with **106** followed by reduction with $NaBH_4$ gives secondary selenoamides in moderate to high yields.

$$(39)$$

Selenium isologues of enolate ions (the eneselenolates **107**) are generated from aliphatic selenoamides and LDA (Eq. 40). Alkylation of eneselenolates usually takes place at the selenium atom to form keteneaminoselenoacetals, which are highly susceptible to hydrolysis, giving ordinary amides. When allylic halides are used as carbon electrophiles, γ,δ-unsaturated selenoamides **109** are obtained in good yields [72]. Keteneaminoselenoacetals **108** may be formed initially, and then undergo seleno-Claisen rearrangement to furnish **109**. Despite the fact that a selenocarbonyl group is less stable than a thiocarbonyl group, the seleno-Claisen rearrangement appears to proceed with an efficiency similar to that of thio-Claisen rearrangement. In terms of the stereochemical outcome of the reaction, the eneselenolates **107** are generated as single isomers, similar to ordinary amides and thioamides. Michael addition reactions of eneselenolates **107** with α,β-unsaturated esters and ketones have been carried out [73]. These reactions are very rapid, and high stereoselectivity is observed for the reaction with aliphatic α,β-unsaturated esters leading to **110** (Eq. 41). Aldol-type condensation reactions of aldehydes [74] and ketones [75] with eneselenolates **107** give β-hydroxy selenoamides **111** in moderate to high yields as stable compounds (Eq. 42). The efficiency of the ketone reaction is dependent on the sub-

$$(40)$$

(41)

stituents adjacent to the carbonyl group. The stereoselectivity of the reaction also depends on the carbonyl compounds used and the substituents at the nitrogen atoms of the selenoamides. Generating the eneselenolate from α-silyl selenoamide **112** and then reacting it with aldehydes produces α,β-unsaturated selenoamides **113** (Eq. 43) [76]. The reaction shows high *E*-selectivity, which is in sharp contrast to the reaction of α-silyl thioacetamide, which shows no stereoselectivity. Deprotonation from α,β-unsaturated selenoamide **114** takes place at the carbon atom γ to the selenocarbonyl group to form dieneselenolates **115** (Eq. 44) [77]. Allylation of **115** gives the selenoamide **116**, where the allyl group is introduced to the carbon atom α to the selenocarbonyl group.

(42)

(43)

(44)

Like the reaction in Eq. 29, the selenoamide dianion **117** is generated by reacting *N*-benzyl selenobenzamide with 2 equiv of BuLi (Eq. 45) [78]. Alkylation of **117** takes place selectively at the carbon atom adjacent to the nitrogen atom to the selenocarbonyl group. Alkyl halides, acid chlorides, and oxiranes are used as carbon electrophiles. Trimethylsilyl chloride performs silylation of **117**, giving the selenoamide **118**.

(45)

3.1.4
Miscellaneous

The lithium alkyneselenolates formed by reacting terminal acetylenes and BuLi with elemental selenium are good precursers for selenoamides. Propargylation of lithium alkyneselenolate **119** followed by reaction with secondary amines gives α,β,γ,δ-unsaturated selenoamide **120** in moderate yields (Eq. 46) [79]. Four-component coupling reactions of terminal acetylenes, elemental selenium, amines, and allylc bromides provide an efficient method of synthesizing γ,δ-unsaturated selenoamides **121** (Eq. 47) [80]. **119** is formed initially, which is allylated to form the alkynyl allyl selenides **122**, which then undergo [3,3] sigmatropic rearrangement to form the selenoketene intermediates **123**. Finally, addition of amines to **123** results in the formation of **121**. It is important that the allylic bromides and amines are added successively to **119**, in order to react the amines with **123** as soon as **123** is generated. Otherwise, yields of **121** drop, probably due to the gradual decomposition of **123**. Various terminal acetylenes, secondary and tertiary amines are used to produce the corresponding selenoamides in good to high yields. The use of crotyl bromide gives diastereoisomeric mixtures with no selectivity.

$$\text{PhC}{\equiv}\text{CH} \xrightarrow[\text{THF}]{\text{BuLi/Se}} \left[\text{PhC}{\equiv}\text{CSeLi}\right] \xrightarrow{\text{EtC}{\equiv}\text{CCH}_2\text{Cl}} \xrightarrow{\text{HNR}_2} \mathbf{120}$$

$$\mathbf{119}$$

(46)

$$\text{RC}{\equiv}\text{CH} \xrightarrow[\text{THF}]{\text{BuLi/Se}} \left[\text{RC}{\equiv}\text{CSeLi}\right] \xrightarrow{} \xrightarrow{\text{HNR''}_2} \mathbf{121}$$

$$\mathbf{119}$$

(47)

$$\text{RC}{\equiv}\text{CSe}\ \mathbf{122} \qquad \mathbf{123}$$

4
Telluroamides

Only limited examples of telluroamides have been reported. This is mainly due to the labilities of telluroamides, particularly toward oxygen. Even when telluroamides are formed in situ, they gradually decompose during purification. Synthetic methods that produce telluroamides along with easily separable co-products are necessary. Reagents **124** and **126**, prepared from diisobutylaluminum hydride [65] or lithium aluminum hydride [60b] with elemental tellurium, have been developed for this purpose. The telluroformamides **125** can be obtained in 25~66% yields by reacting formamides with reagent **124** and then isolating them by column chromatography on silica gel (Eq. 48). Tellurolysis of the selenoiminium salt **127** with the reagent **126** proceeds smoothly below room temperature to form the telluroamide **128** (Eq. 49). This reaction is applicable to the synthesis of telluroformamides and aromatic telluroamides [81]. Aliphatic telluroamides are also formed, but isolated in quite low yields unless two alkyl groups are introduced to the carbon atom α to the tellurocarbonyl group.

$$i\text{-Bu}_2\text{AlH} \quad + \quad \text{Te} \xrightarrow[\substack{\text{toluene} \\ 120 \sim 130\,°C}]{} \left[\substack{(i\text{-Bu}_2\text{AlTe})_n \\ (i\text{-Bu}_2\text{AlTe})_2 \\ \mathbf{124}} \right] \xrightarrow{\underset{\text{H}}{\overset{\text{O}}{\|}}\text{NR'}_2} \underset{\text{H}\quad\text{NR'}_2}{\overset{\text{Te}}{\|}} \quad \mathbf{125} \tag{48}$$

$$\text{LiAlH}_4 \quad + \quad \text{Te} \longrightarrow \mathbf{126} \xrightarrow{\substack{\text{MeSe}\ ^-\text{OTf} \\ R\diagup\!\!\overset{+}{N}\diagdown\ \mathbf{127}}} \underset{R}{\overset{\text{Te}}{\|}}\!\diagup\!\overset{}{N}\diagdown \quad \mathbf{128} \tag{49}$$

$$R = H,\ Ph,\ CH_2\!=\!CHCH_2CH(Ph) \quad \mathbf{128}$$

5
Perspectives

In this chapter we have reviewed methods of synthesizing chalcogen isologues of amides that have been developed over the last few years. Dramatic progress in the chemistry of thioamides looks set to continue. Thioamide moieties are becoming important functional groups in the material and biological sciences. Therefore, we need more efficient, selective, and environmentally benign processes to produce them. For example, the site-selective thionation of amide moieties of oligo- and poly- peptides is demanded. The main drawback with the conversion of amides to thioamides is the formation of equimolar amounts of co-products, such as phosphorus compounds. In an ideal process, the direct replacement of an amide oxygen atom with elemental sulfur would occur, but this is yet to be realized. Future prospects for selenoamides are similar to those for thioamides. As the biological importance of selenium-containing heterocycles and selenoamides becomes more apparent, more synthetic methods for selenoamides will be developed. The use of selenoamides as metal ligands has so far only been studied to a minor extent. Because of the affinity of selenium toward heavy metals, metal complexes formed with selenoamides are expected to show unique properties. Finally, in order to develop the telluroamide chemistry, synthetic reactions performed under mild conditions are desired. Nevertheless, recent results indicate that even aliphatic telluroamides can be made stable enough to handle by introducing appropriate substituents. The fundamental properties of telluroamides have also not yet been fully studied. These topics will be investigated in the future.

References

1. Wiberg KB, Rush, DL (2002) J Org Chem 67:826
2. (a) Sakamoto M, Takahashi M, Kamiya K, Arai W, Yamaguchi K, Mino T, Watanabe S, Fujita T (1998) J Chem Soc Perkin T 1 3731; (b) Sakamoto M, Takahashi M, Arai, W, Kamiya K, Mino T, Watanabe S, Fujita T (1999) J Chem Soc Perkin T 1 3633; (c) Sakamoto M, Takahashi M, Arai W, Mino T, Yamaguchi K, Watanabe S, Fujita T (2000) Tetrahedron 56:6795; (d) Hosoya T, Ohhara T, Uekusa H, Ohashi Y (2002) B Chem Soc Jpn 75:2147; (e) Sakamoto M, Nishio T (2003) Heterocycles 59:399
3. (a) Miwa JH, Patel AK, Vivatrat N, Popek SM, Meyer AM (2001) Org Lett 3:3373; (b) Miwa JH, Pallivathucal L, Gowda S, Lee KE (2002) 4:4655; (c) Tran TT, Zeng J, Treutlein H, Burgess AW (2002) J Am Chem Soc 124:5222
4. Lee HJ, Choi, YS, Lee KB, Park J, Yoon CJ (2002) J Phys Chem A 106:7010
5. Stauffer SR, Sun J, Katzenellenbogen BS, Katzenellenbogen JA (1999) Bioorg Med Chem 8:1293
6. Yu K-L, Torri AF, Luo G, Cianci C, Grant-Young K, Danetz S, Tiley L, Krystal M, Meanwell NA (2002) Bioorg Med Chem 12:3379
7. Krinková J, Dolezal M, Hartl J, Buchta V, Pour M (2002) Farmaco 57:71
8. Dilworth JR, Wheatley N (2000) Coord Chem Rev 199:89
9. Jagodzinski TS (2003) Chem Rev 103:197
10. Manuel J, Fernández G, Mellet CO (2000) Adv Carbohyd Chem Bi 55:35
11. (a) Metzner P (1999) Top Cur Chem 204:127; (b) Castro EA (1999) Chem Rev 99:3505
12. (a) Kanbara T, Kawai Y, Hasegawa K, Morita H, Yamamoto T (2001) J Polym Sci 39:3739; (b) Kagaya S, Sato E, Masore I, Hasegawa K, Kanbara T (2003) Chem Lett 32:622
13. Moghaddam FM, Ghaffarzadeh M (2001) Synth Commun 31:317
14. Zbruyev OI, Stiasni N, Kappe CO (2003) J Comb Chem 5:145
15. Aghapoor K, Darabi HR, Tabar-Heydar K (2002) Phosphorus Sulfur 177:1183
16. Moghaddam FM, Hojabri L, Dohendou M (2003) Synthetic Commun 33:4279
17. Sifferlen T, Rueping M, Gademann K, Jaun B, Seebach D (1999) Helv Chim Acta 82:2067
18. Krumme D, Tschesche H (1999) Tetrahedron 55:3007
19. Miyakoshi K, Oshita J, Kitahara T (2001) Tetrahedron 57:3355
20. Boger DL, Santillán A Jr, Searcey M, Jin Q (1999) J Org Chem 64:5241
21. Arévalo MJ, Avalos M, Babiano R, Cabanillas A, Cintas P, Jiménez JL, Palacios JC (2000) Tetrahedron Asymmetr 11:1985
22. Wang L, Phanstiel IV O (2000) J Org Chem 65:1442
23. Wojczewski C, Schwarzer K, Engels JW (2000) Helv Chim Acta 83:1268
24. Fruit C, Turck A, Plé N, Mojovic L, Quéguiner G (2002) Tetrahedron 58:2743
25. Nishio T, Sekiguchi H (1999) Tetrahedron 55:5017
26. Curphey TJ (2002) J Org Chem 67:6461
27. (a) Varma RS, Kumar D (1999) Org Lett 1:697; (b) Olsson R, Hansen HC, Andersson C-M (2000) Tetrahedron Lett 41:7947
28. (a) Pons J-F, Mishir Q, Nouvet A, Brookfield F (2000) Tetrahedron Lett 41:4965; (b) Coats SJ, Link JS, Hlasta DJ (2003) Org Lett 5:721
29. (a) Ley SV, Leach AG, Storer RI (2001) J Chem Soc Perkin T 1 358; (b) Kellogg RM (2001) Chemtracts: Org Chem 14:646

30. Shalaby MA, Rapoport H (1999) J Org Chem 64:1065
31. (a) Charette AB, Chua P (1998) Tetrahedron Lett 39:245; (b) Charette AB, Grenon M (2003) J Org Chem 68:5792
32. Prabhu KR, Devan N, Chandrasekaran S (2002) Synlett 1762
33. Liboska R, Zyka D, Bobek M (2002) Synthesis 1649
34. Aki S, Fujioka T, Ishigami M, Minamikawa J (2002) Bioorg Med Chem Lett 12:2317
35. Messeri T, Sternbach DD, Tomkinson NCO (1998) Tetrahedron Lett 39:1673
36. Tojo T, Spears GW, Tsuji K, Nishimura H, Ogino T, Seki N, Sugiyama A, Matsuo M (2002) Bioorg Med Chem Lett 12:2427
37. (a) Alayrac C, Nowaczyk S, Lemarié M, Metzner P (1999) Synthesis 669; (b) Alayrac C, Metzner (2000) Tetrahedron Lett 41:2537
38. Sanderson JM, Singh P, Fishwick CWG, Findlay JBC (2000) J Chem Soc Perkin T 1 3227
39. Yeo SK, Choi BG, Kim JD, Lee JH (2002) B Korean Chem Soc 23:1029
40. (a) Doszczak L, Rachon J (2000) Chem Commun 2093; (b) Doszczak L, Rachon J (2002) J Chem Soc Perkin T 1 1271
41. Yokoyama Y, Mochida K (1999) Synthesis 1319
42. Ubiali D, Pagani G, Pregnolato M, Piersimoni C, Munoz JLP, Gascón AR, Terreni M (2002) Bioorg Med Chem Lett 12:2541
43. Sosnicki JG, Jagodzinski TS, Hansen PE (2001) Tetrahedron 57:8705
44. Murguía MC, Ricci CG, Cabrera MI, Luna JA, Grau RJ (2001) J Mol Catal A 165:113
45. Nowaczyk S, Alayrac C, Reboul V, Metzner P, Averbuch-Pouchot M-T (2001) J Org Chem 66:7841
46. Majumdar KC, Ghosh S, Ghosh M (2003) Tetrahedron 59:7251
47. He S, Kozmin SA, Rawal VH (2000) J Am Chem Soc 122:190
48. (a) Lemieux RM, Meyers AI (1998) J Am Chem Soc 120:5453; (b) Lemieux RM, Devine PN, Mechelke MF, Meyers AI (1999) J Org Chem 64:3585; (c) Watson DJ, Lawrence CM, Meyers AI (2000) Tetrahedron Lett 41:815
49. Dantale S, Reboul V, Metzner P, Philouze C (2002) Chem Eur J 8:632
50. Ach D, Reboul V, Metzner P (2002) Eur J Org Chem 2573
51. Murai T, Aso H, Tatematsu Y, Itoh Y, Niwa H, Kato S (2003) J Org Chem 68:8514
52. Keating TA, Armstrong RW (1998) J Org Chem 63:867
53. Besson T, Rees CW, Thiéry V (1999) Synthesis 1345
54. Lehmann J, Linden A, Heimgartner (1999) Tetrahedron 55:5376
55. (a) Petrov ML, Abramov MA (1998) Phosphorus Sulfur 134/135:331; (b) Koketsu M, Senda T, Yoshimura K, Ishihara H (1999) J Chem Soc Perkin T 1 453; (c) Koketsu M, Hiramatsu S, Ishihara H (1999) Chem Lett 485; (d) Zhang P-F, Chen Z-C (2000) Synthesis 1219; (e) Koketsu M, Takenaka Y, Ishihara H (2001) Synthesis 731; (f) Zhang P-F, Chen Z-C (2001) J Heterocyclic Chem 38:503; (g) Koketsu M, Takenaka Y, Hiramatsu S, Ishihara H (2001) Heterocycles 55:1181; Zhao HR, Yu QS (2002) Chinese Chem Lett 13:729; (h) Koketsu M, Ishihara H (2003) Curr Org Chem 7;175; (i) Huang X, Chen J (2003) Synthetic Commun 33:2823
56. Koketsu M, Suzuki N, Ishihara H (1999) J Org Chem 64:6473
57. (a) Zhao HR, Zhao XJ, Huang X (2001) Chinese Chem Lett 12:873; (b) Zhao HR, Zhao XJ, Huang X (2002) Synthetic Commun 32:3383
58. Bailly F, Azaroual N, Bernier J-L (2003) Bioorg Med Chem 11:4623
59. Murai T, Ezaka T, Kato S (1998) Tetrahedron Lett 39:4329
60. (a) Murai T, Mutoh Y, Kato S (2001) Org Lett 3:1993; (b) Mutoh Y, Murai T (2003) Org Lett 5:1361
61. Murai T, Kato S (2000) Top Curr Chem 208:177
62. Yoshifuji M, Higeta N, An D-L, Toyota K (1998) Chem Lett 17

63. Bhattacharyya P, Wollins J D (2001) Tetrahedron Lett 42:5949
64. Bethke J, Karaghiosoff K, Wessjohann LA (2003) Tetrahedron Lett 44:6911
65. (a) Li G-M, Zingaro RA (1998) J Chem Soc Perkin T 1 647; (b) Li G-M, Reibenspies JH, Derecskei-Kovacs A, Zingaro RA (1999) Polyhedron 18:3391
66. (a) Ishihara H, Koketsu M, Fukuta Y, Nada F (2001) J Am Chem Soc 123:8408; (b) Koketsu M, Okayama Y, Aoki H, Ishihara H (2002) Heteroatom Chem 13:195
67. Saravanan V, Mukherjee C, Das S, Chandrasekaran S (2004) Tetrahedron Lett 45:681
68. Hartmann H, Heyde C, Zug I (2000) Synthesis 805
69. Ishihara H, Yoshimura K, Kouketsu M (1998) Chem Lett 1287
70. Koketsu M, Takenaka Y, Ishihara H (2003) Heteroatom Chem 14:106
71. (a) Zhao HR, Yu Q-S (2002) Chinese J Chem 20:891; (b) Zhao HR (2002) Chinese Chem Lett 13:402
72. Murai T (2004) Mini Rev Org Chem 1(3):279
73. Murai T, Suzuki A, Ezaka T, Kato S (2000) Org Lett 2:311
74. Murai T, Suzuki A, Kato S (2001) J Chem Soc Perkin T 1 2711
75. Murai T, Ishizuka M, Suzuki A, Kato S (2003) Tetrahedron Lett 44:1343
76. Murai T, Fujishima A, Iwamoto C, Kato S (2003) J Org Chem 68:7979
77. Murai T, Suzuki A, Takagi M, Kato S (2001) Phosphorus Sulfur 172:101
78. Mura T, Aso H, Kato S (2002) Org Lett 4:1407
79. Koketsu M, Kanoh M, Itoh E, Ishihara H (2001) J Org Chem 66:4099
80. Murai T, Ezaka T, Kato S (1998) B Chem Soc Jpn 71:1193
81. Mutoh Y, Murai T, Yamago S (2004) J Am Chem Soc 126:16697

Author Index Volume 251

Subject Index